IMISCOE Research Series

This series is the official book series of IMISCOE, the largest network of excellence on migration and diversity in the world. It comprises publications which present empirical and theoretical research on different aspects of international migration. The authors are all specialists, and the publications a rich source of information for researchers and others involved in international migration studies.

The series is published under the editorial supervision of the IMISCOE Editorial Committee which includes leading scholars from all over Europe. The series, which contains more than eighty titles already, is internationally peer reviewed which ensures that the book published in this series continue to present excellent academic standards and scholarly quality. Most of the books are available open access.

More information about this series at http://www.springer.com/series/13502

Zvezda Vankova

Circular Migration and the Rights of Migrant Workers in Central and Eastern Europe

The EU Promise of a Triple Win Solution

Springer

Zvezda Vankova
Faculty of Law
Lund University
Lund, Sweden

ISSN 2364-4087 ISSN 2364-4095 (electronic)
IMISCOE Research Series
ISBN 978-3-030-52688-7 ISBN 978-3-030-52689-4 (eBook)
https://doi.org/10.1007/978-3-030-52689-4

This Springer imprint is published by the registered company Springer Nature Switzerland AG
The registered company address is: Gewerbestrasse 11, 6330 Cham, Switzerland

To my family/ На моето семейство

Acknowledgements

This book is based on the doctoral dissertation 'Circular Migration from the Eastern neighbourhood to the EU: The rights of migrant workers in Bulgaria and Poland' drafted between 2013 and 2017 at Maastricht University. The dissertation received financial support as part of the Transnational Migration, Citizenship and the Circulation of Rights and Responsibilities (TRANSMIC) project funded by the European Union under the FP7-PEOPLE-2013-ITN call of the Marie Curie Actions – Initial Training Networks funding scheme (project number – 608417) – and coordinated by Maastricht University. The publication of this book has been funded by a Rubicon grant (project number 019.191SG.008) of the Netherlands Organisation of Scientific Research (NWO), and the Open Access fees have been covered by the IMISCOE network, the Law Faculty of Maastricht University, and Lund University.

So many people from so many different places contributed to the successful accomplishment of this ambitious project, which has travelled throughout Europe like a circular migrant with a constantly changing trajectory. I will try to acknowledge all of them here.

I would like to thank my first supervisor, Prof. Hildegard Schneider, who gave me the chance to pursue a PhD with her. I will never forget our meeting at the hotel cafe facing the Gare Central in Brussels where she offered me the opportunity to start a PhD at Maastricht University as an external candidate and later to work on my dream project as part of TRANSMIC. Encouragement, inspiration, patience, and trust are the qualities that make Hildegard such a great supervisor! I would like to also thank her and René for opening the doors of their home to me so many times and for all the inspiring conversations that were held over a glass of wine.

I owe a big thanks to my second supervisor, Prof. Maarten Vink, who joined my PhD ride a bit later. I am particularly indebted to him because of the structure he brought to my PhD dissertation, for the numerous comments and suggestions for improvements, as well as the faith he had in me. The empirical part of this study would not have been of such quality without his active feedback and constant encouragement.

My heartfelt gratitude also goes to my PhD Assessment Committee who read and assessed my hefty tome: Prof. Bruno de Witte, Prof. Kees Groenendijk, Prof. Sergio Carrera, Prof. Ronald Skeldon, and Prof. Herwig Verschueren.

I would like to give my sincere thanks to Prof. Bruno de Witte who was my supervisor during my research stay at the European University Institute in Florence and contributed to the successful accomplishment of my PhD with inspiring ideas and suggestions for useful literature as well as for introducing me to wonderful scholars at the EUI and beyond.

Throughout the years, Prof. Kees Groenendijk has become my professional mentor and even somewhat of a 'shadow' supervisor, providing countless comments, feedback, and encouragement in numerous countries and on various occasions. I have learned so much from his encyclopaedic knowledge which has helped me overcome various frustrations with European migration law. He is the person who showed me how academic research can also be policy-relevant and that I can continue to do what I love in my new academic path, namely legal empirical research.

I am greatly indebted to Prof. Sergio Carrera for the support he has offered to my PhD since the very beginning when he opened the door to the 'Brussels affairs' for me, helped me with contacts for my interviews, and provided me with feedback throughout the whole project. At the end of the PhD, Sergio also made it possible to use some of my research findings and contribute to two important policy reports for the European Parliament and the Council of Europe.

I was honoured to have Prof. Ronald Skeldon on my Assessment Committee. I would like to thank him for our conversations that brought me back in time when circular migration was first introduced as an academic concept. His comments and inspiration have helped me immensely to finalise this piece of research successfully.

Prof. Herwig Verschueren is the person who made me realise that the only way this study could be close to exhaustive is by including social security coordination as one of the policy areas under consideration. His support throughout my social security academic journey was extremely valuable, and I would also like to thank him for supporting me and Pauline Melin implement our social security coordination workshop with leading scholars in this field in Brussels in 2017.

I would also like to thank all migrants who found the time to participate in this study and allowed me to hear their circular migration stories. My sincere gratitude extends also to the rest of my interviewees who helped me add some flesh to the concept of circular migration.

I want to express my sincerest thanks to all my colleagues at Maastricht University, especially Natasja Reslow for all her support, and Alexander Hoogenboom, Prof. Anne-Pieter van der Mei, and Lavinia Kortese for their feedback. Special thanks also goes to Anja Wiesbrock, who acted as my second supervisor when I started work on my PhD as an external candidate, for her feedback, advice, and friendship.

I am very grateful to TRANSMIC and the European Commission for funding this wonderful network, without which this study would never have come to life. First of all, I would like to express my gratitude to Dr Katharina Eisele for writing such a wonderful project together with Prof. Hildegard Schneider and for her constant support and feedback during the PhD process. Dear Katharina, thank you also for being my paranymph!

Big thanks goes to Prof. Patrick Wautelet and Prof. Patrícia Jerónimo Vink for their comments, training, encouragement, and vivid conversations. I would also like to thank Diana Schabregs for being the most wonderful and supportive financial officer I have ever met in my professional life and for providing me with so much valuable advice that made the financial management of my PhD project very smooth. Big thanks to Niels Philipsen who managed this complicated project and for bearing with us throughout all those stressful moments.

This project would not have been so great without the TRANSMICers. Cordial thanks to all of you: Ali, Clotilde, Ester, Fanny, Hanwei, Marieke, Olga, Polina, and Chun and especially to Leonhard and Pauline for the immense support with comments, feedback, and lovely chats when I needed them the most. Luuk, thanks for being such a wonderful colleague and friend.

My research stay at the Centre for Migration Research at Warsaw University would not have been possible without Prof. Pawel Kaczmarczyk and Dr Joanna Nestorowizc, to whom I am deeply grateful. Big thanks also goes to Monika Szulecka, Prof. Witold Klaus, Dr Marta Kindler, and Dr Anna Kordasiewicz for their feedback and support during and beyond my field research. I would also like to express my heartfelt gratitude to Karolina Grot who assisted me enormously to get in touch with all the different stakeholders, get my hands on statistical data, and who made my stay in Poland very special indeed. Big thanks to Valeria Kochyna for helping me find focus group participants and to my colleagues and friends, Dr Kamila Fiałkowska and Dr Jowanka Jakubek-Lalik.

I am indebted to so many people in Bulgaria! First and foremost, I would like to thank Bistra, Panayot, and the rest of my Multi-Kulti Collective colleagues who supported me immensely throughout my field research in Bulgaria. I am grateful to Bistra for helping me with the transcriptions and serving as a co-interviewer for most of the focus groups, as well as for being such a great friend. Many thanks also to the Red Cross and the IOM office in Bulgaria for assisting me during my field research. Special thanks to Valeria Ilareva for all her intellectual support in helping me grow as a migration expert. I also owe my heartfelt gratitude to Iliana Savova for supporting my migration law journey in Bulgaria with her expertise throughout the years. I would also like to thank Anna Ganeva, Radostina Pavlova, Irina Bulanova, Kina Subeva, Vera Zaharieva, Mariyana Stoyanova, Hristo Simeonov, Emilya Bratanova, and Atanaska Todorova for their assistance. My sincere thanks to my former colleagues at the Open Society Institute in Sofia and especially to Ivanka Ivanova, Assya Kavrakova, and Alexey Pamporov. Many thanks also to Dr Eugenia Markova for her feedback and the introduction to academic life.

I wish to express my gratitude to the ICMPD office in Vienna for hosting me there during my PhD secondment and for giving me access to a lot of information that immensely improved my research, and especially to Veronika Bilger and Katharina Schaur who made my stay in Vienna such a great experience. Many thanks also to Dr Sergo Mananashvili for his valuable comments.

Special thanks also to the Danish Institute for Human Rights for my research stay during a very intensive part of my PhD project, namely the finalisation of my first draft. I would like to thank Eva Ersbøl for making this happen and for her valuable support. I do not think I would have stayed sane without Philip and Katrine Halskov Holten who opened their home to me and kept me going with their Danish *hygge*.

Many thanks also to my Migration Policy Group colleagues and more specifically to Thomas Huddleston for reminding me what a good second case study Poland would be. I wish to also thank Prof. Andrea Rea and Prof. Dirk Jacobs for hosting me at the GERME and providing comments during the initial phase of my PhD project.

I would also like to thank Prof. Godfried Engbersen, Prof. Marta Bivand Erdal, Prof. Jørgen Carling, Prof. Valentina Mazzucato, and Dr Özge Bilgili for showing me the relevance of my research to transnationalism and for their encouragement. I am indebted to Barbara Wegelin for being a wonderful friend and colleague throughout my PhD journey and to Nicholas Blake for his inspiration. I owe special thanks also to Dr Marion Panizzon and Prof. Rainer Bauböck for their feedback and encouragement.

On a more personal note, I would like to thank all my precious friends who gave me their unconditional love throughout this demanding PhD project: Katya, Ivelina, Mira, Denitsa, Anita, Velisalva, Dessi, Kamelia, Gery, Alexander, Sarah, Catharina, Anne-Linde, Vera, Anne Friel, Özge, Leila and Philip, Vanessa, Bene, Elvira, Francesca, Gina, Eva, and Anne Staver.

I would also like to thank my 'Bulgarian' and 'Belgian' families for bearing with me, for understanding and being patient with my weird working hours during holidays and weekends, and my ups and downs during the writing process. Many thanks to my mother and father, to my wonderful sister and her husband, to my granny and my aunt who all contributed to the successful accomplishment of this PhD by giving me continuous confidence boosts. I am also indebted for life to Kety and Jonas for always being there for me during the PhD process and beyond. Thank you Kety for inspiring me to pursue a PhD and for serving as one of my paranymphs.

Last but not least, I would like to express my gratitude to several people who held my hand through the academic journey leading to the publication of this book. Special thanks to Prof. Anna Triandafyllidou and the IMISCOE Editorial Committee for supporting this book project and for the valuable feedback of the IMISCOE reviewers. My sincere thanks go also to Monika Szulecka, Prof. Witold Klaus, Prof. Marek Okólski, Dr Karolina Podstawa, and Dr Mikołaj Pawlak for supporting me through the update of the Polish part of the book. I owe my heartfelt gratitude to my colleagues in Bulgaria who provided me with comments on the updated version of the Bulgarian parts: Iliana Savova, Vladimir Rankov, Petar Achikov, and Ilin Dimitrov. Many thanks also to Prof. Diego Acosta Arcarazo, Lavinia Kortese, Dr Jowanka Jakubek-Lalik, and to Dr Katharina Eisele for their feedback. I am grateful to Ivelina Metchkarova and Diane Shugart for proofreading this book. Cordial thanks also to my colleagues at the Law faculties of Maastricht and Lund Universities for their support with this book project! And finally, on a more personal note, I would like to express my deepest gratitude to Claudio Scalamonti without whose emotional support and backing I would not have accomplished the publication of this book.

September 2019 Zvezda Vankova
Copenhagen

Contents

Abbreviations

ACP	African, Caribbean and Pacific Group of States
AFRB	Act on Foreigners in the Republic of Bulgaria
AHES	Act on Higher Education and Science
ALMLM	Act on Labour Migration and Labour Mobility
ARPQ	Act on the Recognition of Professional Qualifications
BCD	Blue Card Directive
CAMM	Common Agendas on Migration and Mobility
CEE	Central and Eastern Europe(an)
CIS	Commonwealth of Independent States
CJEU	Court of Justice of the European Union
CRC	Convention on the Rights of the Child
EAB	Ethics Advisory Board
ECDPM	European Centre for Development Policy Management
ECMW	European Convention on the Legal Status of Migrant Workers
ECtHR	European Court of Human Rights
EFTA	European Free Trade Association
EMN	European Migration Network
ENIC	European Network of Information Centres
ETF	European Training Foundation
FRD	Family Reunification Directive
GAM	Global Approach to Migration
GAMM	Global Approach to Migration and Mobility
GATS	General Agreement on Trade in Services
GCIM	Global Commission on International Migration
GFMD	Global Forum on Migration and Development
ICCPR	International Covenant on Civil and Political Rights
ICHD	International Centre for Human Development
ICMPD	International Centre for Migration Policy Development
ICRMW	International Convention on the Protection of the Rights of All Migrant Workers and Members of Their Families
ICT	Intra-Corporate Transferees

ICTD	Intra-Corporate Transferees' Directive
ILO	International Labour Organization
IOM	International Organization for Migration
JHA	Justice and Home Affairs
LRC	Lisbon Recognition Convention
LTRD	EU Long-Term Residence Directive
MIPEX	Migrant Integration Policy Index
MISMES	Migrant Support Measures from an Employment and Skills Perspective
MPI	Migration Policy Institute
NARIC	National Academic Recognition Information Centre
NAWA	National Agency for Academic Exchange
NCID	National Centre for Information and Documentation
SRD	Students and Researchers' Directive
SWD	Seasonal Workers' Directive
TEC	Treaty Establishing the European Community
TFEU	Treaty on the Functioning of the European Union
TRANSMIC	Transnational Migration, Citizenship and the Circulation of Rights and Responsibilities
UDHR	Universal Declaration of Human Rights
UNECE	United Nations Economic Commission for Europe
VC	EU Visa Code
WHO	World Health Organization

List of Figures

List of Tables

Chapter 1
Introduction

1.1 Context

The policy idea of facilitating circular migration entered the European Union's agenda more than a decade ago[1] as part of a worldwide buzz among international organisations that it could provide a 'triple win solution' that would benefit all: the countries of origin and destination as well as the migrant workers themselves.[2] It became clear that the European Commission wanted to foster this type of migration in such a way as to allow some degree of legal mobility for migrants back and forth between two countries.[3] For Member States, this 'triple win solution' would provide a tool that resonates with their reluctance to open more channels for legal migration, permanent settlement, and pathways to naturalisation, and it would also reduce any irregular overstaying.[4] Since circular migration is allegedly of temporary nature, states would be able to satisfy their labour market needs[5] and at the same time disengage from the integration challenges associated with permanent migration.[6]

In similar vein, countries of origin would supposedly benefit from the social and economic development that circular migration is claimed to facilitate through a steady flow of remittances,[7] skills and knowledge transfers, as well as brain circulation thus mitigating the negative effects of a brain drain.[8] Lastly, the advocates of circular migration claimed that it could potentially bring advantages for the migrant workers because, amongst other things, this type of migration is emerging as a

[1] European Commission (2005), p. 7.

[2] European Commission (2005), p. 25; Newland et al. (2008), p. 2.

[3] European Commission (2007), p. 8.

[4] Vankova (2016), p. 332.

[5] Zimmerman (2014), p. 2.

[6] Vankova (2016), p. 333.

[7] Hugo (2009); Constant and Zimmerman (2011).

[8] European Commission (2005), p. 7; Zimmerman (2014), p. 3.

Z. Vankova, *Circular Migration and the Rights of Migrant Workers in Central and Eastern Europe*, IMISCOE Research Series,
https://doi.org/10.1007/978-3-030-52689-4_1

natural preference for many migrants, especially when they are encouraged to circulate as a result of flexible policy frameworks.[9]

Many scholars have contested this new 'triple win solution' and its alleged positive outcomes for migrant workers, arguing that one could see it is a revival of the guest-working schemes associated with the circulation of 'labour units'[10] rather than free choice in the migration decision.[11] Judging by policy developments in recent years, one can see that there is little enthusiasm among governments for creating rights-based labour migration schemes; obstacles to accessing long-term residence and family reunification are still a reality for many migrant workers, especially those engaged in low-skilled occupations.[12] Furthermore, as Wickramasekara has argued, the claim that circular migration is a 'natural preference' for many migrants is unsubstantiated.[13] Even though crossing borders to live elsewhere is becoming a lifestyle of its own,[14] this trend very often concerns highly skilled 'global citizens' for whom special regimes are designed and does not necessary entail circularity.

Against the backdrop of the 2015 'European refugee and migrant crisis', several organisations have argued that more legal channels into the EU must be made available[15] and some have stressed that stock needs to be taken of the pathways for legal mobility that are already in existence.[16] More than 10 years after the European Commission began promoting the facilitation of circular migration, there is no clear answer as to whether this type of migration has the potential to provide a legal pathway for migrants into the EU as part of a rights-based policy solution that is beneficial for the migrant worker.[17]

Thus far, much of the literature on circular migration within the context of the EU has focused on conceptualising what is meant by the term[18] and discussing the

[9] Hugo (2013), p. 3; Newland et al. (2008).

[10] Castles and Ozkul (2014); Skeldon (2012); Wickramasekara (2011); Schneider and Wiesbrock (2011).

[11] Vankova (2016), p. 333.

[12] Ibid.

[13] Wickramasekara (2011), pp. 21–3.

[14] Faist et al. (2013), p. 7.

[15] See for instance Jackson (2015).

[16] See for instance Collett et al. (2016).

[17] Vankova (2016), p. 333.

[18] The Euro-Mediterranean Consortium for Applied Research on International Migration (CARIM), which was managed by the Robert Schuman Centre for Advanced Studies of the European University Institute, Italy, produced research examining the demographic, legal, and socio-political aspects of circular migration in the Euro-Mediterranean context for the European Commission. See for instance Fargues (2008). Moreover, the CARIM East project covered explanatory notes that examined the demographic, legal, and socio-political aspects of circular migration between Eastern Europe and the European Union and within the post-Soviet space. Available at: http://www.carim-east.eu/database/legal-module/?ls=4&ind=exnocm&lang= last (accessed 25 Sept 2017).

critical issues related to this type of migration,[19] as well as analysing the existing patterns between the EU and its neighbours,[20] rather than discussing the implementation of the EU's circular migration approach and its consequences for migrant workers.[21] Little is known about the ways in which the existing supranational and national structures and the normative frameworks they create influence the possibility for migrants to circulate within and across transnational spaces. At the same time, however, the current policy turn to circular migration policies is largely driven by the relatively recent recognition of the importance of migrant transnational practices.[22] As Vertovec has emphasised, the study of transnational migration would benefit from examining how transnational social structures and practices have emerged in the light of opportunity structures – such as visa, residency, citizenship, pension, and health care provisions – in both the sending and receiving states and how they influence migrants' own desires and strategies to conduct their transnational lives.[23]

This study aims to contribute to filling this gap by assessing the implementation of EU policies and legal instruments designed to foster circular migration and, additionally, how they affect migrant workers' rights in the context of circularity. The study focuses geographically on the Eastern neighbourhood, which comprises countries in Central and Eastern Europe (CEE) attracting migrant workers from the former Soviet Union republics. This region is an interesting case for research because it is understudied both in terms of issues related to legislation and policy of these new countries of immigration as well as in terms of implementation of the EU's circular migration approach at the national level. Furthermore, the CEE comprises both EU and non-EU countries and is characterised by an increasing cross-border migration after the latest EU enlargement.

1.2 Understanding the Term 'Circular Migration'

1.2.1 Definitions of Circular Migration

In order to assess circular migration policies, it is important to first define what is meant by this term. Previous literature review of this topic demonstrates that 'circular migration' was first referred to in the academic literature on urbanisation, rural

[19] Castles (2006); Castles and Ozkul (2014); Vertovec (2007); Pastore (2008); Skeldon (2012).

[20] Triandafyllidou (2013b).

[21] Carrera and i Sagrera (2009); Angenendt (2007).

[22] Vertovec (2007), p. 3.

[23] Vertovec (2003), pp. 653–4.

development, and internal migration in developing countries[24] in the late 1960s and early 1970s as a spontaneous pattern of mobility.[25] Zelynski defined circulation in the context of migration as 'a great variety of movements usually short term, repetitive and cyclical in nature but all having in common the lack of any declared intention of a permanent or long lasting change of residence'.[26] Later, the concept's introduction into the international policy-making discourse marked a shift in the use and understanding of circular migration – from a pattern of spontaneous movement to a potential migration policy tool.[27]

Hitherto, however, neither academics nor policy researchers have been able to arrive at one universally accepted definition of the notion of circular migration. It was not until the first Global Forum on Migration and Development in 2007 that a working definition was provided at the international level: 'Circular Migration is the fluid movement of people between countries, including temporary or more permanent movement which, when it occurs voluntarily and is linked to the labour needs of countries of origin and destination, can be beneficial to all involved'.[28] This definition, developed by the Migration Policy Institute (MPI), recognised that circular migration can take different forms and that it could be either temporary or permanent. It also included an aspirational policy element related to 'mutual benefit and voluntariness'.[29]

Skeldon further argues that circular migration is a subset of return and temporary migration, where the migrant 'engages in a regular and repetitive series of outward and return movements between an origin and a destination or destinations', and is 'free' to return at any time.[30] The author emphasises that circular migration is repetitive, regular, and involves more than one return. However, his key point is that circular migration occurs at its best when individuals are entitled to free movement across international boundaries and when it is in line with the idea of 'voluntariness' that has been developed by Newland et al.[31]

Furthermore, according to Skeldon, it is contradictory to talk about managing circular migration because this will turn it into a temporary migration programme.[32] In other words, facilitating spontaneous circular migration that already occurs is a more successful policy option than managing migration by placing restrictions on

[24] Prothero and Chapman (1985), pp. 1–26; Newland (2009), p. 5.

[25] Vankova (2016), p. 334.

[26] Zelynski (1971), pp. 225–6.

[27] Vankova (2016), p. 334.

[28] Newland and Agunias (2007), p. 4.

[29] Newland (2009), p. 7.

[30] Skeldon (2012), pp. 46–7.

[31] Newland et al. (2008), p. 7.

[32] Skeldon (2012), p. 53.

labour and human rights, length of stay, or change of employer. As demonstrated above, this is a distinctive feature in the definitions that have been given by some scholars and policymakers.[33] Cassarino is equally critical in this regard, arguing that circular migration programs build upon past temporary schemes aiming to manage labour migration, whose adoption is linked to new 'security-driven safeguards'.[34] These authors believe that the concept of managed circular migration does not differ greatly from the old guest-working models that were developed after World War II that emphasised restrictions and security in order to manage migration. For instance, according to the German model, which was notable for both its scale and its administrative framework, guest workers were to be used as 'temporary labour units'[35] to be returned when no longer needed by employers.[36] This allowed employers to have full control over the worker's status, from recruitment to deportation.

Hugo, who was among the first scholars to start using the term, stresses the elements of repetition and regularity in his definition: 'circular migration refers to repeated migration experiences between an origin and destination involving more than one migration and return'.[37] He argues that circular migration is best understood as an occurring and spontaneous migration pattern that should be facilitated and encouraged through policy measures by both destination countries and countries of origin; moreover, where permanent settlement occurred, it would not hinder circulation in the long term.[38] Such is the policy example from Sweden, which built on the idea of facilitating circulation of settled migrants back to their countries of origin in order to promote development as part of its migration policy.[39]

Many authors have also emphasised the economic characteristics of circular migration along with the element of repetition and temporariness.[40] Fargues takes a rather normative approach and underlines the essential policy ingredients: circular migration is characterised as 'temporary, renewable, circulatory, legal, respectful of

[33] See also Triandafyllidou (2013a), p. 5.

[34] Cassarino (2013), p. 23.

[35] Martin and Miller (1980), p. 316.

[36] Huhn (2011), p. 24.

[37] Hugo (2013), p. 2.

[38] Ibid.

[39] Government Bill 2013/14:213 approved by the Riksdag on 18 June 2014 allowed permanent residence permit holders who reside outside of Sweden to keep their permits for up to 2 years provided they have notified the Swedish Migration Board. It also allows for labour migrants with temporary residence permits to spend certain periods of time outside Sweden and still be able to qualify for a permanent residence permit if they have been working in Sweden for at least 4 years of the past 7 years.

[40] Newland et al. (2008); Cassarino (2008).

the migrants right, and managed in such way as to optimise labour markets at both ends, in sending and receiving countries'.[41] As to the notion of 'circulatory', the author means freedom of movement between countries of origin and destination, as per the definitions given by Skeldon and Newland et al. Fargues also emphasises the importance of migrants' rights and suggests that another important aspect of a circular migration policy could be the enhancement of skills and skill transfers.[42]

To summarise, the general understanding of circular migration is that it entails repetitive, recurrent, and temporary cross-border movement for short or long periods of residence in a country of destination and then a return to the country of origin. It is of crucial importance to stress that authors such as Skeldon, Hugo, and Newland et al. do not perceive permanent migration or settlement as counter to circular migration. Rather, they have a broad understanding of this type of migration as a fluid movement. A good example illustrating this is the case of 'circulatory transmigration' of Chinese migrants into New Zealand over the past few decades, where migrant families circulate according to the specific needs of their members at various stages of their life cycle.[43]

1.2.2 *Circular Migration in the EU Context*

In 2005, the European Commission initially defined circular migration as a spontaneous pattern: '[…] migration, in which migrants tend to go back and forth between the source country and the destination country […]'.[44] The 2007 Commission Communication conceptualised it further as 'a form of migration that is managed in a way allowing some degree of mobility back and forth between two countries', marking a shift in the Commission's approach towards the policy idea of circular migration management. In 2010, the European Migration Network was commissioned to carry out a study on circular migration.[45] Instead of applying the definition promulgated by the Commission, it proposed its own definition: 'a repetition of legal migration by the same person between two or more countries'.[46] This

[41] Fargues (2008), p. 11.

[42] Ibid, p. 2.

[43] For more details see Ip (2011).

[44] European Commission (2005), Annex 5, p. 25.

[45] This study aimed 'to illustrate different policy preferences and approaches to temporary and circular migration, and to provide evidence of their characteristics, as well as to identify lessons learned, best practices and possible policy options, which could be further explored at national and EU political levels'. Thus, the EMN study also responded to 'the request from the Council, through its Council Conclusions and the Stockholm Programme, regarding further exploration and development of circular migration as an integral part of EU migration policy'. In European Migration Network (2011), p. 1.

[46] Ibid., p.14.

definition stressed that circular migration takes place through legal channels and that circularity may occur between more than two countries.

This ambivalence in defining circular migration at the EU level required data collection through interviews in order to further clarify the meaning of the term.[47] One of the interviewees in this study, who participated in the introduction of this concept in the EU's policy agenda, shared that the concept was deliberately left broad so as to capture a wider policy context and instruments based on different Treaty articles.[48] The term 'mobility' was not only used to forge an explicit link to the EU visa policy but also to denote the circular 'back and forth' movement as well as 'to bring in the concept of dynamics'.[49] Another interviewed official emphasised that different stakeholders would associate circular migration with something that they were familiar with, such as for example, guest working schemes and seasonal work.[50]

An interviewed Council official said that, according to his personal understanding, circular migration referred to a scheme of temporary migration.[51] It was mostly used as a tool to achieve 'a status of protected people' so that migrants could keep their work permits and use them again when they go back to the country of destination.[52] Circular migration was also seen as an incentive against 'illegality'. However, for the interviewee, the biggest question was whether one could control circular migration patterns, which by their very nature were spontaneous acts. He associated managed circular migration with the creation of schemes that seemed complicated and beset with problems.[53] Another interviewee working at an international organisation shared the view that EU policy discussions on circular migration were, indeed, focused on establishing temporary migration schemes.[54]

A European Commission official confirmed that '[…] we want to have well-managed migration and therefore when we talk about circular migration, we have to put in a management scheme'.[55] On a personal level, however, the official believed in the model that facilitated spontaneous circular migration, even though this model could not be promoted among Member States because 'spontaneous' sounded like 'uncontrolled'.[56] According to the interviewee: '[…] for this reason, people have preference for a more organised system, even though in reality, it hardly happens'.[57]

[47] For more details on the study's methodology see Chap. 2.

[48] Interview # 29 with European Commission official, Belgium, May 2017, Annex I.

[49] Interview # 29 with European Commission official, Belgium, May 2017, Annex I.

[50] Interview # 11 with European Commission official, Belgium, May 2013, Annex I.

[51] Interview # 8 with EU Council official, Belgium, February 2013, Annex I.

[52] Interview # 8 with EU Council official, Belgium, February 2013, Annex I.

[53] Interview # 8 with EU Council official, Belgium, February 2013, Annex I.

[54] Interview # 28 with representative of international organisation, Austria, March 2017, Annex I.

[55] Interview # 9 with European Commission official, Belgium, March 2013, Annex I.

[56] Interview # 9 with European Commission official, Belgium, March 2013, Annex I.

[57] Interview # 9 with European Commission official, Belgium, March 2013 Annex I.

One of the interviewed academics underlined that this was a concept 'especially established to look nice on paper: it allows the policy makers to show the voters that it is something that it is manageable, so it is a tool.[58] If you have a tool, it means that you can manage a phenomenon'. According to this interviewee, there were Member States such as Germany, Austria, the Netherlands, and France where one could not 'tell the voters that you are going to have temporary migration scheme, but when they were given this beautiful package of circular migration, they adopted it and moved it up on the European agenda'.[59] In addition, this academic noted, it was also developed to attract funding from the donors: 'the donors would never give money for temporary migration schemes but it is a different thing for circular migration. In the end, it is exactly the same'.[60]

As this section demonstrates, there is no uniform understanding of the policy concept of circular migration in the EU context either. The understanding of this notion by EU actors differs in terms of the approach to managing this type of migration – by establishing temporary schemes or facilitating spontaneous patterns – as well as concerning the aims of these measures and the spectrum of mobility – between two or more countries. This poses a conceptual challenge and for the purposes of this study requires the adoption of a working definition that would guide the assessment of the implementation of the EU's approach to circular migration. In order to do that, one needs to further distinguish circular migration from other types of migration and also look into the different patterns of circular migration.

1.2.3 Towards a Working Definition of Circular Migration

Typologies of patterns of circular migration naturally vary depending on each author's approach to defining circular migration. Newland et al., for instance, outline three patterns of circular migration, differentiating between the character of the movement and the level of the migrant's skills: seasonal; non-seasonal and low-wage; and, mobility of professionals, knowledge workers, and transnational entrepreneurs.[61] In his research on circular migration in the Euro-Mediterranean area, Cassarino provides an alternative typology of circular migration patterns, which are shaped by the migrant's mobility strategy (migrant's agency) as well as by changing circumstances and structural factors such as state migration and border policies.[62] According to this typology, patterns of circular migration can be hindered, embedded, and regulated.

58 Interview # 12 with academic, Italy, May 2013, Annex I.

59 Interview # 12 with academic, Italy, May 2013, Annex I.

60 Interview # 12 with academic, Italy, May 2013, Annex I.

61 Newland et al. (2008), p. 3.

62 Cassarino (2008), pp. 2–7.

Both typologies capture the intrinsic characteristics of circular migration which can only demonstrate the variety of circular movements if they are examined together. Patterns of circular migration can be differentiated based on the seasonality (or not) of the movement, migrants' skill levels, and whether the circular migration pattern is regulated or embedded (non-regulated). The only category that does not fit within this joint model is the hindered pattern, which as Wickramasekara stresses, is not a separate category.[63] The hindered pattern does not actually describe a different type of movement, but rather it presents the factors that could drive circular movement to an end – and which are applicable to both embedded and regulated types. Only by joining the two typologies can one arrive at a variety of circular patterns that, in turn, give a better picture of the diversity that is redolent of circular migration.

Nevertheless, as Skeldon underlines, 'the identification of a constant, clearly identifiable form of migration that can be called "circular" is problematic', especially across borders.[64] The author believes that circular movements are prone to change their duration and composition over time.[65] Therefore, another fact that poses additional challenges to the conceptualisation of circular migration is the difficulty of separating it from other forms of migration, such as temporary and return migration.[66] Furthermore, 'leakage', as Skeldon refers to permanent settlement in the country of destination, is an inherent characteristic of circular migration.[67] Despite these challenges, and for the sake of clarity when arriving at a working definition for this study, it is important to attempt to distinguish between the typical features of circular migration and other types of migration and, furthermore, to demonstrate how they relate to each other.

Circular vs. Permanent Migration Permanent migration involves migration to another country for long-term or permanent settlement. This type of migration, however, does not exclude circularity. Circular migration can 'leak' into permanent migration and resume again at some point since there are long-term or permanently settled migrants who engage in circularity to their county of destination. In order to illustrate the blurred line between these types of migration, scholars have coined the terms 'intermittent circulars'[68] and 'circular transmigrants' in an attempt to distinguish the varying trajectories and experiences of engagement that could span a lifetime and should be seen as holistic circular migration projects.[69] Actually, more and more scholars consider that in the age of globalisation, which provides for so much accessible communication and transport, this naturally means that 'all migration is to some extent circular'.[70]

---

63 Wickramasekara (2011), p. 10.

64 Skeldon (2012), p. 55.

65 Ibid.

66 Vankova (2016), p. 335.

67 Skeldon (2012), p. 53.

68 See Triandafyllidou (2013a).

69 See Górny (2017).

70 Pastore (2008), p. 3.

Circular vs. Temporary Migration The terms temporary and circular migration are frequently used interchangeably[71] in policy circles. It is, therefore, important to note salient differences. The main distinction between these two types of migration is that temporary migration 'involves a one-time only temporary stay and eventual return which closes the migration cycle'[72] whereas circular migration involves recurrent temporary movements after the initial return. In line with that premise, Newland et al. argue that 'circular migration is distinct from temporary migration in that circular migration denotes a migrant's continuous engagement in both home and adopted countries'.[73] Therefore, circular migration is also seen as being closely connected to transnationalism, featuring migration as well as return to the country of origin.[74] For instance, a study of Moroccan immigrants in the Netherlands demonstrates that their integration in the host country does not lead to less engagement with their country of origin.[75] They continue to remit as well as transfer skills and knowledge (including through return visits) when they have the capacity to do so and exhibit higher levels of attachment to their country of origin.

Furthermore, both terms relate to temporary movement and stay, thus raising the following question: how long is temporary? According to the working definition given by the EURA-NET project that focuses on the impact of temporary migration, this type of movement can last between 3 months and up to 5 years.[76] The variations in the range of 'human mobility experiences' within this spectrum depend on the different categories of migrants.[77] For instance, persons who stay in the EU for a period shorter than 3 months are involved in short-term circularity, whereas at the other end of the spectrum – migrants spending more than 5 years outside their home country – are considered to be long-term residents settled more permanently abroad.[78] Hence, temporary migration excludes any permanent stays.

In contrast to those temporary migration stays that do not last for more than 5 years, the circular migration cycle has no formal beginning or end. It could begin for less than 3 months in the case of seasonal migration or continue after 5-year-long periods, when a third-country national qualifies for a long-term residence status, and even last until the person becomes eligible for naturalisation and is granted citizenship. Therefore, circular migration can encompass both temporary and longer-term stays. The distinctive feature is that circular migrants spend significant periods of their lives in both their countries of origin and destination.[79]

[71] Or as Triandafyllidou notes in a 'slash fashion "temporary/circular"'. In Triandafyllidou (2013a), p. 4.

[72] European Migration Network (2011), p. 21.

[73] Newland et al. (2008), p. 2.

[74] Triandafyllidou (2013a), p. 5; Vertovec (2007), pp. 2–3.

[75] Bilgili (2014), p. 183.

[76] Carrera et al. (2014), p. 18.

[77] Aksakal and Schmidt-Verkerk (2015), p. 5.

[78] Ibid.

[79] Hugo (2013), p. 2.

Circular vs. Return Migration Usually the term 'return migration' refers to a one-off movement back to the country of origin. Nevertheless, as was the case with the example of permanent migration, return does not exclude circularity. There are migrants who live abroad and permanently return to their countries of origin, but there are also migrants who return temporarily and continue to circulate. Thus, the main difference with circular migration is that one-off return migration closes the migratory cycle and circular migration 'implies more than just a single out-and-return movement'.[80]

Circular vs. Labour Migration Circular migration occurs primarily for economic reasons such as higher wages, economic survival, and better working conditions, and can thus be considered as being predominantly labour migration.[81] As mentioned above, circular migrants can have different skill levels and work in a variety of different sectors. Moreover, it is important to distinguish it from cross-border commuting as another type of economic circulation that is not considered circular migration. Commuting involves a return to the country of origin within the same day or within the working week and therefore does not involve long stays abroad.

Circular vs. Other Types of Migration That Are Circular in Nature It is important to differentiate circular migration from shuttle and pendular migration,[82] which are also referred to as incomplete migration.[83] These types of migration emerged in the post-1989 period between the CEE countries. They possess a 'quasi-migratory character' and denote the movement of 'incomplete migrants' that does not last for more than 3 months after which they return home, having usually entered as 'false tourists' on a short-term visa and informally engaged in petty trade commerce or interregional trading.[84] For example, this was the strategy used predominantly by Ukrainians in Poland.[85] However, the main difference with circular migration is that this type of movement falls within the ambit of short-term mobility and involves migrants working irregularly on the basis of a tourist visa. Finally, 'liquid' migration'[86] is a concept applied to intra-EU mobility and not easily applicable to third-country nationals facing legal barriers.[87]

Circular Migration vs. Mobility Traditionally, the term 'mobility' in the EU context was reserved for the free movement of persons. After 2006, however, with the

---

[80] Skeldon (2012), p. 46.

[81] Triandafyllidou (2013a), p. 12.

[82] This was first introduced by Morokvasic (1992), pp. 31–36. See also Iglicka (2000), pp. 61–73.

[83] Okólski (2004), p. 44; Kaczmarczyk and Okólski (2005), pp. 18–20.

[84] Okólski (2004), p. 44. This strategy was also used in the domestic sector. See for instance, Marchetti (2013).

[85] Iglicka and Gmaj (2010).

[86] Engebersen et al. (2010).

[87] Górny (2017), p. 3.

publication of the Commission Communication 'Global Approach to Migration one year on',[88] the term 'mobility' began being used in the context of the external dimension of the EU's migration policy as well, denoting cross-border mobility. As Mananshivili stresses 'the boundaries of this concept have not been demarcated yet, and accordingly, its precise legal definition is so far absent at the EU level'.[89] Since this is an implementation study, the usage of both terms in the context of this book follows the EU's usage.

Despite the conceptual challenges, this section of the book has outlined some of the key characteristics of circular migration that will form the basis of a working definition: it is legal labour migration occurring through legal channels; it is repeated migration, involving more than one outward movement and return; and it is temporary migration encompassing both temporary and long-term stays. To sum up, this study adopts the following working definition: circular migration involves legal temporary migration of economically active migrants, moving repeatedly between their countries of destination and their countries of origin.[90] This definition is left broad on purpose, following Skeldon's understanding that a clear-cut identification of circular migration is difficult as it changes over time and is prone to leakages into permanent migration that are not considered opposite but rather interconnected types of migration.

1.3 Analytical Tools

This book takes as a starting point the rights-based benchmark framework for assessment of circular migration policies that was developed for this study.[91] Previous work on this topic suggests that the key policy areas and problematic issues that need to be considered if this type of circular migration is to be facilitated are entry and re-entry conditions, work authorisation, residence status, and social security coordination.[92] In addition, this study focuses on two policy areas that are considered secondary – entry conditions for family members and recognition of qualifications – but which could influence the willingness of migrants to engage in circular migration. They are illustrative of the diversity of issues associated with this type of migration. As already argued elsewhere, family reunification might not be an issue for a seasonal worker engaged in circular migration for less than 3 months.[93]

[88] European Commission (2006).

[89] Neither relevant EU legal acts nor the EMN glossary entail the explanation of what might be meant under 'mobility'. In S. Mananshvili, 'What is EU's understanding of Mobility?' (unpublished note).

[90] Vankova (2016), p. 336.

[91] Ibid., pp. 332–52.

[92] Ibid., pp. 336–9.

[93] Vankova (2016), p. 338.

By contrast, it could become an insurmountable obstacle for a worker who embarks on a 1-year contract and wishes to bring family members along. Finally, it is important to examine the issues arising from rules on recognition of qualifications; advocates of circular migration claim that it enables 'brain circulation and skill transfer, yet there is evidence that this is not always the case.[94]

The developed analytical framework aims to assess whether these policy areas provide a 'win' for the migrant worker within the context of the 'triple win solution' that circular migration ostensibly offers. The premises thereof are, firstly, whether the migrant has a certain degree of voluntarism and 'free' movement or generally a free choice in the migration decision. This condition differentiates circular migration policies from general time-bound migration policies that are redolent of the guest-worker models.[95] In other words, if policymakers want to implement effective circular migration policies, they need to design them in such a way so as to accommodate the migrants' transnational links with both the country of origin and destination, as well as to allow the possibility for migrants to determine their own trajectory.[96] Secondly, what needs to be assessed is whether these policies adequately protect the migrant workers' fundamental rights and rights that allow them to benefit from the circulation, such as export of social security benefits when they return home and provisions to ensure that their qualifications can be recognised.

In line with these premises, a two-level benchmark[97] framework for assessing circular migration policies has been developed.[98] On the primary level, international and European standards developed in the field of human rights, migration, labour, and social security law are identified as benchmarks in the study's key policy areas. Only those provisions that cover the two premises of voluntarism and protection of migrants' rights are included within the benchmark framework, meaning that they are employed as aspirational standards against which circular migration policies can be assessed.

On the secondary level, the benchmark framework includes policy instruments that can help in the implementation of these benchmarks. They are identified as being conducive to circular migration management on the basis of a literature review of the lessons that have been learned from the application of similar time-bound labour migration policies, such as the experience with guest-worker schemes as well as good practices that have been identified among the emerging new generation of circular migration programmes.[99]

Each of the chapters containing analysis of the rights of migrants in the six policy areas at stake with circular migration commences with a presentation of the

---

[94] Hooper and Sumption (2016), pp. 20–1.

[95] Vankova (2016), p. 339.

[96] Newland et al. (2008), p. 1; Newland and Agunias (2007), p. 11.

[97] The English Oxford Dictionary defines 'benchmark' as 'a standard or point of reference against which things may be compared'.

[98] Vankova (2016), pp. 339–50.

[99] Ibid.

benchmarks that are applied as analytical tools to assess existing policies. A summary of the benchmarks can also be found in Annex V.

1.4 Focus

With regard to the presented definition of circular migration and the policy issues at stake with this type of migration, this study discusses the formulation of the legal and policy instruments of EU migration policy that aim to foster circular migration or which incorporate elements of circular migration (referred to hereinafter as the EU's approach to circular migration) as well as their implementation at the national level in CEE. It focuses on the rights of migrant workers in six policy areas – namely entry and re-entry conditions, work authorisation and residence status, social security coordination, entry and residence conditions for family members, and recognition of qualifications – in order to assess whether the EU is fostering rights-based circular migration.

This book examines what has been achieved in relation to the two categories of circular labour migration that were targeted by the EU: temporary engagement of EU settled third-country nationals returning to their countries of origin and temporary opportunities for entry and re-entry for persons residing in a third country for the purposes of working in the EU.[100] The first category covers third-country nationals who reside in one of the Member States and who travel back and forth to their country of origin in order to engage in some form of professional activity. This category refers mainly to diaspora members who decide to circulate.

The second category incorporates those third-country nationals who reside outside of the EU but temporarily engage in a professional activity within a Member State and afterwards return to reside in their country of origin. According to the European Commission, circularity in this case is to be achieved on the basis of simplified re-entry procedures to and from the Member States where the migrants were temporarily engaged professionally.[101] Unlike the common understanding that circular migration concerns predominantly low-educated migrants who rely on their vocational skills,[102] this EU category covers a broad range of migrant categories engaged in employment, research, study, intercultural exchanges, and voluntary work. As demonstrated above, in most cases migrants engage in work-related circular migration. Therefore, this study focuses on economically-driven circular migration and excludes migration where the main purpose is study, intercultural exchange, or volunteering.

These two categories of circular migration are to be facilitated at the European level based on a two-fold approach. Firstly, the Commission planned to promote it through a legislative framework by using existing legal migration instruments and

[100] European Commission (2007), pp. 8–9.

[101] Ibid.

[102] Constant and Zimmerman (2011), pp. 495–515.

introducing special measures in future legislative acts. Secondly, it has been incorporated as a policy instrument within the context of the Global Approach to Migration and Mobility (GAMM), which is the overarching framework of the EU external migration and asylum policy. The European Commission has planned to facilitate the development of circular migration schemes with third countries within the framework of the GAMM. This raises the question of what kind of migrant categories this two-fold approach covers.

The first part of the EU's approach to circular migration covers the legal migration directives adopted before the circular migration concept was introduced in 2005, such as the EU Long-term Residence Directive,[103] as well as the Researchers'[104] and Students' Directives[105] which were repealed and replaced by the new Students' and Researchers' Directive.[106] It also includes legal instruments that were adopted after 2005 and featured in the gradual development of the circular migration approach at the EU level: the Blue Card Directive,[107] the Seasonal Workers' Directive,[108] the Single Permit Directive,[109] and the Intra-corporate Transferee's Directive.[110] Thus, this legal approach to circular migration encompasses the sectoral legal migration framework that has already been promulgated at the EU level and which regulates the conditions of entry and residence for different categories of immigrants, such as highly-skilled workers, seasonal workers, students and researchers, family migrants, and EU long-term residence holders. The second part of the EU's approach to circular migration, within the framework of the GAMM, does not limit the spectrum of migrant categories that can participate in circular migration schemes.

---

[103] Council Directive 2003/109/EC of 25 November 2003 concerning the status of third-country nationals who are long-term residents [2004] OJ L 16.

[104] Council Directive 2005/71/EC of 12 October 2005 on a specific procedure for admitting third-country nationals for the purposes of scientific research [2005] OJ L 289.

[105] Council Directive 2004/114/EC of 13 December 2004 on the conditions of admission of third-country nationals for the purposes of studies, pupil exchange, unremunerated training or voluntary service [2004] OJ L 375/12.

[106] Directive (EU) 2016/801 of the European Parliament and of the Council of 11 May 2016 on the conditions of entry and residence of third-country nationals for the purposes of research, studies, training, voluntary service, pupil exchange schemes or educational projects and au pairing [2016] OJ L 132.

[107] Council Directive 2009/50/EC of 25 May 2009 on the conditions of entry and residence of third-country nationals for the purposes of highly qualified employment [2009] OJ L 155/17.

[108] Directive 2014/36/EU of the European Parliament and of the Council of 26 February 2014 on the conditions of entry and stay of third-country nationals for the purpose of employment as seasonal workers [2014] OJ L 94.

[109] Directive 2011/98/EU of the European Parliament and of the Council of 13 December 2011 on a single application procedure for a single permit for third-country nationals to reside and work in the territory of a Member State and on a common set of rights for third-country workers legally residing in a Member [2011] OJ L 343.

[110] Directive 2014/66/EU of the European Parliament and of the Council of 15 May 2014 on the conditions of entry and residence of third-country nationals in the framework of an intra-corporate transfer [2014] OJ L 157.

In order to analyse the whole spectrum of economically-active migrants who can engage in circular migration, this study focuses on migrants in both low-skilled (e.g., seasonal) and highly-skilled occupations (e.g., Blue Card holders, researchers, intra-corporate transferees). As outlined above, circular migration can also include migrants possessing a more permanent status. Indeed, the EU's approach to circular migration aims to also facilitate such migration for third-country nationals who have settled in a Member State and are willing to go back to their countries of origin. Hence, it is crucial to include EU long-term residence holders within the scope of this study. Any such analysis would be incomplete if it were to exclude the family members of third-country nationals who move between two countries for different periods since this type of migration can have an impact on the family life of the migrants. What should be kept in mind, however, is that in reality these categories are not constant and there are shifts from one form of migration status to another throughout the lives of migrants as they fall in love, reunite with families, settle, or decide to restart their engagement in circular migration. Therefore, this study also attempts to take account of these shifts in migration statuses.

To conclude, the purpose of this study is to answer the core research questions: how has the EU's approach to circular migration been implemented through its legal and policy instruments and does it provide for rights-based circularity for migrant workers in the CEE context? In order to answer that question, this study covers within its scope economically-active third-country nationals in both low- and high-skilled occupations with temporary and permanent statuses as well as their family members who are engaged in circular migration to and from the EU.

1.5 Structure

This book is organised into nine chapters. Chapter 2 presents the employed methodological framework of this interdisciplinary research. It discusses the case study selection by outlining the differences and similarities in terms of contextual factors and legal and policy issues. This chapter concludes with a focus on the data sources used and the ethical considerations at the heart of this study.

Chapters 3 and 4 place circular migration into its European legal and policy context by focusing on the process of formulation and implementation of the EU's circular migration approach respectively. Chapter 3 starts by examining the genesis of the circular migration notion in the EU's migration policy and outlines the formulation of a two-fold approach towards its facilitation at the EU level, incorporating legal migration directives as well as policy initiatives developed under the auspices of the GAMM. Chapter 4, then, brings together the legal and policy instruments developed as part of the approach at the EU level in order to assess its implementation and establish whether it provides rights-based outcomes for migrant workers according to the developed benchmark framework for analysis.

Chapters 5, 6, 7, and 8 focus on national variances of instruments conducive to circular migration as part of this implementation study on the EU's approach. Chapter 5 is an introduction to the national approaches to circular migration facilitation developed in Bulgaria and Poland, which are chosen as case studies of this book. Chapter 6 brings together the developed EU and national instruments in Bulgaria and Poland conducive to circular migration and assesses their implementation against the background of the developed benchmarks concerning entry and re-entry conditions for migrant workers. Chapter 7 focuses on EU and national instruments in the policy areas of work authorisation, residence status, and social security coordination which are considered key policy areas that need to be taken into account if this type of migration is to be facilitated. Finally, Chap. 8 assesses entry and residence conditions for family members and recognition of qualifications, which despite being secondary policy areas could still influence the willingness of migrants to engage in circular migration.

The above four chapters are based on legal analysis of national and transposed EU instruments, updated as of September 2019. In addition to this layer of legal analysis, these chapters incorporate insights into the implementation dynamics of the EU and national instruments conducive to circular migration on the basis of data collected through focus groups with migrants from Ukraine and Russia in 2016 (before the introduction of a visa-free regime with Ukraine); interviews conducted with stakeholders in the period 2013–2017; data on permits retrieved from the national administrations of the two countries chosen for case studies as of mid-2019 (where possible); and, available data in recent studies.

The book culminates with Chap. 9, which returns to the study's main research question, namely: how has the EU's approach to circular migration been implemented through its legal and policy instruments and does it provide for rights-based circularity for migrant workers in the CEE context. It concludes by proposing policy recommendations to this end.

References

Aksakal, M. & Schmidt-Verkerk, K. (2015). Conceptual framework on temporary migration. Report on conceptual clarifications for the EURA-NET project.

Angenendt, S. (2007). Circular migration: A sustainable concept for migration policy?. SWP Comments 11, Stiftung Wissenschaft and Politik, Berlin.

Bilgili, Ö. (2014). Simultaneity in transnational migration research: Links between migrants' host and home country orientation. PhD thesis, Boekenplan, Maastricht.

Carrera, S., Eisele, K. & Guild, E. (2014). European and international perspectives and standards on temporary migration. In P. Pitkanen & S. Carrera (Eds.), *Transnational migration in transition: State of the art report on temporary migration.* Collected working papers from the EURA-NET project. Tampere: University of Tampere.

Carrera, S. & Hernández i Sagrera, R. (2009). The externalisation of the EU's labour immigration policy. Towards mobility or insecurity partnerships?. Working document 321. Brussels: CEPS.

Cassarino, J. P. (2008). Patterns of circular migration in the Euro-Mediterranean area. Implications for policy-making. *CARIM Analytic and Synthetic Notes* 29, Circular Migration Series, Political and Social Module.

Cassarino, J. P. (2013). The drive for securitized temporariness. In A. Triandafyllidou (Ed.), *Circular migration between Europe and its neighbourhood: Choice or necessity?* (pp. 22–41). Oxford: Oxford University Press.

Castles, S. (2006). Guestworkers in Europe: A resurrection? *International Migration Review, 40*(4), 741–766.

Castles, S., & Ozkul, D. (2014). Circular migration: Triple win, or a new label for temporary migration? In G. Battistella (Ed.), *Global and Asian perspectives on international migration. Global Migration Issues* (pp. 27–49). Cham: Springer.

Collett, E., Clewett, P., & Fratzke, S. (2016). *No way out? Making additional migration channels work for refugees*. Brussels: Migration Policy Institute Europe.

Constant, A., & Zimmermann, K. (2011). Circular and repeat migration: Counts of exits and years away from the host country. *Population Research and Policy Review, 30*(4), 495–515.

Engbersen, G., Snel, E., & De Boom, J. (2010). "A van full of Poles": Liquid migration from Central and Eastern Europe. In R. Black, G. Engbersen, M. Okólski, & C. Pantîru (Eds.), *A continent moving West? EU enlargement and labour migration from Central and Eastern Europe* (pp. 115–140). Amsterdam: Amsterdam University Press.

European Commission. (2005). Communication on migration and development: Some concrete orientations, COM (2005) 390 final, Brussels, 1 September 2005.

European Commission. (2006). Communication on the Global Approach to Migration one year on: Towards a comprehensive European migration policy, COM (2006) 735, Brussels, 30 November 2006.

European Commission. (2007). Communication on circular migration and mobility partnerships between the European Union and third countries, COM (2007) 248 final, Brussels, 16 May 2007.

European Migration Network. (2011). Temporary and circular migration: Empirical evidence, current policy practice and future options in EU member states. Synthesis Report, Brussels.

Faist, T., Fauser, M., & Reisenauer, E. (2013). *Transnational migration*. Cambridge: Polity Press.

Fargues, F. (2008). Circular migration: Is it relevant for the south and east of the Mediterranean? *CARIM Analytic and Synthetic Notes, 40*.

Górny, A. (2017). All circular but different: Variation in patterns of Ukraine-to-Poland migration. *Population, Space and Place, 23*(8), e2074.

Hooper, K., & Sumption, M. (2016). *Reaching a 'Fair' deal on talent. Emigration, circulation and human capital in countries of origin*. Washington DC: Migration Policy Institute.

Hugo, G. (2009). Circular migration and development: An Asia-Pacific perspective. In O. Hofírek, R. Klvaňová, & M. Nekorjak (Eds.), *Boundaries in motion: Rethinking contemporary migration events* (pp. 165–180). Brno: Centre for the Study of Democracy and Culture.

Hugo, G. (2013). What we know about circular migration and enhanced mobility. *Policy Brief* No 7. Migration Policy Institute, Washington DC, September 2013.

Huhn, K. (2011). Arbeitsplaz Deutschland, Heimat Türkei? Die Anwerbung von Arbeitskräften aus der Türkei im Kontext der bundesdeutschen Ausländerbeschäftigungspolitik. Ein Policy Paper mit Empfehlungen für die künftige Gestaltung der Zuwanderung im Auftrag der Bertelsmann Stiftung.

Iglicka, K. (2000). Mechanisms of migration from Poland before and during the transitional period. *Journal of Ethnic and Migration Studies, 26*(1), 61–73.

Iglicka, K & Gmaj, K. (2010). Circular migration patterns. Migration between Ukraine and Poland. METOIKOS Project Robert Schuman Centre for Advanced Studies Research Paper. Florence: European University Institute.

Ip, M. (2011). Rethinking contemporary Chinese circulatory transmigration: The New Zealand case. In *Transmigration and the new Chinese* (pp. 21–56). Hong Kong: Hong Kong Institute for the Humanities and Social Sciences, University of Hong Kong.

Jackson, G. (2015). UN's François Crépeau on the refugee crisis: "Instead of resisting migration, let's organise it". The Guardian, 22 April 2015. https://www.theguardian.com/world/2015/apr/22/uns-francois-crepeau-on-the-refugee-crisis-instead-of-resisting-migration-lets-organise-it?CMP=share_btn_tw Accessed 16 August 2016.

Kaczmarczyk, P. & Okólski, M. (2005). *International migration in Central and Eastern Europe: Current and future trends*. Report for the United Nations Expert Group meeting on International Migration and Development UN/POP/PD/2005/12.

Marchetti, S. (2013). Dreaming circularity? Eastern European women and job sharing in paid home care. *Journal of Immigrant and Refugee Studies, 11*(4), 347–363.

Martin, P. L., & Miller, M. J. (1980). Guestworkers: Lessons from Western Europe. *Industrial and Labour Relations Review, 33*(3), 315–330.

Morokvasic, M. (1992). Une migration pendulaire: les Polonais en Allemagne. *Hommes et Migrations, 1155*, 31–36.

Newland, K. (2009). Circular migration and human development. *UNDP Human Development Reports, 42*.

Newland, K. & Agunias, D. (2007). How can circular migration and sustainable return serve as development tools? Background paper for Roundtable 1.4, Global Forum on Migration and Development, Brussels.

Newland, K., Mendoza, D. R., & Terrazas, A. (2008). Learning by doing: Experiences of circular migration. *Migration Policy Institute Insight,* Washington DC, September, 2008.

Okólski, M. (2004). The effects of political and economic transition on international migration in Central and Eastern Europe. In J. E. Taylor & D. S. Massey (Eds.), *International migration. Prospects and policies in a global market* (pp. 35–58). Oxford: Oxford University Press.

Pastore, F. (2008). Circular migration. Background note for the meeting of experts on legal migration, Rabat, 3–4 March.

Prothero, R., & Chapman, M. (Eds.). (1985). *Circulation in Third World countries*. London: Routledge & Kegan Paul.

Schneider, H., & Wiesbrock, A. (2011). Circular migration: A triple win situation? Wishful thinking or a serious option for suitable migration policy? In D. Schiek, U. Liebert, & H. Schneider (Eds.), *European economic and social constitutionalism after the Treaty of Lisbon*. Cambridge: Cambridge University Press.

Skeldon, R. (2012). Going round in circles: Circular migration, poverty alleviation and marginality. *International Migration, 50*(3), 43–60.

Triandafyllidou, A. (2013a). Circular migration: Introductory remarks. In A. Triandafyllidou (Ed.), *Circular migration between Europe and its neighbourhood: Choice or necessity?* (pp. 1–21). Oxford: Oxford University Press.

Triandafyllidou, A. (Ed.). (2013b). *Circular migration between Europe and its neighbourhood: Choice or necessity?* Oxford: Oxford University Press.

Vankova, Z. (2016). EU circular migration policies: Dead or alive? Developing a rights-based benchmark framework for policy assessment. *Journal of Immigration, Asylum and Nationality Law, 30*(4), 332–352.

Vertovec, S. (2003). Migration and other modes of transnationalism: Towards conceptual cross-fertilization. *International Migration Review, 37*(3), 641–665.

Vertovec, S. (2007). *Circular migration: The way forward in global policy?* International Migration Institute Working Papers, No. 4.

Wickramasekara, P. (2011). Circular migration: A triple win or a dead end. *Discussion Paper No 15*. The Global Union Research Network.

Zelinsky, W. (1971). The hypothesis of the mobility transition. *Geographical Review, 61*(2), 219–249.

Zimmermann, K. (2014). Circular migration. Why restricting labour mobility can be counterproductive. *IZA World of Labor, 1*. https://doi.org/10.15185/izawol.1.

Chapter 2
Methodological Framework

This chapter aims to shed light on the methodological framework employed in this interdisciplinary research. It presents the case study selection by outlining the differences and similarities in terms of contextual factors and legal and policy issues. This chapter concludes with a focus on the data sources used and the ethical considerations at the heart of this study.

2.1 Interdisciplinary Research Design

This book combines international, European, and national law and implementation evaluation as part of empirical legal research studies[1] in order to answer the question of how the EU's circular migration approach has been implemented and how this has affected the rights of migrant workers. It focuses on the process of implementation as an object of legal evaluation and not on the impact and effectiveness of the approach under investigation.[2] In line with that focus, the study first examines how the circular migration approach has been translated into policy and then traces the policy formulation process that has taken place at the EU level. Secondly, it focuses on how this policy is implemented at both the EU and national levels by analysing policy outputs: the activities established to execute the policy (e.g., laws, decisions, and so on). Finally, the study also examines the outcomes of the circular migration policy in order to assess the practical challenges that are related to the rights of migrant workers in the context of circular migration. Tracing the policy formulation and implementation thereof requires a three-level analysis: policy formulation at the

[1] See Leeuw and Schmeets (2016), p. 6; p. 35.

[2] See Ibid., p. 46.

Z. Vankova, *Circular Migration and the Rights of Migrant Workers in Central and Eastern Europe*, IMISCOE Research Series,
https://doi.org/10.1007/978-3-030-52689-4_2

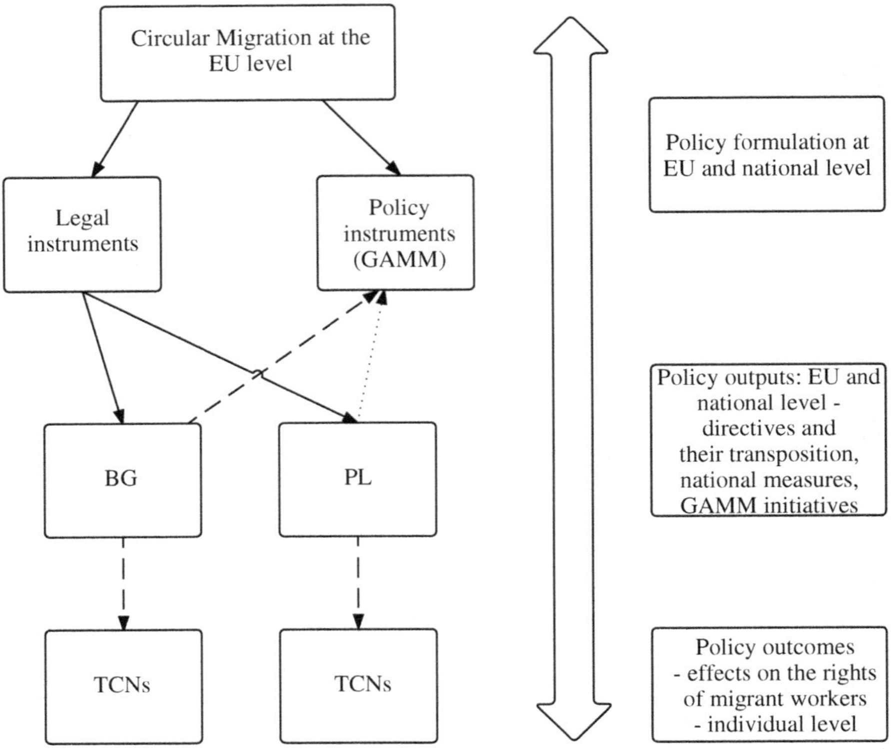

Fig. 2.1 Three-level framework for analysis – formulation and implementation of EU's circular migration approach through legal and policy channels. Source: Author's own elaboration

EU (and national) level, policy outputs developed at the EU and national levels, and policy outcomes measured at the individual level (see Fig. 2.1).

As discussed, the circular migration approach at the EU level has been formulated very broadly, consisting of both legal and policy channels for implementation that also allows, to a certain extent, policy co-formulation from Member States at the national level. This occurs mainly through the policy formulation process within the Global Approach to Migration and Mobility (GAMM), whereby countries can initiate circular migration projects. As a result of these two channels for implementing the circular migration policy, there are two main types of outputs that can be identified: legal instruments developed at the EU as well as at the national level and the policy instruments developed as part of the GAMM.

The legal outputs developed at the EU level require implementation, in most cases, at the national level within a margin of appreciation that is given to the Member States, which can additionally have the effect of reshaping these outputs. Furthermore, those cases where EU law has not been correctly transposed also need to be considered. In addition, the legal outputs developed at the national level, such as bilateral agreements, will also be examined in this study. Finally, different policy outputs – such as circular migration initiatives and bilateral social security agreements – can be identified as a result of Member States' participation in the GAMM, such as for example in Mobility Partnerships, which are one of the 'principle bilateral frameworks for facilitating policy dialogue and operational cooperation with partner countries'.[3]

In order to evaluate the implementation of the formulated policies and established outputs at both the EU and national levels, this study also examines a third level – the individual level – where the outcomes of circular migration policy can be assessed. Taking the policy outcomes into consideration makes it possible to analyse whether migrant workers benefit in practice from rights-based circular migration.

2.2 Comparative Case Study Approach

This study employs a comparative case study research design as part of the legal empirical research methodology. The case study approach is considered the most suitable because it allows for the developed policy outputs at the national level to be examined in their specific legal systems or contexts, and it also provides for a 'detailed consideration of contextual factors'.[4] Furthermore, by employing a comparative strategy to the case study approach, one can examine the differences in the transposition of the EU legal instruments that have a bearing on circular migration, the establishment of national instruments as part of the legal frameworks of each country, and the implementation of circular migration through the GAMM policy channels.

According to Yin, the case study method is 'an empirical study that investigates a contemporary phenomenon in depth within its real-world context, especially when the boundaries between the phenomenon and context may not be clearly evident'.[5] Another advantage of the case study method is that it is able to 'test views directly in relation to phenomena as they unfold in practice'.[6] It can be employed as an umbrella strategy for conducting research that is capable of utilising a range of different research methods and data sources.[7]

The selection of the case is the most critical step in undertaking any case study research.[8] There are single case studies that are typically used for anthropological studies and multiple case studies that are more often used in the ambit of

[3] European Commission (2014), p. 2. The other principal framework is the Common Agendas on Migration and Mobility (CAMMs). Two CAMMs with Ethiopia and Nigeria have been signed so far. Therefore, this instrument is left outside the scope of this study.

[4] George and Bennett (2005), p. 19.

[5] Yin (2014), p. 16.

[6] Flyvbjerg (2006), p. 235.

[7] Webley (2010), p. 940.

[8] Stake (1994), p. 243.

sociological research.[9] This book is focused on a multiple-case study comparison, employing a 'most similar system' design,[10] which is a method that focuses the analysis on comparable cases within the same geographical-cultural area and allows identification of factors that help understand differences in outcomes.

Comparative legal enquiries usually consist of three major steps: a selection of what is to be compared, a description of the law and its context, and an analysis thereof.[11] The first step begins with a determination of the basis of comparison, containing the objects of comparative research and the sources that are consulted.[12] Even though researchers conducting comparative legal enquiries are generally advised to look for both similarities and differences[13] at the selection stage, the researcher must strive to have at least a minimum similarity in order to provide a basis for the comparative analysis.[14] Thus, selecting countries from the same legal family and which – in the context of this study – share a similar communist past, are part of the latest EU enlargement, are active in the Eastern Partnership, and attract migrants from the same regions were among the criteria that were used to choose the countries for the purposes of this case study. These common characteristics thereby allow the researcher to account for whether similar laws and policies have similar or different outcomes when they are applied in the same contexts.

The next section describes the case selection in greater detail. It shows that the two countries analysed in this study share common characteristics thus allowing the researcher to test the findings derived from the country comparisons. Even if the results from this study cannot be formally generalised in the end, it follows that purely descriptive case studies still have the potential of contributing to the 'process of knowledge accumulation' in the field of legal empirical research and can 'certainly be of value in this process and has often helped cut a path toward scientific innovation'.[15]

2.2.1 Case Study Selection: Bulgaria and Poland

Bulgaria and Poland have been selected from the Central and Eastern European (CEE) countries as the two cases that will be compared in this study. They share similarities and differences which, as mentioned above, makes them useful subjects for comparative research. Firstly, due to their geographical proximity and their ethnic and historical ties, both countries attract migrants from the former Soviet Union

[9] Webley (2010), p. 940.

[10] See Lijphart (1975), p. 159.

[11] Dannemann (2006), p. 406.

[12] Ibid., p. 407.

[13] See Ibid., p. 419.

[14] Zweigert and Kötz (1998), pp. 34–35.

[15] Flyvbjerg (2006), p. 227.

republics that are situated in the Eastern European neighbourhood as well as from Russia. Therefore, this case selection allows the study to focus geographically on the implementation of the EU's circular migration approach with regard to the Eastern Partnership countries. Secondly, they share similar contextual factors: they are former communist countries with some experience in temporary migration; they are new EU Member States with a geopolitical interest in the Eastern Partnership; and, they participate in the GAMM instruments with these countries. Finally, they diverge in their legal and policy contexts; both experienced different EU accession processes, which resulted in distinctive national migration legislation and differences in the transposition of relevant EU legislation.

2.2.1.1 Contextual Factors: Similarities and Differences

Bulgaria and Poland share a similar communist history of immigration when inward and outward flows were heavily controlled until 1989. Temporary labour migration took place mainly within the Soviet bloc or according to bilateral agreements with other Soviet republics. Poland had a continuous excess of labour supply and Poles were involved in regular migration related to temporary employment in neighbouring Czechoslovakia and Hungary, as well as engaged in petty trade conducted under the guise of tourism. This was a type of circular mobility that became a widespread phenomenon in the 1980s and which was known as 'incomplete migration'.[16] Later, in the 1990s, this type of movement was referred to as 'shuttle mobility' and began taking place from former Soviet republics, especially Ukraine, Belarus, and Russia.

Temporary migration opportunities were also present in communist Bulgaria. The country concluded bilateral agreements with other socialist countries for the exchange of workers. For example, in the 1980s, many Bulgarians worked in the construction and timber industries of the Komi Autonomous Soviet Socialist Republic under an exchange agreement signed with the Soviet Union. Furthermore, some Vietnamese construction workers were sent to Bulgaria under agreements that were concluded in the 1980s.

Additionally, both Bulgaria and Poland are slowly evolving from countries of emigration and transit to new countries of immigration. As such, in both countries foreign-born individuals comprise 2 per cent of the population.[17] As of 1 January 2018, there are 156,505 foreign-born residents in Bulgaria (both EU and non-EU migrants) and 695,850 in Poland.[18] However, it should be stressed that this is an estimated number and in the case of Poland, for instance, excludes migrants working on a visa. Therefore, these numbers are used only for indicative purposes.

---

[16] See Sect. 1.2.3 of Chap. 1.

[17] OECD, International Migration Outlook 2019. https://www.oecd.org/migration/international-migration-outlook-1999124x.htm Accessed 22 October 2019.

[18] Eurostat data on Foreign-born population. http://ec.europa.eu/eurostat/tgm/table.do?tab=table&plugin=1&language=en&pcode=tps00178 Accessed 22 October 2019.

However, the two countries differ with regard to the main countries of origin of immigrants. In Bulgaria, according to the latest OECD data from 2017, the main countries of origin of non-EU country newcomers are Turkey, Russia, Syria, and Ukraine.[19] In Poland, as of 2017, Ukrainians formed more than 60 per cent of the total foreigner population, followed by immigrants from Belarus, China, and India.[20] Until 2016, Russians were the largest group of third-country nationals with first residence permits in Bulgaria while, in the past couple of years, Ukrainians have emerged as the third-largest group (see Table 2.1).[21] In Poland, Ukrainians are in the majority, followed by Belarusians as the second-largest group; Russians were in the top five until 2016 (see Table 2.2).

The reasons for immigration of the first permit holders also differ between the two countries. Bulgaria attracts mainly family migration, as well as migrants for other purposes such as pensioners and refugees; Poland attracts primarily migration for employment-related reasons (see Figs. 2.2 and 2.3). A total of 87.4 per cent of

Table 2.1 Top five of first residence permits in Bulgaria per country of citizenship (2015–2018). Source: Eurostat

Citizen of	2015	Citizen of	2016	Citizen of	2017	Citizen of	2018
Russia	2.782	Turkey	2.838	Turkey	4,060	Turkey	4,553
Turkey	2.558	Russia	1.509	Russia	2,137	Russia	2,004
Ukraine	1.488	Ukraine	1.086	Ukraine	1,115	Ukraine	1,095
North Macedonia	300	North Macedonia	348	North Macedonia	664	North Macedonia	840
Kazakhstan	234	Serbia	206	Serbia	379	Serbia	536
Total	9.595	Total	7.942	Total	10,958	Total	11,864

Table 2.2 Top five of first residence permits in Poland per country of citizenship (2015–2018). Source: Eurostat

Citizen of	2015	Citizen of	2016	Citizen of	2017	Citizen of	2018
Ukraine	430.081	Ukraine	512.552	Ukraine	585,439	Ukraine	413,449
Belarus	75.394	Belarus	28.165	Belarus	42,756	Belarus	126,576
Moldova	7.987	Moldova	7.613	Moldova	7,803	Turkey	13,264
Turkey	4.226	India	5.473	India	6,244	India	10,999
Russia	3.932	Turkey	5.133	Turkey	6,094	Russia	8,786
Total	541.583	Total	585.969	Total	683,228	Total	635,335

[19] OECD, International Migration Outlook 2019. https://www.oecd-ilibrary.org/sites/f5a90a0d-en/index.html?itemId=/content/component/f5a90a0d-en&mimeType=text/html Accessed 22 October 2019.

[20] OECD, International Migration Outlook 2019. https://www.oecd-ilibrary.org/sites/c9a909d8-en/index.html?itemId=/content/component/c9a909d8-en&mimeType=text/html Accessed 22 October 2019.

[21] Eurostat data on first permits by reason, length of validity and citizenship as of 22 October 2019. http://appsso.eurostat.ec.europa.eu/nui/show.do?dataset=migr_resfirst&lang=en Accessed 22 October 2019.

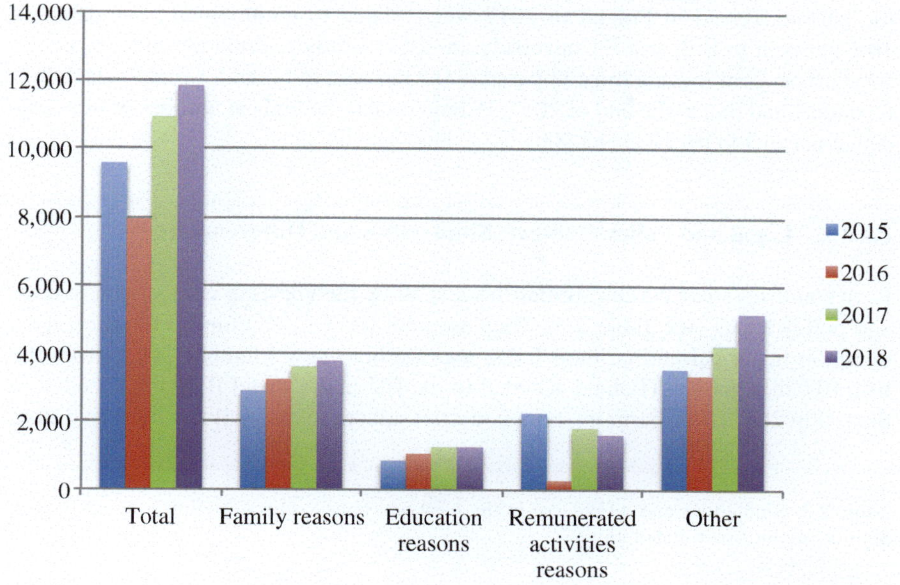

Fig. 2.2 First permits by reason, Bulgaria (2016–2018). Source: Eurostat

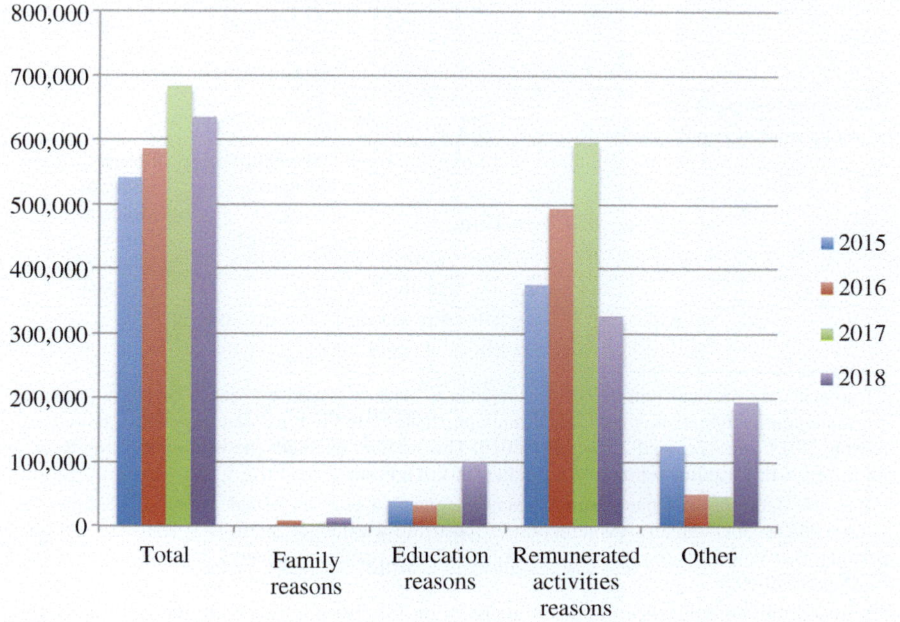

Fig. 2.3 First permits by reason, Poland (2015–2018). Source: Eurostat

the permits issued in Poland in 2017 were related to employment, compared to 16.6 per cent in Bulgaria.[22] Conversely, in 2017, family reunion permits accounted for 33.0 per cent in Bulgaria compared to 0.5 per cent in Poland.[23] Finally, it should be underlined that at the end of 2017, Poland issued the highest number of first residence permits in the EU (683,000).[24]

2.2.1.2 Legal and Policy Context: Similarities and Differences

Both countries share certain similarities that allow for a basis of comparison in legal and policy terms (see Table 2.3). They are both former communist countries, have common historical and cultural links, and are both new Member States from the fifth EU enlargement (Poland acceded to the EU in 2004 and Bulgaria acceded to the EU in 2007). Furthermore, both countries adhere to the civil law legal tradition,

Table 2.3 Similarities and differences in the legal and policy context of Bulgaria and Poland. Source: Author's own elaboration

Legal/Policy issue	Bulgaria	Poland
Legal family	✓	✓
New Member States	✓	✓
Participation in Mobility Partnership with Eastern partnership countries	Moldova (2008) Georgia (2009) Armenia (2011) Azerbaijan (2013) Belarus (2016)	Moldova (2008) Georgia (2009) Armenia (2011) Azerbaijan (2013) Belarus (2016)
Part of the Schengen area	☒	✓
Circular migration policy/scheme	✓ Armenia Moldova Georgia (pending)	✓ Belarus Georgia Moldova Russia Ukraine Armenia

[22] Eurostat data on total number of first residence permits issued by reason in 2017. https://ec.europa.eu/eurostat/statistics-explained/index.php?title=File:First_residence_permits_issued_by_reason,_2017.png Accessed 22 October 2019. This number excludes the workers coming through the simplified procedure whom work on the basis of a visa.

[23] Eurostat data on residence permits – statistics on first permits issued during 2017. https://ec.europa.eu/eurostat/statistics-explained/index.php?title=Residence_permits_-_statistics_on_first_permits_issued_during_the_year&oldid=427234#First_residence_permits_by_reason Accessed 22 October 2019.

[24] Eurostat data on residence permits – statistics on first permits issued during 2017. https://ec.europa.eu/eurostat/statistics-explained/index.php?title=Residence_permits_-_statistics_on_first_permits_issued_during_the_year&oldid=427234#First_residence_permits_by_reason Accessed 22 October 2019.

but remnants of their former socialist legal tradition are still present in their respective legal frameworks.[25]

As new Member States, Poland and Bulgaria were obliged to apply all of the Justice and Home Affairs (JHA) legal instruments without the possibility of opting out. Therefore all these instruments, including those incorporating circular migration elements, had to be implemented at the national level. Also, both countries have chosen to participate in Mobility Partnerships with Eastern Partnership countries.[26]

However, neither Bulgaria nor Poland acceded to the ILO Migration for Employment Convention (No. 97),[27] ILO Migrant Workers Convention (No. 143)[28] and International Convention on the Protection of the Rights of All Migrant Workers and Members of Their Families (ICRMW)[29] that contain underlying principles for national laws and regulations concerning labour migration and the protection of migrant workers.[30] They are also among the instruments forming part of the study's rights-based benchmark framework for assessment of circular migration policies.[31]

Bulgaria and Poland both underwent a transition from communism to democracy and when they started their accession periods, the EU rules 'had to compete with a layer of old regulations inherited from the communist regimes'.[32] These processes are part of the legal context of both countries and must naturally be taken into consideration in the comparative case study analysis. However, even though both coun-

---

[25] See for instance Manko (2013).

[26] For more details see Chap. 3, Sect. 3.3.

[27] Convention concerning Migration for Employment (Revised 1949), C097, adopted at 32nd ILC session on 01 July 1949, Geneva, entry into force on 22 January 1952. It has 50 ratifications, among which are ten EU Member States.

[28] Convention concerning Migrations in Abusive Conditions and the Promotion of Equality of Opportunity and Treatment of Migrant Workers, C143, adopted at 60th ILC session on 24 June 1975, Geneva, entry into force on 09 December 1978. It has 25 ratifications, among which are five EU Member States.

[29] International Convention on the Protection of the Rights of All Migrant Workers and Members of Their Families, G.A. res. 45/158, 18 December 1990, entered into force on 1 July 2003. It has 55 ratifications. No EU Member States have signed this convention.

[30] Principle 9 of the ILO Multilateral Framework on Labour Migration. Non-binding principles and guidelines for a rights-based approach to labour migration, adopted by Tripartite Meeting of Experts, which convened in Geneva from 31 October to 2 November 2005.

[31] For more details on the benchmark framework, see Chap. 1, Sect. 1.3 and Annex V. Other instruments that are forming part of the study's benchmark framework and have not been signed by Bulgaria and Poland are: Equality of Treatment (Social Security) Convention, C118, adopted at 46th ILC session on 28 June 1962, entry into force on 25 April 1964; Maintenance of Social Security Rights Convention, C157, adopted at 68th ILC session on 21 June 1982, entry into force on 11 September 1986; European Agreement on Regulations governing the Movement of Persons between Member States of the Council of Europe, CETS No.025, 13 December 1957, entry into Force 1 January 1958; European Convention on the Legal Status of Migrant Workers, ETS No. 93, 24 November 1977, entry into force on 1 May 1983; European Convention on Establishment, ETS No.019, 13 December 1955, entry in force 23 February 1965; European Code of Social Security, CETS No.048, 16 April 1964, entry in force 17 March 1968.

[32] Toshkov (2012), p. 4.

tries are new EU Member States, they had different EU accession periods, which resulted in distinctive national migration legislation and different transposition of EU legislation.

Moreover, the EU migration policy is an area of shared competence and Member States have a certain discretion, which is an important part of the legal context and must therefore also be considered. As a result, Poland, for example, has developed a scheme that facilitates circular migration through its national legislation and under the aegis of the GAMM, whilst Bulgaria promotes this type of migration on the basis of bilateral agreements with third countries. Another difference in the policy contexts is that Poland has become a member of the Schengen area and fully applies the Schengen *acquis*. Bulgaria, however, is still a candidate country and is awaiting the political approval that is required in order to become a member of the Schengen area.

In order to examine the challenges that are related to migrant workers' rights in the context of circular migration and ensure comparability of the data, one should focus on migrants with similar characteristics. Since the geographical focus of the study is on migrants circulating from the Eastern neighbourhood countries to the EU, the first logical step is to determine the largest immigrant groups in Bulgaria and Poland, respectively. According to the most recent available data, in the case of Bulgaria, the main countries of origin of first residence permits holders from the Eastern neighbourhood countries are Russia and Ukraine. In the case of Poland, these are Ukraine followed by Belarus, Moldova, and Russia (see Table 2.2).

Apart from being the largest immigrant group granted first residence permits in the EU 28,[33] Ukrainians are, as the data show, among the largest immigrant groups from the Eastern Partnership countries in both Poland and Bulgaria; therefore it is logical to identify them as one of the immigrant groups suitable for the aims of this research. Russians are the largest immigrant group in Bulgaria, and Russia was in the top five countries of origin of immigrants in Poland in 2015 and 2018. Furthermore, Russians are in the top ten of immigrant groups granted first residence permits in the EU 28.[34] Even though Russia is not one of the Eastern Partnership countries, it is an important partner in the external dimension of EU migration policy and more specifically within the context of the GAMM (at least until the annexation of Crimea, when all migration-related EU–Russian dialogues were suspended). Therefore, Russians are identified as the second suitable group for the purposes of ascertaining the research aims of this study.

---

[33] Eurostat data on developments for the number of first residence permits issued by country of citizenship, EU-28, 2015–2017. https://ec.europa.eu/eurostat/statistics-explained/index.php?title=File:Figure_5_Developments_for_the_number_of_first_residence_permits_issued_by_country_of_citizenship,_EU-28,_2015-2017_(1_000_persons).png Accessed 22 October 2019.

[34] Eurostat data on developments for the number of first residence permits issued by country of citizenship, EU-28, 2015–2017. https://ec.europa.eu/eurostat/statistics-explained/index.php?title=File:Figure_5_Developments_for_the_number_of_first_residence_permits_issued_by_country_of_citizenship,_EU-28,_2015-2017_(1_000_persons).png Accessed 22 October 2019.

2.2.2 *Data Sources*

The comparative case study approach is carried out using legal and policy sources as well as qualitative data gathered from semi-structured interviews and focus groups. The legal and policy sources – in combination with semi-structured interviews with relevant policy actors – are mainly used for the analysis of the formulation and implementation of the EU's circular migration approach through the established outputs at both the EU and the national levels. The semi-structured interviews are used to fill gaps in the interpretation and implementation of legal and policy sources when analysing both the EU's approach and national instruments facilitating circular migration. The aim of the focus groups is to examine the perceptions of migrant workers and thus the outcomes at the individual level.

2.2.2.1 Legal Sources

This study is based on an analysis of primary and secondary European legislation: EU treaties, EU regulations, and EU legal migration directives as well as national legal acts on foreigners and labour migration. Universal and regional human rights instruments are also used to construct a rights-based analytical framework that supports the assessment of the developed circular migration policies.[35] Furthermore, national legal acts are analysed to track the implementation of the EU's approach to circular migration at the national level. This includes taking stock of the transposition of EU directives in the area of legal migration and implementation of other sources of EU law (e.g. on visa and social security matters) that incorporate elements of circular migration into the national laws on foreigners, migration, and employment in Bulgaria and Poland, respectively. The case law of Court of Justice of the European Union (CJEU) has also served as relevant legal source. However, no national court cases were identified as having a direct bearing to the topic of circular migration. An important caveat to keep in mind is that the review of national court cases has not been exhaustive due to language barriers and time limitations. The legal sources have been updated to reflect any important amendments that have occurred up to September 2019.

2.2.2.2 Policy and Academic Sources

Policy documents at the international, European, and national levels in Bulgaria, Poland, and the Eastern Partnership countries that participate in Mobility Partnerships are also used as data sources. Firstly, EU policy sources such as the Communications of the European Commission are used to conceptualise the term circular migration and trace its two-fold implementation at the EU level. Secondly,

---

[35] For more details, see Chap. 1, Sect. 1.3, Annex V and Vankova (2016).

data consisting of academic publications and NGO reports are employed in order to complement the conceptualisation of the EU's approach to circular migration that is based on EU legal and policy data sources. This concept has attracted considerable scholarly and policy attention in the years following its introduction, and these data sources are able to provide valuable insights into the understanding of its meaning, its key conceptual elements, and the policy routes to its implementation.

Policy and academic literature[36] on the topic is also used to trace the implementation of the EU's approach. The reports of Brussels-based NGOs and think tanks provide valuable insight into the problems associated with the concept.[37] Furthermore, the reports produced by the European Migration Network and the European Training Foundation are among the limited data sources that provide useful insights into the challenges related to the concept's implementation at the national level.[38]

Most of the EU-related documents necessary for this study are available online through the websites of EU institutions and the European Migration Network. The only issue with regards to EU documents is securing access to the so-called scoreboards that help further operationalise the annexes of the Mobility Partnerships, listing – among other things – the state of play of the different projects, the leading partners, and their budgets. The scoreboards were accessed through official requests for information that were filed to the European Commission or through requests to experts from the respective national administrations. The book contains information retrieved through the latest available scoreboards obtained in September 2019. Several official requests for information were filed with the Bulgarian and Polish administrations in order to access statistical data related to the implementation of different instruments analysed in this study.

2.2.2.3 Semi-structured Interviews

In order to complement the information on the implementation of the EU's circular migration approach that was gathered through legal and policy sources, this research also employed in-depth interviews with policymakers and NGOs at the EU and national levels in Bulgaria, Poland, and some of the Eastern Partnership countries participating in Mobility Partnerships. This method is extensively used in empirical legal research because interviews are an effective means of gathering data on other individuals' perceptions and views as well as on the reasoning behind the responses.[39] The study employs semi-structured in-depth interviews, allowing for

[36] For instance, F. Fargues (2008). Also the CARIM East explanatory notes examining the demographic, legal and socio-political aspects of circular migration between Eastern Europe and European Union and within the post-Soviet space. http://www.carim-east.eu/database/legal-module/?ls=4&ind=exnocm&lang= Accessed 25 September 2017.

[37] Carrera and Hernández i Sagrera (2009); McLoughlin and Münz (2011).

[38] See for instance, European Migration Network (2011); Kalantaryan (2015).

[39] Webley (2010), p. 937.

some set questions formulated on the basis of the analysis of the other data sources but also leaving open-ended questions that enable the interviewees to reflect on the topic on the basis of their experience and understandings.[40]

The main challenge in relation to interviewing is how to avoid biases from both the interviewer and interviewee.[41] Therefore, different authors recommend the use of various techniques in order to ensure researchers maintain the reliability and validity of their research. Some techniques used in this research were: keeping focus on the core questions and themes of the interview guide; remaining relevant and directed and thus maintaining control; avoiding misleading questions or socially acceptable bias, as well as value judgements, in order to maintain objectivity; and ensuring that the interviewee had correctly understood the questions and terms used.[42] Additionally, the interview guide was piloted in the first phase of the field research in 2013.

Several semi-structured interview guides were used: one for policymakers and stakeholders at the EU and national levels who were engaged in the formulation or implementation of the EU's circular migration approach and several specific guides for interviews with national experts working in one of the six specific policy areas under consideration. In a few cases, additional semi-structured guides were used when a given policy actor was approached with a specific set of questions for a second interview, with the aim of gathering additional information.

Firstly, the interview method was used to provide insight into policy formulation at the EU level. Several EU institutions are involved in policy formulation of the circular migration concept. However, given the complicated legislative and decision-making process at the EU level, it is not always easy to identify the people who are involved in the development of the concept. Furthermore, since the concept was introduced more than 10 years ago, some of the individuals have either transferred to another Directorate General of the European Commission or left their employment at the EU institutions altogether. Therefore, the only way to create a sample was to use a snowball method.[43] The first interview for this research was carried out at the Centre for European Policy Studies in 2013 and was used as an entry point in the 'circular migration policy-making world'.[44]

Thus, by asking every respondent to recommend other potential interviewees, 18 initial interviews were conducted in the period 2013–2014 with officials from EU institutions and representatives of Brussels-based NGOs and think tanks (see Annex I). These interviews allowed the researcher to acquire an initial understanding of the concept and its use. This phase also provided the author with the opportunity to pilot the developed semi-structured interview guide, which was adjusted slightly

[40] Ibid.

[41] Brewerton and Millward (2001), p. 74.

[42] See ibid., p. 71.

[43] Goldstein and Getting (2002), p. 671; Weiss (1994), p. 25.

[44] These interviews were conducted in 2013, outside of the TRANSMIC framework, and therefore did not follow the requirement for registering the informed consent of the interviewees.

following the initial interviews. During a secondment at one of the headquarters of International Centre for Migration Policy Development (ICMPD) throughout the period between January–March 2017, an additional round of ten interviews with representatives of this international organisation, as well as other relevant policy actors from the Eastern Partnership countries, were conducted (see Annex I). Additionally, during the last phase of writing this study, two more interviews with European Commission officials were conducted.

Secondly, in order to gain an understanding of how policy formulation is transformed into policy outputs, a total of 45 (23 in Bulgaria and 22 in Poland) interviews with national stakeholders from the relevant authorities in Bulgaria and Poland, as well as NGO representatives, were conducted on the basis of snowball sampling (see Annexes II and III). These interviews were of crucial importance for grasping the co-formulation and implementation of the circular migration approach that has been developed at the national level since very often the policies or legal frameworks that are developed at the national level do not explicitly refer to the term 'circular migration'. The interviews with policymakers also constituted a source of information about the attitudes toward this approach and provided an explanation of some of the policy decisions in this regard. Finally, interviews with lawyers and national experts from academia, NGOs, and think tanks were conducted to ascertain a critical view of the implementation (or lack thereof) of the circular migration approach and the consequences on the rights of the migrant workers (see Annexes II and III).

2.2.2.4 Focus Groups

The focus group (or group interview) entails a collective interview process.[45] Its purpose is to understand how people think or feel about a certain issue, to gather opinions,[46] and provide opportunities for the participants to share their experiences.[47] There are several reasons why focus groups were chosen as a method for examining the policy outcomes at the individual level rather than by recourse to individual interviews. First, the main feature distinguishing focus groups from individual interviews is that they produce what is referred to as 'interactive data'.[48] Focus groups are a unique method because they allow data to be gathered from both the individual and from the individual as part of a group.[49] This process is driven by the communication between participants with specific shared characteristics that relate to the topic of the focus group which allows the research to ascertain a richer knowledge of the subject that is being discussed because the topics and opinions

---

[45] Ayala (2012), p. 123.

[46] Krueger and Casey (2015), p. 2.

[47] See for instance Massey (2011), p.22.

[48] Hydén and Bülow (2003), p. 306.

[49] Massey (2011), p. 21.

'unfold and are negotiated in the focus group's discussion'.[50] It also provides the opportunity for direct comparisons among the experiences and views of the participants, rather than analysing differences of the interviewees on the basis of aggregated individual data.[51]

Second, this method is often used to determine an individual's reaction to the introduction of a policy or a policy change that affects a population in order to provide a policy evaluation.[52] Focus groups were therefore considered to be a suitable method for the purposes of this research because they bring the perspective of migrants on the challenges related to their circulation and rights in the six policy areas identified as pertinent to circular migration on the basis of group interactions. Thus, this method allows for an examination of policy outcomes at the individual level. Third, focus groups complement the other methods used in this research and build on the findings derived from the legal and policy analysis and the data gathered from the interviews with policymakers, thus allowing for triangulation by adding 'the human element of the voices of multiple subjects'.[53]

Finally, there were certain pragmatic considerations as a result of this method. The combination of focus groups with interviews 'has the advantage of getting reactions from a relatively wide range of participants in a relatively short time'.[54] Therefore, focus groups were considered to be an efficient research method, especially given the limited duration of the periods of field research that were possible under this study.

Some of the critiques of the use of focus groups include the argument that focus group participants could make up the answers, that the dominant participants could influence the results, and that this method may produce unreliable and trivial results.[55] As with any other research methods, the reliability question and the quality of the results produced concerns the sampling approach and the recruitment strategy used as well as the number and size of the groups. The justification of all the steps that were taken in designing this research is presented below. Another important factor for the successful implementation of the focus groups concerns the role and skills of the moderator, who in this study served as 'a levelling force that allows participants to reflect on various arguments without pressure'.[56] In order to prevent participants from fabricating answers, the research employed a strategy of presenting, in detail, the rules of the method and asking additional questions during the interviewing process in order to eliminate any doubt. Furthermore, these problems are considered minimised when multiple strategies of inquiry are employed.[57]

---

[50] Hydén and Bülow (2003), p. 307.

[51] Morgan (1996), p. 139.

[52] Krueger and Casey (2015), p. 8; Frey and Fontana (1991), p. 176.

[53] Frey and Fontana (1991), p. 178.

[54] Morgan (1996), p. 134.

[55] Krueger and Casey (2015), pp. 14–16.

[56] Ibid., p. 16.

[57] Ibid., p. 14.

The aim of the focus group is to gather the opinions of individuals who have something in common. The first such feature of migrants that needs to be considered is their country of origin. As already mentioned, Ukraine and Russia were identified as the most suitable countries of origin with regards to this study's aim. Another important feature that the focus group participants had to share was being economically active in line with the definition adopted by this study. As the respective national laws of Poland and Bulgaria do not refer to a 'circular migrant status', the aim was to select participants who are labour migrants and employed workers. Focus groups need to be homogeneous, but there must also be sufficient variation therein.[58] Therefore, the focus groups that were conducted with Ukrainians and Russians included both workers in low-skilled (e.g. seasonal labour) and highly-skilled occupations (e.g., Blue Card holders, national permit holders). They also included migrants who had recently retired. This group of participants offered a perspective on the challenges related to pension rights that migrants who are still in employment cannot provide. Therefore, their participation in the focus group was considered important.

Additionally, one of the characteristics of circular migrants is that they circulate repeatedly between their countries of origin and destination. Therefore, the participants recruited for the focus groups had to have returned at least once to their country of origin and have come back to Bulgaria or Poland. Furthermore, since circular migration in the EU context is not exclusively limited to temporary stays and does not exclude circular migration of permanently settled migrants, migrants with either temporary or permanent status were considered suitable for participation in the focus groups.

Thus the selection criteria for the focus groups included: country of origin (Russia and Ukraine); legal status (temporary and permanent residence permit holders); economic status (employed economic migrants and retirees); and, when possible, with circular migration experience (returning to the country of origin for work-related reasons, e.g. return for work under a temporary contract in the country of origin or renewal of documents related to work permits /residence permits).

Potential focus group participants were recruited in several different Bulgarian and Polish cities on the basis of snowball sampling, which is an approach that helps the researcher to locate 'information-rich key informants' or events that serve as a starting point for the development of the sample.[59] The strategies for finding potential participants included identification of informants from the immigrant communities on the basis of meetings with different stakeholders. The informants were asked to think of migrant workers who fit the focus group profile and also distribute a brochure among their community or network. Sometimes informants directly contacted the researcher with suggestions for potential participants and in other cases participants contacted the researcher directly.

Among the informants in Bulgaria were representatives of the International Organisation for Migration (IOM) and its Integration Offices in Sofia and Bourgas,

[58] Krueger and Casey (2015), p. 81.

[59] Patton (2002), p.237.

representatives of the Bulgarian Red Cross and NGOs working in Varna, the network of the 'Multi Kulti Collective Association', contacts received from interviewees, and personal contacts. One member of the 'Multi Kulti Collective Association' supported the recruitment process of focus group participants in Bulgaria. The recruitment strategy for Russian migrants included contacting Russian restaurants in Sofia, the Russian Culture Centre, and the Russian Embassy, among others, as well as all shops from the Russian retailer Berizoka. Ukrainians were recruited through Facebook groups, the Ukrainian association Matti Di, and other informal channels. An information brochure was prepared in line with the developed ethics self-assessment procedure (see Sect. 2.3), which was translated into Ukrainian and Russian and distributed among more than 65 different informants and 17 Facebook groups.

The recruitment process in Poland followed the same strategy. The only difference was that the recruitment of potential focus group participants was supported by the network of interviewers and scholars working at the Centre for Migration Research in Warsaw. At the time of the field research in the period from October–December 2016, the Centre was conducting a survey among Ukrainians in Poland and some focus group participants were recruited through this pool of migrants, adhering to the ethical standards of the Centre's survey. In addition, one of the Centre's interviewers served as a focus group recruiter in the Ukrainian and Russian communities in Warsaw. Representatives of NGOs providing support to migrants in Warsaw were among the informants for this study. Finally, the prepared information brochure was distributed through the Russian and Ukrainian churches in Warsaw, the Russian Cultural Centre *Ośrodek rosyjskiej kultury*, several Russian restaurants, more than 14 Facebook groups, and 12 vk.com groups.

The recruitment of Blue Card holders was organised in a slightly different manner. It required the identification of companies employing Blue Card holders from the Eastern Partnership countries and Russia. Based on information obtained from informants in Bulgaria, it became clear that the Blue Card Directive was used primarily by IT companies. It took more than 2 months of targeted attempts to identify a company that was willing to give access to its employees for the sake of conducting several focus groups. After the field research in Bulgaria was completed, the Bulgarian IT company facilitated access to a Polish enterprise that recruited migrants through the Blue Card mechanism.

The recruitment period lasted 3 months in both countries. In Bulgaria, even though Ukrainians were the smaller immigrant group compared to Russians, they were easier to find and, furthermore, were eager to participate in the study. Finding Russian migrants willing to take part in the focus groups turned out to be a great challenge in both countries. Despite the broad network of informants used to recruit Russian participants, they were often not interested in participating or did not match the required profile. Most of the Russians who agreed to participate were highly-skilled which, according to the study's informants, matches the general profile of Russian migrants in these two countries.

Furthermore, the initial profile that was envisaged in the research design had to be adapted several times. In Bulgaria it was extremely difficult to find 'true' labour migrants who had an employment contract and came to work as employed persons. Most of the migrants reached were students, businessmen, or were in the country through family reunification. In addition, it was impossible to find migrant workers circulating voluntarily between Bulgaria and their country of origin. There is, however, a particular reason for that, which is presented in Chap. 6. Voluntary circular migration seemed to be a typical feature only for students and businessmen. Thus, the final sample for the focus groups in Bulgaria had to be broadened and it ultimately included labour migrants as well as family migrants and migrants with registered companies. Some of them had circulated; some of them did not have this experience. In Poland, the biggest challenge encountered, apart from recruiting Russian migrants, was finding migrants working legally; it emerged that many migrant workers coming through the Polish simplified procedure (referred to as the *Oświadczenie* procedure in this study) entered the country legally but were working irregularly without any employment contract.

Initially the scope of the study's research design included focus groups with seasonal workers. According to the assessment of the Ethics Advisory Body presented below, only legal migrants could be interviewed for the purposes of this study. However, it was impossible, in both Bulgaria and Poland, to find any seasonal workers who were working legally at the time of the field research in the second half of 2016. Furthermore, these types of workers were engaged in different sectors in the countries chosen for case studies; in Bulgaria they worked primarily in tourism and in Poland they worked mainly in agriculture. This required an additional recruitment strategy and an extended field research period that did not fit the study's timeline. Therefore, seasonal workers were outside the scope of the focus groups. To compensate for this empirical gap, additional interviews with employers who recruit seasonal workers were conducted in both Poland and Bulgaria.

A total of nine focus groups were conducted as part of this study in 2016: four focus groups in Bulgaria and five focus groups in Poland; four of the focus groups covered general labour migrants and the rest included only Blue Card holders (see Annex IV).

2.3 Ethical Considerations

This study employs interviews and focus groups with migrants to examine how EU and national legislation is implemented in practice and the challenges migrant workers face. Thus, the project includes human participation, particularly policy actors and migrants, and thus the use of personal data, including information relating to ethnicity.

As this study was part of the TRANSMIC project, its methodology had to be reviewed by an interdisciplinary Ethics Advisory Body (EAB) at Maastricht University. The EAB reviewed the procedure for recruiting participants for the interviews and the focus groups, the procedure for handling any sensitive or personal data collected, and the informed consent procedure. The research activities were designed to ensure respect for people and human dignity, fair distribution of research benefits and burden, and protect the values, rights, and interests of the research participants. In order to achieve this and following the assessment of the EAB, only migrants possessing legal status were recruited for the focus groups. The sampling excluded migrants who were not able to give informed written consent themselves or provide ways of evidencing both the informed consent and the understanding of the risks that the project could bring for them.

Information brochures were translated into Russian and Ukrainian and formed part of the invitation for the recruitment of participants for the focus groups. Migrants gave informed consent orally during the recruitment phase. On the day of the focus group, participants received an informed consent form in a language and terms that were fully understandable to them, respectively in Bulgarian, English, Russian, or Ukrainian. The documents contained the aims, methods (how the data would be used in the analysis), duration, and implications of the research (including the impact on national and EU immigration policies), the nature of the participation (interviews) and any benefits, risks, or burdens that might be involved as a result.

The informed consent form and the detailed information brochures also explicitly stated that participation was voluntary and that each person had the right to refuse to participate and withdraw their participation or data at any time — without any consequences. The participants were informed about their right to ask questions and receive understandable answers before deciding. They were provided with the names and contact details of the principal researcher conducting the research, the supervisors of the principal researcher, and the EAB's complaint procedure. The focus group participants gave their consent in writing by signing the informed consent form or by signing an informed consent list stating that they had been informed orally. The participants had their costs covered (transportation and food) and received a small remuneration in the form of a voucher.[60]

Policymakers and other stakeholders were not covered by the informed consent procedure envisaged for the migrant workers. They were recruited through an invitation containing information about the project, its aims, focus, and the purpose of the interview. Once they agreed to participate (either via email or telephone), an interview was then scheduled. At the beginning of the interview the information about the project was repeated and they were asked whether they agreed to be recorded. Their consent was either registered or they were provided with a simplified informed consent form for stakeholders. Stakeholders could choose to be anonymous and most preferred this option.

The current research involved collecting and processing personal data through interviews and focus groups. The personal data included name, occupation, age, type of work or residence permit or both, citizenship, ethnicity, circular migration history, and telephone number or email. Ethnicity is sensitive personal data that was

[60] The voucher in Bulgaria was for 30 BGN (Apr. 15 EUR) and the voucher in Poland was for 40 PLN (Apr. 9.50 EUR). The field research in Poland was conducted after the research in Bulgaria and the difference between the voucher amounts is because this was the standard fee offered to respondents by the Centre of Migration Research, Warsaw.

collected because some of the policies existing in the countries chosen for the case studies give preferential treatment to migrants with a particular ethnic background (for instance the '*бесарабски българи*' [besarabski Bulgari] who are considered to be Ukrainians of Bulgarian origin), which is important with regards to the project findings.

During the informed consent procedure, both focus group participants and interviewees were informed that their data would be anonymised, protected during the project, and destroyed at the end of the research. The data was safely stored in password-protected storage devices and encrypted files on the Maastricht University servers. Only the principal researcher, the research assistants, and supervisors had access to the data. The procedures on destruction or re-use followed the guidelines that were given by the EAB.

References

Ayala, L. S. (2012) Interviewing techniques for migrant minority groups. In C. Vargas-Silva (Ed.), *Handbook of Research Methods in Migration*. Cheltenham: Edward Elgar Publishing, pp. 117–136.

Brewerton, P. M. and L. J. Millward (2001) *Organizational Research Methods: A Guide for Students and Researchers*. Thousand Oaks: Sage Publications.

Carrera, S. and R. Hernández i Sagrera (2009) The Externalisation of the EU's Labour immigration policy. Towards mobility or insecurity partnerships? CEPS Working Documents No. 321. Centre for European Policy Studies, Brussels.

Dannemann, G. (2006) Comparative law: Study of Similarities or Differences? In M. Reimann and R. Zimmermann (eds), *The Oxford Handbook on Comparative Law*. Oxford: Oxford University Press, pp. 383–419.

European Commission (2014) Report on the Implementation of the Global Approach to Migration and Mobility 2012–2013, COM (2014) 96 final, Brussels, 21 February 2014.

European Migration Network (2011) *Temporary and Circular Migration: empirical evidence, current policy practice and future options in EU Member States. Synthesis Report*, Brussels.

Kalantaryan, S. (2015) Inventory of Migrant Support Measures from an Employment and Skills Perspective. Armenia. European Training Foundation.

Fargues, F. (2008) Circular Migration: Is it relevant for the south and east of the Mediterranean? *CARIM Analytic and Synthetic Notes* 40.

Flyvbjerg, B. (2006) Five Misunderstandings About Case-Study Research. *Qualitative Inquiry* 12 (2): 219–245.

Frey, H. and A. Fontana (1991) The Group Interview in Social Research. *Social Science Journal* 28 (2): 175–187.

George, A. L. and A. Bennett (2005) *Case Studies and Theory Development in the Social Science*. Cambridge, MA: MIT Press.

Goldstein, K. and K. Getting (2002) In the Door: Sampling and Completing Elite Interviews. *Political Science and Politics* 35 (4): 669–672.

Hydén, L. C. and P. H. Bülow (2003) Who's talking: drawing conclusions from focus groups—some methodological considerations, *International Journal of Social Reasearch Methdology*, 6 (4): 305–321.

Krueger, R. A. and M. A. Casey (2015) *Focus Groups. A Practical Guide for Applied Research* (5th ed.) SAGE Publishing.

Leeuw, F. L. and H. Schmeets (2016) *Empirical Legal Research. A Guidance Book for Lawyers, Legislators and Regulators*. Cheltenham: Edward Elgar Publishing.

Lijphart, A. (1975) The Comparable-cases Strategy in Comparative Research. *Comparative Political Studies* 8 (2): 158–177.

Manko, R. (2013). Survival of the Socialist Legal Tradition? A Polish Perspective. *Comparative Law Review* 4 (2): 1–28.

Massey, O. (2011) A proposed model for the analysis and interpretation of focus groups in evaluation research. *Evaluation and Program Planning* 34 (1): 21–28.

McLoughlin, S. and R. Münz (2011) Temporary and circular migration: opportunities and challenges. Working Paper No. 35. European Policy Centre, Brussels.

Morgan, D. L. (1996) Focus Groups, Annual Review of Sociology, 22.

Patton, M. Q. (2002) Qualitative Evaluation and Research Methods. 3rd edition. Thousand Oaks: Sage Publications.

Stake, R. E. (1994) Case Studies. In N. K. Denzin and Y. S. Lincoln (eds), *Handbook of Qualitative Research*. Thousand Oaks: Sage Publications, pp. 236–247.

Toshkov, D. (2012) Compliance with EU law in Central and Eastern Europe: The Disaster that Didn't Happen [Unpublished].

Webley, L. (2010) Qualitative Approaches to Empirical Legal Research. In P. Cane and H. Kritzer (eds), *The Oxford Handbook of Empirical Legal Research*. Oxford: Oxford University Press, pp. 926–950.

Weiss, R. S. (1994) *Learning from Strangers. The Art and Method of Qualitative Interview Studies*. New York: The Free Press.

Yin, R. K. (2014) *Case study research: Design and methods* (5th ed.). Thousand Oaks, CA: Sage Publications.

Zweigert, K. and H. Kötz (1998) *An Introduction to Comparative Law* (3rd ed.). Oxford: Oxford University Press.

Chapter 3
The Formulation of the EU's Approach to Circular Migration

This chapter presents the process of the formulation of the circular migration approach at the European level. It examines the genesis of the circular migration concept in the EU's migration policy by discussing the main policy developments and legal acts adopted since the Tampere programme as well as by analysing the interviews concluded with EU and international actors that have had a bearing on the topic. In doing so it outlines the formulation of a two-fold approach towards the facilitation of circular migration at the EU level, incorporating legal migration directives and policy initiatives developed under the auspices of the Global Approach to Migration and Development (GAMM).

3.1 The Genesis of the Circular Migration Concept in EU's Migration Policy

The development of the EU's approach to circular migration can be traced back to the entry into force of the Treaty of Amsterdam, when the EU was endowed with competences in the field of migration under Title IV 'Visas, asylum, immigration and other policies related to free movement of persons'.[1] The newly-inserted Title IV gave the EU the competence to adopt measures in the ambit of internal and external border controls; short-term visas; asylum; and immigration policy, covering conditions of entry and residence, family reunion, measures concerning the rights and conditions under which legally-residing third-country nationals may reside in other Member States, as well as competence in the field of irregular migration and repatriation.[2]

[1] Vankova (2018), p. 188.

[2] Articles 61–63 TEC. For more details, see Hailbronner (2000), pp. 35–122.

Z. Vankova, *Circular Migration and the Rights of Migrant Workers in Central and Eastern Europe*, IMISCOE Research Series,
https://doi.org/10.1007/978-3-030-52689-4_3

This provided a legal basis for the EU to start adopting secondary legislation in all of these respective policy fields and work towards the development of an Area of Justice, Freedom and Security within the EU, as well as securing a more robust role for the European Commission, not only in proposing policy and legislation but also in negotiating agreements with third countries.[3] The first multi-annual programme for the development of the EU's Justice and Home Affairs policies was adopted by the European Council in Tampere on 15–16 October 1999. Covering a 5-year period, the Tampere programme set out the political guidelines for the establishment, inter alia, of a common EU migration and asylum policy.[4]

Making use of the competences granted to it by the Treaty of Amsterdam to legislate on the immigration of third-country nationals, four instruments pertaining to the admission of third-country nationals were gradually proposed by the European Commission between 1999 and 2002: on the right to family reunification; the status and rights of long-term residents; on admission for paid employment and self-employed activities; and, finally on the admission conditions for the purposes of studies, vocational training, or voluntary service.[5] In 2004, in order to complement the proposal for the Students' Directive, the Commission also presented a proposal on the admission and mobility of researchers. Despite the 'ambitious' agenda on legal migration that followed from the European Council in Tampere, however, it took three-and-a-half years of difficult negotiations before the Family Reunification and the EU Long-term Residence Directives were finally adopted, followed by the adoption of the Students' and Researchers' Directives, respectively.[6]

Yet, due to the limitations of the institutional settings under Title IV of the Treaty of Amsterdam and the strong resistance exhibited by the Member States, the negotiations behind the proposal that focused on the admission for paid employment and self-employed activities were put on hold, and it was subsequently withdrawn.[7] This proposal, however, should be briefly reviewed because it illustrates the EU's first attempt to legislate on the possibility for circular migration through 'secure legal status for temporary workers who intend to return to their countries of origin, while at the same time providing a pathway leading eventually to a permanent status for those who wish to stay and who meet certain criteria'.[8] Furthermore, it contained provisions in the policy areas that needed to be addressed with regard to circular migration.

---

[3] Lavenex and Stucky (2011), p. 118.

[4] The rest of the policy domains are a genuine European area of justice; a union-wide fight against crime and stronger external action. In European Council (1999). These issues will not be discussed as they fall outside the focus of this study. For more details on the implementation of the Tampere Presidency Conclusions, see Wiesbrock (2010), pp. 717–31; De Somer et al. (2019) and Carrera et al. (2020).

[5] Papagianni (2006), pp. 157–158.

[6] Ibid., p. 159.

[7] European Commission (2005a).

[8] Interview # 30 with European Commission official, Belgium, November 2017, Annex I. European Commission (2000), p. 17. Also included in the Explanatory Memorandum in European Commission (2001).

3.1.1 *The 2001 Proposal for a Council Directive on Admission for Paid Employment and Self-Employed Economic Activities*[9]

As Papagianni notes, the proposal was based on two main categories of principles: the establishment of a transparent, flexible, and clear mechanism to recruit workers both quickly and successfully, and respect for the domestic labour market needs of the Member States.[10] Article 11 of the Commission proposal listed the rights conferred on economic migrants including, *inter alia*, the right to entry and re-entry after a temporary absence from the territory of the Member State that issued the permit.[11] The Commission stressed in its Explanatory Memorandum that 'this proposed Directive shall ensure that migrants are not cut off from their country of origin and that they have possibilities of going back as the situation develops in the country of origin'.[12]

Article 11 (3) of the proposal also legislated for the right to request and obtain the payment of the contributions made by economic migrants and their employers into public pension schemes during the period of validity of their permits under certain conditions. According to the Explanatory Memorandum, return migration would have been discouraged '(…) if third-country nationals were too "lose" the payments they made into public pension schemes in a Member State upon return to a third country'. Therefore, this provision served as 'supplementary protection addressing those cases in which the concerned third-country national has neither acquired a right to an EU pension to be paid now or in the future in a third country, nor a possibility to transfer his/her EU pension rights into a scheme of the third country where he/she resides'.[13]

What also needs to be mentioned is that the proposal envisaged that economic migrants should enjoy 'the same treatment in substance as citizens of the Union' at least with regards to certain basic rights, such as working conditions, access to vocational training, recognition of qualifications, and social security including healthcare.[14] According to the Explanatory Memorandum, this was aligned with the catalogue of rights that were present in the Commission's proposal for an EU Long-term Residence Directive, but was also designed in line with the principle that the rights of third-country nationals should be incremental with their length of stay. In addition, Article 10 (3) stipulated that unemployment per se was not to constitute a sufficient reason for revoking the permit, unless the period of unemployment exceeded a specified period of time. Nevertheless, economic migrants were

---

[9] European Commission (2001).

[10] Papagianni (2006), p. 170.

[11] Article 11 (1) (a) and (b) of the Proposal. In European Commission (2001).

[12] Article 11 (3) of the Explanatory Memorandum. In European Commission (2001).

[13] Article 11 (3) of the Explanatory Memorandum. In European Commission (2001).

[14] Article 11 (1) of the Explanatory Memorandum. In European Commission (2001).

restricted to the exercise of specific professional activities or fields of activities for an initial period of 3 years.[15]

The Commission's proposal provided for specific rules in relation to certain categories of migrants such as seasonal workers, intra-corporate transferees, and trainees. This proposal's failure gradually led to the establishment of a sectoral approach to labour migration addressing different categories of migrants at the EU level and, eventually, to the adoption of the Single Permit Directive. These instruments are discussed later in this chapter.

3.1.2 *The Notion of Partnerships with Countries of Origin*

The development of the EU circular migration approach is also intertwined with the notion of partnerships with countries of origin.[16] The European Council in Tampere identified partnerships with countries of origin as being one of the essential elements of a future comprehensive approach to migration, with the potential to address political, human rights, and development issues in third countries.[17] Additionally, more efficient management of migration flows was highlighted as another integral part of the future common EU migration and asylum policy.[18] The Council underlined that there should be efforts to conclude readmission agreements with third countries, as well as to promote voluntary return, both considered key elements in this development. Legal migration was one of the aspects acknowledged by the European Council, stressing the need for approximation of national laws on the conditions for the admission and residence of third-country nationals.[19] Partnerships with countries of origin and readmission agreements were identified once again as being among the essential policy measures in the Council's report on 'European Union priorities and policy objectives for external relations in the field of justice and home affairs'.[20]

The Communication of the Commission on a Community Immigration Policy that was issued 1 year later presented the idea of creating legal migration channels to the EU for third-country nationals.[21] As Lavenex noted, however, 'the comprehensive partnership approach reached its rhetorical peak at Tampere'; in subsequent years, this gave way to the notion of partnership with a one-sided focus on migration control and readmission.[22]

---

[15] Article 8 of the Proposal. In European Commission (2001).

[16] Carrera and Hernández i Sagrera (2009), p. 4.

[17] European Council (1999), paras. 10–12.

[18] Ibid, paras. 22–27.

[19] European Council (1999), para. 20.

[20] Council of the European Union (2000); see also previous draft, document number 7512/00 LIMIITE JAI 33 of 24 May 2000.

[21] European Commission (2000), p. 17.

[22] Lavenex and Stucky (2011), p. 119.

The Laeken Presidency Conclusions from December 2001 reconfirmed the need for integrating migration policy into the European Union's foreign policy, particularly through the conclusion of readmission agreements with third countries.[23] The conclusions from the Seville European Council took this one step further by placing a special emphasis on the link between the EU's relations with third countries and combating 'illegal migration'.[24] The European Council underlined that '(...) any future cooperation, association or equivalent agreement which the European Union or the European Community concluded with any country should include a clause on joint management of migration flows and on compulsory readmission in the event of illegal immigration'.[25] As Carrera and Hernandez I Sagrera stress, the Spanish presidency marked the beginning of a 'securitarian approach to human mobility' in the development of the external dimension of the EU's immigration policy.[26] However, that trend had actually commenced in 1985 following the initiation of the Schengen cooperation, when the set of rules adopted aimed primarily at preventing the entry of undesired migrants.[27] The prevalence of measures targeting visas, irregular migrants, border control, and expulsion continued after 1999.

In December 2002, the European Commission published a Communication titled 'Integrating Migration Issues in the European Union's Relation with Third Countries'.[28] This was the Commission's answer to the request from the Heads of States and Governments to apply a targeted approach and integrate immigration policy into the Union's relations with third countries by using all of the appropriate EU external relations instruments. One of the purposes of the Communication was to 'put the migration issue back in its broader context' within the political dialogue part of the Association and Co-operation Agreements negotiated and signed between the EU and third countries.[29] The Commission exhibited a shift in its approach by stressing that the focus of the dialogues with third countries would be broadened to address not only readmission and irregular migration, but also the legal migration channels. More specifically, the agreements would also include: measures targeting the root causes of migration; legal migration management, including ways of regulating the demand and supply of low skilled labour through temporary working permits; the integration of legal migrants; and, the facilitation of 'brain circulation' for legally residing migrants in the EU who wished to contribute to the development of their countries of origin.[30]

In line with the securitarian approach promulgated by the Seville European Council, the Commission proposed in its Communication that Article 13 of the

---

23 European Council (2001), para. 40.

24 European Council (2002), p. 8–9, para. 33–36.

25 European Council (2002), p. 8, para. 33.

26 Carrera and Hernández i Sagrera (2009), p. 5.

27 Gronendijk (2012), p. 12.

28 European Commission (2002), p. 4.

29 Ibid., p. 23.

30 Ibid., p. 23.

Cotonou Agreement between the EU and African, Caribbean and Pacific Group of States (ACP States) should serve as a flexible model for migration clauses that are to be negotiated in future agreements with other third countries.[31] Article 13 of the Cotonou Agreement contains provisions pertaining to cooperation on migration issues, including, inter alia, ensuring respect for human rights and the elimination of all forms of discrimination, fair treatment, and integration of legally residing third-country nationals.[32]

However, Article 13 also provides for combatting and prevention of irregular migration, as well as a standard readmission clause containing a commitment for each of the ACP States to 'accept the return and readmission of any of its nationals who are illegally present in the territory of a Member State of the European Union, at that Member State's request and without further formalities'.[33] In addition, Article 13 includes a commitment to negotiating readmission agreements, if this is requested by one of the parties. All of this showed that even though the Commission advocated for 'a broader context', its rationale actually remained security-driven. This was also evident from the way that the European Commission justified the shift in its approach, which is reproduced as follows:

> 'Experience so far has taught that the time needed to negotiate a readmission agreement, which is seen as being in the sole interest of the Community, should not be underestimated and no quick results should be expected. They can only succeed if they are part of a broader co-operation agenda, which takes duly into account the problems encountered by partner countries to effectively address migration issues. This is the reason why the Commission considers that the issue of "leverage" – i.e. providing incentives to obtain the co-operation of third countries in the negotiation and conclusion of readmission agreements with the European Community – should be envisaged on a country by country basis, in the context of the global policy, cooperation and programming dialogues with the third countries concerned'.[34]

Part 11 'Readmission agreements' of the Communication 'Integrating Migration Issues in the European Union's Relation with Third Countries' revealed that all measures, except those in the fields of readmission and irregular migration, were outside 'the sole interest of the Community' and were used as 'leverage' in order to ensure cooperation with third countries.[35] The ultimate goal of the partnership with third countries was to negotiate and sign readmission agreements. Therefore, the Commission also envisaged the possibility to ensure accompanying support in the form of technical and financial assistance for the better management of migration

---

[31] Partnership agreement between the members of the African, Caribbean and Pacific Group of States of the one part, and the European Community and its Member States, of the other part, signed in Cotonou on 23 June 2000 [2000] OJ L 317.

[32] For more details, see Koeb and Hohmeister (2010).

[33] Ibid, p. 25.

[34] Ibid, p. 25.

[35] In European Commission (2002).

flows[36] as well as compensatory measures for cooperating countries such as a more generous visa policy or increased quotas for migrant workers.[37]

Even though the Communication did not explicitly mention the term circular migration, it addressed the brain drain phenomenon and the need to facilitate brain circulation within the EU in Part 4.2, titled 'Brain circulation'. The Commission stressed the 'win-win scenarios' for all parties – the migrant as well as the sending and the receiving states – that were possible when migrants maintained links with their country of origin and returned voluntarily on either a permanent or temporary basis.[38]

On the basis of the Communication 'Integrating Migration Issues in the European Union's Relation with Third Countries', the Council adopted conclusions on migration and development, acknowledging the 'development of a real partnership with third countries' in the 'broader context' set by the Commission in May 2003 as a key element of a future successful migration policy.[39] The Council formulated five key principles for action on migration and development, covering, *inter alia*, the Union's objectives to tackle the root causes of irregular migration, to combat the smuggling and trafficking of human beings, to improve the control and management of migration flows, and to address the root causes of migration as part of a comprehensive approach for dialogue and action with third countries.[40] The Commission was invited to advance the migration agenda within the context of current and future Association, Cooperation, or equivalent Agreements, as well as to continue to invest in efforts for the conclusion of readmission agreements for compulsory readmission in the event of irregular migration.[41] The European Commission was also requested to formulate ways to regulate 'demand and supply and organising access of labour, e.g. through temporary residence-work permits' and to facilitate the brain circulation of highly-skilled migrants.[42]

Five years after the adoption of the Tampere programme in 1999, the European Council approved a second multi-annual programme on the Union's policy priorities in the Area of Justice and Home Affairs for the 2005–2009 period. The Hague Programme once again emphasised the importance of partnership with third countries.[43] In light of the terrorist attacks in the United States on 11 September 2001 and in Madrid on 11 March 2004, and the strengthened securitarian approach, the

[36] After Tampere, the European Commission started to provide support to third countries through programmes specifically dedicated to border management, the fight against irregular migration, and migration management. These include, for example, MEDA for the Mediterranean region, CARDS for the Western Balkans, TACIS in Eastern Europe and Central Asia, etc. In European Commission (2002), p. 18.

[37] In European Commission (2002), p. 18.

[38] Ibid., p. 16.

[39] Council of the European Union (2003), p. 4.

[40] Ibid., p. 5.

[41] Ibid., p. 9.

[42] Ibid., p. 8.

[43] European Council (2005a), p. 11.

European Council demanded the establishment of a removal and repatriation policy that was based on common standards.[44]

To sum up, the circular migration concept emerged in the EU's policy agenda as a 'brain circulation' tool that was gaining prominence as part of the migration-development nexus discourse.[45] Initially adopted as one of the instruments of the external dimension of the EU's migration policy, it was soon coupled with the tools that were needed for the fight against irregular migration, border control, and readmission and it gradually became an integral part of the EU's securitarian approach towards migration. Eventually, these two parallel policy tracks merged, thereby forming security 'channels' for migration.

3.2 Introduction of the Circular Migration Approach in the EU's Migration Policy

The European Commission introduced the term circular migration into European policymaking circles through its Communication on Migration and Development in 2005.[46] One Commission official, who was involved in this process, shared that a number of academics and business sector representatives proposed to the Commission that they should utilise this concept.[47] The main motivation behind engaging with it was the fact that the labour market was changing – there were more temporary jobs requiring very specific skills and different means of production – and thus the argument was that the traditional understanding of migration leading to settlement as a main goal was not capable of capturing these new realities.[48] Additional reasons for introducing this concept were to counteract the 'brain drain' from third countries' health sectors and also because circular migration was seen as means of allowing 'skills to move back and forth'.[49] Another official with the institutional memory of the introduction of this concept shared with the author that the Commission honed in on the 'famous triple-win idea' and used it to make a case for legal migration in general and to '(…) make it easier for the Member States to step in (…)'.[50] In addition, the idea was also to use circular migration under the General Agreement on Trade in Services (GATS).[51] Yet, the Directorate-General for Economic and Financial Affairs was not open to accepting this new concept.

[44] Ibid., p. 13.

[45] See for instance Global Commission on International Migration (2005), p. 17, p. 31.

[46] European Commission (2005b).

[47] Interview #29 with European Commission official, Belgium, May 2017, Annex I.

[48] Interview #29 with European Commission official, Belgium, May 2017, Annex I.

[49] Interview #29 with European Commission official, Belgium, May 2017, Annex I.

[50] Interview #9 with European Commission official, Belgium, March 2013, Annex I.

[51] Interview #29 with European Commission official, Belgium, May 2017, Annex I.

Therefore, it had to make its way by being included in the different Communications that were related to development and migration.[52]

The Commission defined it as '(…) migration, in which migrants tend to go back and forth between the source country and the destination country (...)'.[53] The term was used as an umbrella for return and temporary migration, which according to the European Commission had the potential to maximise the benefits of migration on the development of the countries of origin, mitigating the impact of brain drain and fostering brain circulation. The Commission's working definition was rather broad, covering seasonal and temporary migrants as well as highly-skilled migrants (e.g., researchers) residing both in the country of origin and in the host country (e.g. diaspora members). One interviewee who witnessed the introduction of this concept said: '(…) the interpretation of the Spanish was: we talk about seasonal workers, for whom there are instruments and which is nothing new, which could be better organized and Europeanized (...). The Swedish interpretation sometimes went so far by saying: if the kids of migrants go back, we should also see that as circular migration. We talked from something ultra-short, people going back after 3 months, to something that is interpreted extremely wide: everyone with migrant roots keeping a relationship with their country of origin going back'.[54]

In its Communication, the European Commission also underlined the obstacles and prerequisites that had to be in place in order to realise the potential of this type of migration for the purposes of development. In order to encourage migrants to engage in circular migration and travel back and forth between their country of origin and their country of destination, the Commission emphasised the need to grant returnees a multi-entry visa allowing them to go back to their former country of residence, as well as maintaining the validity of the returning migrants' residence permits once back in their countries of origin.[55]

Furthermore, according to the Communication, Member States could encourage circular migration '(…) by giving a priority for further temporary employment to workers who have already worked under such schemes and have returned at the end of their contract'.[56] It could also be facilitated by 'rewarding' participating migrants by reimbursing their pension contributions at the end of their contracts. In addition to financial incentives, public authorities in the Member States could: offer secondment opportunities to institutions in developing countries for migrants or diaspora members who wish to engage in such activities; remove the legal obstacles to unpaid sabbatical leave; and, encourage businesses to enable their foreign employees to take unpaid leave for engaging in such activities.

Other possible EU measures for facilitating circular migration envisaged in the Communication was the establishment of a general framework for the entry and

---

[52] Interview #29 with European Commission official, Belgium, May 2017, Annex I.

[53] European Commission, (2005b), Annex 5, p. 25.

[54] Interview #22 with representative of international organisation, Austria, March 2017, Annex I.

[55] European Commission (2005b). Annex 5, p. 27.

[56] Ibid., p. 26.

short-term stay of seasonal migrants[57] as well as facilitating conditions for issuing uniform short-stay visas for researchers from third countries travelling within the EU for the purposes of conducting their scientific research.[58] The European Commission also presented the possibility of proposing measures regarding the transferability of pension rights and social security schemes benefits, the recognition of qualifications and mechanisms to facilitate voluntary returns, and the successful reintegration of researchers or other professionals who have worked in the EU.

One interviewee stressed that these were references to the rights under EU law from which circular migrants should benefit.[59] It was also clear that this required the establishment of certain 'regimes'. Yet, the Commission wanted to first see whether Member States were willing to engage with these types of policies and then subsequently focus on the rights dimension.[60] One interviewed representative of an international organisation underlined that many of the rights-related questions were not resolved at the European level mainly because of the lack of a completely harmonised interpretation.[61]

A broad interest in circular migration and its benefits was taken one step further in the Policy Plan on Legal Migration[62] that was presented in 2005 by the European Commission, which proposed specific policy instruments.[63] The Policy plan was elaborated on the basis of the Green Paper on the EU's approach to managing economic migration[64] and the analysis of the contributions received. Its purpose was to serve as a roadmap for the remaining period of the Hague Programme (2006–2009), listing the actions and legislative initiatives that would lead to the coherent development of an EU legal migration policy.[65] The Commission envisaged a general framework directive that would regulate the rights of all third-country nationals in legal employment who were already present in a Member State but who had not yet obtained a EU long-term residence status, as well as the following four specific instruments:

1. a Proposal for a directive on the conditions of entry and residence of highly-skilled workers;
2. a Proposal for a directive on the conditions of entry and residence of seasonal workers;

[57] Refer to European Commission (2005c).

[58] The Commission already proposed in March 2004 a Recommendation concerning the facilitation of issuing conditions for uniform short stay visas for researchers from third countries travelling within the EU for the purpose of carrying out scientific research.

[59] Interview #29 with European Commission official, Belgium, May 2017, Annex I.

[60] Interview #29 with European Commission official, Belgium, May 2017, Annex I.

[61] Interview #22 with representative of international organisation, Austria, March 2017, Annex I.

[62] European Commission (2005d).

[63] Vertovec (2007), p. 4.

[64] European Commission (2005c).

[65] European Commission (2005d), p. 3.

3. a Proposal for a directive on the procedures regulating the entry into, the temporary stay and residence of Intra-Corporate Transferees (ICT); and
4. a Proposal for a directive on the conditions of entry and residence of remunerated trainees.

The proposed instruments did not make any explicit reference to circular migration. However, in a separate section called ‘Cooperation with countries of origin’, the Commission proposed feasibility studies for separate measures to support circular and return migration, and stressed that arrangements on managed temporary and circular migration would be included in some of the specific instruments.[66] Among the envisaged measures that aimed to facilitate circular migration were the provisions for long-term multi-entry visas for returning migrants; a possibility for former migrants to be given priority for obtaining new residence permits for further temporary employment under a simplified procedure; and the creation of an EU database of third-country nationals who left the EU upon the expiration of their temporary residence or work permit.

Moreover, the Communication referred to the EU Long-term Residence Directive as a good example that offered ‘interesting possibilities’ regarding circular and return migration. According to the Commission, the Directive offered a possibility for Member States to allow returning migrants to retain this status for longer than the 1-year period that was provided for under Article 9 thereof. Therefore the Roadmap for the proposed measures in the Policy Plan envisaged an analysis of the transposition and implementation of Article 9 of the Directive, which was due to be implemented in 2007.[67] Among the other Roadmap activities, there were also feasibility studies carried out in relation to long-term multi-entry visas and on the effective implementation of circular migration; financial support under EU instruments for pilot projects for the creation of training structures in the countries of origin; and presentation of proposals tangibly supporting circular and return migration based on preceding studies and feasibility analyses.

3.3 Circular Migration as Part of the Global Approach to Migration and Mobility (GAMM)

The deaths of hundreds of African migrants trying to enter the Spanish enclaves of Ceuta and Melilla in North Africa in 2005 triggered a new impetus for a policy response at the EU level in relation to cooperation with third countries.[68] At an informal meeting in Hampton Court in October 2005, the EU Heads of State and Government agreed on a comprehensive approach to tackle migration.[69] The Council

[66] European Commission (2005d), p.11.

[67] Ibid., p.13.

[68] Eisele (2014), p. 90.

[69] Ibid.

further stressed the importance of creating partnerships with third countries and the benefits of a 'comprehensive and balanced approach'.[70] It concluded that such an approach would 'enhance the benefits of migration for both third countries and the EU, as well as migrants themselves, whilst ensuring co-ordinated action against illegal migration, trafficking in human beings and people smuggling'.[71] The existing agreements with third countries – the Cotonou Agreement, the Stabilisation and Association Agreements, the Neighbourhood Action Plans, and the Euro-Mediterranean Association Agreements, as well as the dialogue within the Barcelona Process – were identified as the basis of this future comprehensive approach.[72]

In a follow-up Communication, the Commission also recognised the need for a coherent overall approach to migration issues. It committed itself to focusing on all aspects of migration and intensifying financial assistance within its existing institutional framework, covering the areas of Development, External Relations, European Neighbourhood Policy, Freedom, Security and Justice, and Employment.[73] Acknowledging that migration was a global phenomenon, it proposed priority actions aimed at increasing operational coordination between Member States and strengthening the dialogue with neighbouring countries and countries of origin, with a specific focus on the Mediterranean region and African countries.

The European Council officially adopted this approach, which started being referred to as the 'Global Approach to Migration' (GAM) in December 2005.[74] According to the Council Conclusions, the new approach consisted of policy measures that were designed to combat irregular immigration, ensure safe return, strengthen durable solutions for refugees, and build the capacity for better managed migration, including 'through maximising the benefits to all partners of legal migration'.[75] However, beyond that statement, it did not envisage any concrete labour migration measures and thus there was no mention of circular migration.[76]

Circular migration based on EU partnerships with third countries became an attractive policy tool within the Franco-German initiative for a 'New European Migration Policy' initiated in 2006.[77] This approach was identified by Nicolas Sarkozy and Wolfgang Schäuble as a strategy that would reduce and control irregular migration on the basis of quotas for temporary labour migration into certain occupations and that would be accompanied by measures for readmission in cases where migrants do not want to return voluntarily.[78] Therefore, the Commission was

[70] Council of the European Union (2005), pp. 4–7.

[71] Ibid.

[72] Ibid.

[73] European Commission (2005e), p.2.

[74] European Council (2005b).

[75] European Council (2005b), para. 3.

[76] Carrera and Hernández i Sagrera (2009), p. 11.

[77] Angenendt (2007), p. 1.

[78] "German-French Initiative for a New European Migration Policy", a strategy paper presented by Nicolas Sarkozy and Wolfgang Schäuble to a meeting of interior ministers from France, Germany, Great Britain, Italy, Spain and Poland in Stratford-upon-Avon, England, 26 October 2006, p. 4.

invited to continue to negotiate readmission agreements, with a view to securing quotas and permits for temporary workers. Moreover, in order to address the need for enhanced cooperation with third countries, the European Commission also had to present to the Council a plan for developing partnerships between the Member States and main countries of origin based on a 'European treaty'.

By the end of 2006, the European Commission reported on the first year of the implementation of the Global Approach, which at this stage only focused on Africa and the Mediterranean region.[79] The Commission proposed the inclusion of two new policy areas in the GAM that were not covered in 2005 – legal migration and integration measures – in order to make 'the European Union's approach truly comprehensive'.[80] It acknowledged that legal migration had to form part of both external and internal EU policies, and it consequently proposed two concrete policy measures: the creation of migration centres and mobility packages.[81]

The aim of the migration centres was to provide information on employment opportunities within the EU and facilitate development of different skills that would increase the chances of third-country nationals finding legal employment in Europe. The Commission proposed to establish these centres with EU funding in partnering third countries and claimed that they could also contribute to facilitating the management of seasonal workers as well as the exchange of students and researchers.[82]

The Communication linked circular migration to the newly-introduced Mobility Packages. In order to facilitate circular migration as a migration management tool, the Commission envisaged the adoption of measures relating to administrative capacity-building in third countries.[83] The boosted administrative capacity would enable third countries to meet 'certain conditions' with regards to cooperation on irregular migration and establish effective mechanisms for readmission that would allow them to start negotiations for Mobility Packages with the EU. The aim of the Mobility Packages was to serve as a framework for managing legal migration that was based on the 'possibilities offered by the Member States and the European Community, while fully respecting the division of competences as provided by the Treaty'.[84] The Commission also stressed that in the context of this framework, the conclusion of readmission agreements, strengthening cooperation on irregular immigration, and border management could be prerequisites for visa facilitation with the partnering third countries.

In December 2006, following the proposals of the Commission, the European Council[85] identified circular migration as one of the guiding principles in the devel-

---

[79] The geographical focus of this study is on the CEE and Eastern Partnership countries, thus the GAM's provision on cooperation with African countries of origin is not presented here.

[80] European Commission (2006), p. 2.

[81] European Commission (2006), p.7.

[82] Ibid.

[83] Ibid.

[84] Ibid.

[85] European Council (2006), point 24, lett. a, p. 9.

opment of the EU's policy on legal migration.[86] The Commission was invited to present proposals by June 2007 on legal migration focused on the development of a balanced partnership with third countries and on the ways and means of facilitating circular and temporary migration.[87] The European Council also decided to extend the GAM's geographical focus to cover the neighbouring Eastern and South-Eastern regions.

3.4 Towards an EU Two-Fold Approach to Facilitation of Circular Migration

In response to the invitation of the European Council, the Commission published a Communication in May 2007 that focused entirely on circular migration and Mobility Partnerships between the European Union and third countries.[88] The Commission used the working definition of circular migration that it adopted earlier in its Communication on Migration and Development in 2005. The Communication further elaborated upon the working definition by outlining two main forms of circular migration that were relevant in the EU context: temporary engagement of third-country nationals settled in the EU with business, professional, or voluntary activities in their countries of origin; and temporary opportunities for entry and re-entry for persons residing in a third country for the purposes of work and study or training in the EU.

The Commission envisaged two approaches for facilitating circular migration: based on a legislative framework promoting circular migration and through development of circular migration schemes enabling migrants to temporarily perform certain activities in the EU. In the Commission's view both approaches had to be accompanied by the necessary conditions and safeguards to ensure that this kind of migration would remain circular.[89] As Wiesbrock and Schneider underline,[90] the Communication emphasised the temporary character of circular migration that the EU wished to facilitate: 'if not properly designed and managed, migration intended to be circular can easily become permanent and, thus, defeat its objective'.[91] The legislative harmonisation of the already-existing instruments and the introduction of special measures in future legislative acts, announced in the Policy Plan on Legal Migration, were identified as ways to put in place a framework conducive to circular migration.

---

[86] Schneider and Wiesbrock (2009) p. 4.

[87] European Council (2006), point 24, lett. a, p.9.

[88] European Commission (2007).

[89] European Commission (2007), p.8.

[90] Schneider and Wiesbrock (2009), p. 5.

[91] European Commission (2007), p.8.

The directives that the Commission considered adjusting for the sake of fostering circularity included the EU Long-term Residence Directive,[92] the Students' Directive,[93] and the Researchers' Directive.[94] The Policy Plan on Legal Migration already identified the EU Long-term Residence Directive as an instrument that offered 'interesting possibilities' in relation to circular migration, and it was envisaged to carry out an analysis of the transposition and implementation of Article 9 thereof. The Communication reconfirmed the Commission's idea to extend the permitted period of absence from the territory of the EU from 12 consecutive months to 2 or 3 years, after which the EU long-term status could be withdrawn. A similar measure was also envisaged in relation to the Students' and Researchers' Directives – the introduction of a multi-entry residence permit, allowing the holder to be absent from EU territory for long periods without losing his or her residence rights – already provided for in the Communication on Migration and Development in 2005.[95] Some of the special measures foreseen as beneficial to circular migration included fast-track admission procedures for highly-skilled migrants already legally resident in the EU, multi-annual residence/work permits for seasonal workers – a matter already proposed in the 2005 Communication – and facilitated re-entry for former trainees, allowing them to develop their skills.

Ensuring effective return was among the conditions and safeguards that the Commission foresaw as essential in order to make circular migration work. Therefore, Member States were encouraged to make return for migrants an attractive option through the enactment of measures that reward 'bona-fide migrants' as well as measures supporting the reintegration of returning migrants.[96] In cases of an irregular stay after the expiration of a migrant's permit, readmission measures were envisaged that would be facilitated through readmission agreements between the countries of origin and the EU or the Member States. The development of a set of criteria for monitoring future circular migration schemes was also part of the conditions and safeguards put forward by the Commission, along with an assessment of the contribution of the relevant legal instruments that were designed to facilitate circular migration.[97]

The Commission's proposals were welcomed by the Council in December 2007. The adopted Council Conclusions on Mobility Partnerships and circular migration in the framework of the Global Approach to Migration[98] made it clear that circular

---

[92] Council Directive 2003/109/EC of 25 November 2003 concerning the status of third-country nationals who are long-term residents [2004] OJ L 16.

[93] Council Directive 2004/114/EC of 13 December 2004 on the conditions of admission of third-country nationals for the purposes of studies, pupil exchange, unremunerated training or voluntary service [2004] OJ L 375/12, no longer in force.

[94] Council Directive 2005/71/EC of 12 October 2005 on a specific procedure for admitting third-country nationals for the purposes of scientific research [2005] OJ L 289, no longer in force.

[95] European Commission (2005b), p.10.

[96] Schneider and Wiesbrock (2009), p. 6.

[97] European Commission (2007), p. 12.

[98] Council of the European Union (2007).

migration should become an integral part of future Mobility Partnerships. The definition used in the document once again confirmed the temporary character of the migration at stake as follows: '(…) circular migration could be understood as the temporary, legal movement of people between one or more Member States and particular third countries, whereby third country nationals take up legal employment opportunities in the EU or persons legally residing in the EU go to their country of origin'.

The Council proposed a set of 'possible elements which could be addressed when facilitating circular migration' including integration and the accompanying measures that would be available to migrants prior to their arrival in the EU; partnerships between labour market agencies of partner countries and Member States to better match supply and demand; improved mutual recognition of qualifications; measures to ensure return and readmission; an adequate legal framework to promote circular migration, and so on. As Pastore stressed, most of these elements were not specific to the management of circular migration, but rather were part of a very traditional 'linear' migration management approach.[99]

In line with the Policy Plan on Legal Migration, the Commission proposed two legislative measures in 2007: a directive on the admission of highly qualified migrants in the EU, creating the so-called EU Blue Card adopted in 2009, and a directive establishing a single permit and common set of rights for third-country workers legally residing in the EU. This latter measure could not be agreed before the entry into force of the Treaty of Lisbon. As Thym emphasised, unlike the instruments in the field of entry and border controls, political agreement on new legal migration measures was not easy to obtain.[100]

The Treaty of Lisbon's entry into force recalibrated the EU's competence in the field of economic migration and brought about changes to the decision-making process that paved the way for the adoption of the legal instruments envisaged by the Policy Plan on Legal Migration on the basis of Article 79 TFEU. The EU migration policy that was conceived by the EU Treaties was based on 'differentiated and selective admission process' that was provided by statutory rules decided by the EU legislature on the basis of the ordinary legislative procedure (Article 79 (2) TFEU).[101] In addition, this EU migration policy was established as a 'process of legal status change', allowing the EU to provide different permits depending on the circumstances of each individual case, including not allowing an option of renewal as was the case for the Seasonal Workers' Directive.[102] However, the EU's established competence to adopt legal rules in the ambit of economic migration as a matter of principle meant that Member States were free to determine the numbers of third-country national workers coming to the EU (Article 79 (5) TFEU).[103]

[99] Pastore (2008), p. 7.

[100] Thym (2016), p. 272.

[101] Thym (2016), p. 276.

[102] Ibid.

[103] Ibid, p. 278.

Following the entry into force of the Treaty of Lisbon, the European Council adopted the Stockholm Programme in 2009, which provided the Justice and Home Affairs programme to cover the period 2010–2014.[104] The European Council invited the Commission to submit proposals before 2012 'on ways to further explore the concept of circular migration and study ways to facilitate orderly circulation of migrants, either taking place within, or outside, the framework of specific projects or programmes including a wide-ranging study on how relevant policy areas may contribute to and affect the preconditions for increased temporary and circular mobility'.[105] Furthermore, the Justice and Home Affairs Council Conclusions at the end of 2009 stated that the Member States and the Commission were committed 'to further examine issues which may have the potential to facilitate circular migration and voluntary return, such as portability of social rights, migrants' opportunities to return to their countries of origin for longer periods of time without losing their right to residence in countries of destination as well as the promotion of viable livelihood options in countries of origin'.[106] The Council also called for an 'in-depth qualitative and quantitative analysis in order to further explore the concepts of temporary and circular migration'.[107]

In response to this invitation, the European Migration Network (EMN) undertook a study entitled 'Temporary and circular migration: empirical evidence, current policy, practice and future options', which was published in 2011.[108] The study concluded that despite the EU's focus on facilitating this type of migration, the policy and legal developments at the national level were still 'in an embryonic stage' and were considerably diverse; whilst some of the Member States' policies contained elements of circular migration, these were not explicitly acknowledged.[109] It also stressed that the initial evaluations pointed to positive outcomes for participating migrants.

The Treaty of Lisbon also made it possible to reach agreement on the Single Permit Directive in early 2010, after the Council had been deadlocked over this issue for more than 3 years.[110] In 2010, the Commission also proceeded with their proposals on the position of intra-corporate transferees and seasonal workers. The EMN study that evaluated the impact of the Stockholm Programme stressed that the adoption of the Seasonal Workers' Directive was a significant development with regards to circular migration, and that this legal instrument formalises 'incremental work undertaken by Member States in this policy area throughout the period covered by the Stockholm Programme'.[111]

[104] European Council (2010).

[105] Point 6.1.2. of the Stockholm Programme. European Council (2010).

[106] Council of the European Union (2009), p. 2.

[107] Ibid.

[108] European Migration Network (2011).

[109] Ibid., p. 63.

[110] For more details see Peers et al. (2012), p. 223.

[111] European Migration Network (2014), p. 22.

In addition, on the basis of a GAM evaluation, the European Commission promulgated a renewed Global Approach to Migration and Mobility (GAMM) which aimed to make it 'more strategic and more efficient, with stronger links and alignment between relevant EU policy areas and between the external and internal dimensions of those policies'.[112] The inclusion of 'mobility' in the framework's title aimed to explicitly show that using migration and mobility in a safe environment was an essential part of the EU's external relations.[113] Migration was understood as a general category covering both legal channels for migration and combating irregular migration. Mobility, on the other hand, was perceived as part of the general category of migration referring to the possibility to legally enter the territory of the EU and denoting the notion that the EU wanted to increase 'the movement of persons'.[114]

One of the interviewees who was involved in the Task Force on Temporary and Circular Migration coordinated by the European Policy Institute in the period 2010–2011 commented that interest in this concept sharply declined after the Task Force report was published as a result of the 'very difficult political context'.[115] Therefore, it was not surprising that at the dawn of the next phase of the EU's Area of Freedom, Security and Justice for the period 2015–2020, none of the EU's institutional actors made reference to the concept of circular migration in their respective policy agendas and programmes.[116] Nor was it mentioned in the European Agenda on Migration that was adopted as a response to the 2015 'European refugee and migrant crisis' and the tragedies that took place in the Mediterranean Sea.[117] However, it kept being used as an instrument that was part of the framework of the GAMM,[118] including in the latest Mobility Partnership concluded with Belarus. Furthermore, it was also part of the Legal Migration Fitness Check of the European Commission.[119] A representative of an international organisation who was interviewed said that '(…) it lingers within the EU frameworks because it was mentioned as a point that nobody elaborated on and is part of all projects'.[120] However, according to the interviewee, 'it was out'.[121]

---

112 European Commission (2011).

113 Interview # 8 with EU Council official, Belgium, February 2013, Annex I.

114 Interview # 8 with EU Council official, Belgium, February 2013, Annex I.

115 Interview #14 with former think-tank officer, Belgium, August 2013.

116 For more details, see also Carrera and Guild (2014), pp. 19–50.

117 See European Commission (2015).

118 European Commission (2014).

119 European Commission (2016), p. 3 and p. 8.

120 Interview #28 with representative of international organisation, Austria, March 2017, Annex I.

121 Interview #28 with representative of international organisation, Austria, March 2017, Annex I.

3.5 Conclusions

This chapter demonstrated that the circular migration concept entered the EU level through the policy efforts at the international level aimed at maximising the positive effects of the migration-development nexus. At the EU level it was perceived as a policy tool for migration management, which was to be facilitated through legal instruments and through formal, bilateral and multilateral, programmes and projects. The main difference with the international discourse was that at the EU level there was an additional securitarian rationale. Circular migration became one of the instruments of the Directorate General for Migration and Home Affairs and part of its securitarian approach to migration, putting emphasis on return and readmission rather than on facilitation of legal migration.

Despite the fact that the EU has invested efforts in promoting this type of migration since 2001, almost 20 years later it is difficult to identify a cohesive line of policy formulation that pinpoints the EU's approach to circular migration. Scattered among different instruments and characterised by volatility, the EU's approach is difficult to comprehend without putting some flesh on the concept through deeper analysis of its components, which is the task of the next chapter.

References

Angenendt, S. (2007) Circular Migration: A Sustainable Concept for Migration Policy?. *SWP Comments* 11, Stiftung Wissenschaft and Politik, Berlin.

Carrera, S. and E. Guild (2014) The European Council's Guidelines for the Area of Freedom, Security and Justice 2020. Subverting the 'Lisbonisation' of Justice and Home Affairs? *CEPS Essays*, No. 13/14 July 2014, Brussels.

Carrera, S. and R. Hernández i Sagrera (2009) *The Externalisation of the EU's Labour immigration policy. Towards mobility or insecurity partnerships?* CEPS Working Document No. 321, Brussels.

Carrera, S., D. Curtin and A. Geddes (2020) *20 year anniversary of the Tampere Programme. Europeanisation Dynamics of the EU area of Freedom, Security and Justice.* Florence: European University Institute.

Council of the European Union (2000) "A" Item Note from COREPER to the General Affairs Council/ European Council, 7653/00 LIMITE JAI 35 on "European Union priorities and policy objectives for external relations in the field of justice and home affairs", 6 June 2000.

Council of the European Union (2003) Council Conclusions on migration and development, 8927/03, Brussels, 5 May 2003.

Council of the European Union (2005) Council Conclusions on Migration and External Relations, 14769/05, 21 November 2005.

Council of the European Union (2007) Council Conclusions on Mobility Partnerships and circular migration in the framework of the Global approach to migration, 2839th General Affairs Council meeting, Brussels, 10 December 2007.

Council of the European Union (2009) Council Conclusions on Migration for Development, 2979th Justice and Home Affairs Council meeting, Brussels, 30 November 2009.

De Somer, M., P. De Bruycker and J.-L. De Brouwer (eds) (2019) *From Tampere 20 to Tampere 2.0: Towards a New European Consensus on Migration*. European Policy Centre, Brussels.

Eisele, K. (2014) The External Dimension of the EU's Migration Policy. Different Legal Positions of Third-Country Nationals in the EU: A Comparative Perspective. Leiden: Brill /Nijhoff.

European Commission (2000) Communication on a Community immigration policy, COM (2000) 757 final, Brussels, 22 November 2000.

European Commission (2001) Proposal for a Council Directive on the conditions of entry and residence of third-country nationals for the purpose of paid employment and self-employed economic activities, COM (2001) 386, 11 July 2001.

European Commission (2002) Communication on integrating migration issues in the European Union's relations with third countries, COM (2002) 703 final, Brussels, 3 December 2002.

European Commission (2005a) Communication from the Commission to the Council and the European Parliament – Outcome of the screening of legislative proposals pending before the Legislator, COM (2005) 0462 final, Brussels, 27 September 2005.

European Commission (2005b) Communication on Migration and Development: Some concrete orientations, COM (2005) 390 final, Brussels, 1 September 2005.

European Commission (2005c) Green Paper of 11 January 2005 on an EU approach to managing economic migration, COM (2004) 811 final, Brussels, 11 January 2005.

European Commission (2005d) Communication on Policy Plan on Legal Migration, COM (2005) 669 final, Brussels, 21 December 2005.

European Commission (2005e) Communication on Priority actions for responding to the challenges of migration – First follow-up to Hampton Court, COM (2005) 621, Brussels, 30 November 2005.

European Commission (2006) Communication on the Global Approach to Migration one year on: Towards a comprehensive European migration policy, COM (2006) 735, Brussels, 30 November 2006.

European Commission (2007) Communication on circular migration and Mobility Partnerships between the European Union and third countries, COM (2007) 248 final, Brussels, 16 May 2007.

European Commission (2011) Communication on the Global Approach to Migration and Mobility, COM (2011) 743 final, Brussels, 8 November 2011.

European Commission (2014) Report on the implementation of the Global Approach to Migration and Mobility 2012–2013, COM (2014) 96 final, Brussels, 21 February 2014.

European Commission (2015) Communication on a European Agenda on Migration, COM (2015) 240 final, Brussels, 13 May 2015.

European Commission (2016) REFIT- Legal Migration Fitness Check, Indicative Road Map, 1 September 2016.

European Council (1999) Tampere Presidency Conclusions of 15 and 16 October 1999, SN 200/99.

European Council (2001) Laeken Presidency Conclusions, 14 and 15 December 2001, SN 300/1/01 REV 1.

European Council (2002) Seville Presidency Conclusions, 21–11 June 2002, 13463/02.

European Council (2005a) The Hague Programme: strengthening freedom, security and justice in the European Union, OJ C 53/1, 3 March 2005.

European Council (2005b) Brussels Presidency Conclusions of 15 and 16 December 2005, SN 15914/01/05, 30 December 2005, Annex 1: "Global Approach to Migration: Priority Actions focusing on Africa and Mediterranean".

European Council (2006) Presidency Conclusions of 14/15 December 2006, 16879/1/06 REV 1.

European Council (2010) Stockholm Programme: An open and secure Europe serving and protecting the citizens, OJ C 115/1, 4 May 2010.

European Migration Network (2011) Temporary and Circular Migration: empirical evidence, current policy practice and future options in EU Member States. Synthesis Report, Brussels.

European Migration Network (2014) Synthesis Report contributing to evaluation of the Stockholm Programme 2010–2013 from the European Migration Network. A Descriptive Analysis of the Impacts of the Stockholm Programme 2010–2013.

Global Commission on International Migration (2005) Migration in an Interconnected World: New Directions for Action. *Report of the Global Commission on International Migration*.

Gronendijk, K. (2012) Introduction: Migration and Law in Europe. In E. Guild and P. Minderhoud (eds) The First Decade of EU Migration and Asylum Law. Leiden: Brill/Nijhoff, pp. 1-22.
Hailbronner, K. (2000) Immigration and Asylum Law and Policy of the European Union. The Hague: Kluwer Law International.
Koeb, E. and H. Hohmeister (2010) *The revision of Article 13 on Migration of the Cotonou Partnership Agreement. What's at stake for the ACP?* European Centre for Development Policy Management, Maastricht.
Lavenex, S. and R. Stucky (2011) Partnering for migration in EU external relations. In R. Kunz, S. Lavenex and M. Panizzon (eds) Multilayered Migration Governance: The Promise of a Partnership. London: Routledge, pp. 116-142.
Papagianni, G. (2006) Institutional and Policy Dynamics of EU Migration Law. Leiden: Martinus Nijhoff Publishers.
Pastore, F. (2008) Circular Migration. Background note for the Meeting of Experts on Legal Migration, Rabat, 3–4 March.
Peers, S., E. Guild, D. A. Arcarazo, K. Groenendijk and V. Moreno-Lax (eds) (2012) EU Immigration and Asylum Law (Text and Commentary): Second Revised Edition. Volume 2: EU Immigration Law. Leiden: Martinus Nijhoff Publishers.
Schneider, H. and Wiesbrock, A. (2009) *Circular Migration and Mobility Partnerships*. Briefing paper, Directorate-General for Internal Policies, European Parliament.
Thym, D. (2016) Legal Framework for EU Immigration Policy. In K. Hailbronner and D. Thym (eds), EU Immigration and Asylum Law. A Commentary. (2nd edition). Munchen/Oxford/ Baden-Baden: C. H. Beck/Hart/Nomos, pp. 271-299.
Vertovec, S. (2007) Circular Migration: The Way Forward in Global Policy? International Migration Institute Working Papers, No. 4.
Wiesbrock, A. (2010) Legal Migration to the European Union. Leiden: Brill/Nijhoff.
Vankova, Z. (2018) Poland and Bulgaria's bilateral agreements with Eastern partnership countries in the context of circular migration. European Journal of Social Security 20 (2): 188-203.

Chapter 4
The Implementation of the EU's Approach to Circular Migration Through Legal and Policy Routes

This chapter brings together the legal and policy instruments developed as part of the EU's circular migration approach in order to assess its implementation and establish whether it provides rights-based outcomes for migrant workers. In order to do so, it employs a benchmark framework for analysis based on universal and regional international standards and soft law principles as well as policy measures that have been identified as conducive to circular migration (presented in Chap. 1). The benchmark framework covers six policy areas considered inherent to this type of labour migration and which at the same time could help distinguish circular migration from the guest-worker model and other time-bound migration policies: entry and re-entry conditions, work authorisastion, residence status, social security coordination, entry and residence conditions for family members, and recognition of qualifications. To be beneficial for migrant workers, policies need to allow for a certain degree of migrant-led trajectory of movement and provide adequate protection of the rights of migrant workers.[1]

4.1 Legal and Policy Instruments Forming Part of the EU's Approach to Circular Migration

There are several EU legal instruments that must be taken into consideration when analysing the implementation of the EU's approach to circular migration. These are legal instruments that explicitly mention, amongst their aims, the facilitation of

[1] Vankova (2016), p. 339.

Z. Vankova, *Circular Migration and the Rights of Migrant Workers in Central and Eastern Europe*, IMISCOE Research Series,
https://doi.org/10.1007/978-3-030-52689-4_4

circular migration[2]: the Seasonal Workers' Directive[3] and the Blue Card Directive.[4] Additionally, instruments that do not explicitly mention the term circular migration, but contain elements of circular migration that may foster this type of migration: the EU Visa Code,[5] the EU Visa List Regulation,[6] the Visa Facilitation Agreements with the Eastern Partnership countries,[7] the EU Long-term Residence Directive,[8] the Intra-corporate Transferees' Directive,[9] and the Students' and Researchers' Directive.[10] And, finally, instruments that do not refer to circular migration but should be considered because they can contribute to rights-based circulation through provisions on flanking rights thereby allowing for a migrant-led trajectory of circular

[2] Vankova (2018), p. 157.

[3] Directive 2014/36/EU of the European Parliament and of the Council of 26 February 2014 on the conditions of entry and stay of third-country nationals for the purpose of employment as seasonal workers [2014] OJ L 94. See Recital 34 of the Preamble to the Directive containing obligation for Member States to foster circular migration in relation to *bona-fide* seasonal workers.

[4] Council Directive 2009/50/EC of 25 May 2009 on the conditions of entry and residence of third-country nationals for the purposes of highly qualified employment [2009] OJ L 155/17. It explicitly aims to 'encourage geographical and circular migration' of highly skilled third-country nationals. See European Commission (2007) 637, p. 11, p. 16. See also Recitals 20, 21 and 22 of the Preamble to the Directive.

[5] Regulation (EC) No 810/2009 of the European Parliament and of the Council of 13 July 2009 establishing a Community Code on Visas (Visa Code) [2009] OJ L 243/1.

[6] Council Regulation (EC) No 539/2001 of 15 March 2001 listing the third countries whose nationals must be in possession of visas when crossing the external borders and those whose nationals are exempt from that requirement [2001] OJ L 81.

[7] For an overview see Van Elsuwege and Vankova (2020).

[8] Council Directive 2003/109/EC of 25 November 2003 concerning the status of third-country nationals who are long-term residents [2004] OJ L 16.

[9] Directive 2014/66/EU of the European Parliament and of the Council of 15 May 2014 on the conditions of entry and residence of third-country nationals in the framework of an intra-corporate transfer [2014] OJ L 157. It focuses on temporary migration (see Recital 7 of the Preamble). During the negotiations, however, a new provision was added to the text of the Directive, which makes it possible for ICTs to return for a subsequent transfer, at the end of the maximum period of stay that is permitted, which could transform the temporary migration into circular migration. See Council of the European Union (2011a), p. 24.

[10] Directive (EU) 2016/801 of the European Parliament and of the Council of 11 May 2016 on the conditions of entry and residence of third-country nationals for the purposes of research, studies, training, voluntary service, pupil exchange schemes or educational projects and au pairing [2016] OJ L 132. The explanatory memorandum of the Recast Proposal contained a reference to 'brain circulation', stressing that this would be encouraged by allowing third-country nationals to acquire skills and knowledge through a period of training which would benefit both sending and receiving countries. Both the Proposal and the adopted Directive 2016/801 also stress that 'fostering people-to-people contacts and mobility' is an important element of the European Neighbourhood Policy and will contribute to the GAMM and its Mobility Partnerships by potentially providing measures for promoting the mobility of the categories of migrants that are covered by the scope of the Directive. See European Commission (2013), p.3 and Recital 6 of the Preamble to the SRD.

migration: the Single Permit Directive,[11] the Family Reunification Directive,[12] and the EU Long-term Residence Directive. The latter instrument must be understood as having a dual role[13] as it contains a provision that according to the European Commission can facilitate circular migration for settled third-country nationals in the EU and also provides the general rules on accessing the EU long-term residence status.[14]

With regards to the GAMM as a policy route for fostering circular migration, one should focus on the implementation of the Mobility Partnerships[15] with the Eastern Partnership countries.[16] Since 2008, five Mobility Partnerships have been concluded with Eastern Partnership countries: Moldova,[17] Georgia,[18] Armenia,[19] Azerbaijan,[20] and Belarus.[21] As a prerequisite to signing these partnerships and visa facilitation agreements, Moldova, Georgia, Armenia, and Azerbaijan had to conclude readmission agreements that regulate the return of irregular migrants.[22] Belarus was an exception in this regard, as the signing of a Mobility Partnership preceded the conclusion of both a readmission and a visa facilitation agreement.[23]

Circular migration is seen as covering many aspects of the GAMM: legal migration, development, and mobility.[24] Therefore, one of the interviewed Council officials stated that sometimes it was difficult to make a decision as to whether to include circular migration initiatives under the development or the legal migration sections of the documents. He stressed that the concept had received interest from the third-country partners because it did not deprive them of their human resources and at the same time provided them with an opportunity to increase their skills and

[11] Directive 2011/98/EU of the European Parliament and of the Council of 13 December 2011 on a single application procedure for a single permit for third-country nationals to reside and work in the territory of a Member State and on a common set of rights for third-country workers legally residing in a Member [2011] OJ L 343

[12] Council Directive 2003/86/EC of 22 September 2003 on the right to family reunification [2003] OJ L 251.

[13] Vankova (2018), p. 157.

[14] See Chap. 3, Sects. 3.2 and 3.4.

[15] For an overall picture on Mobility Partnerships, see Reslow (2013).

[16] For overview of all instruments, see Van Elsuwege and Vankova (2020).

[17] Referred in the text as EU-Moldova Mobility Partnership. Council of the European Union (2008).

[18] Referred in the text as EU-Georgia Mobility Partnership. Council of the European Union (2009).

[19] Referred in the text as EU-Armenia Mobility Partnership. Council of the European Union (2011b).

[20] Referred in the text as EU-Azerbaijan Mobility Partnership. Council of the European Union (2013).

[21] Referred in the text as EU-Belarus Mobility Partnership. European Commission (2015).

[22] For details see Van Elsuwege and Vankova (2020) and the Treaties Office database of the European External Action Service. http://ec.europa.eu/world/ agreements/.

[23] See ibid.

[24] Interview # 8 with EU Council official, Belgium, February 2013, Annex I.

salary abroad.[25] However, there was no common understanding of circular migration within the framework of the GAMM.

One European Commission official stressed that when third countries were participating in Mobility Partnerships, they kept asking the Commission how to implement circular migration in practice, and were then advised by the Commission to negotiate that with the participating Member States.[26] The Commission was promoting circular migration as a potential additional initiative within the GAMM, but did not play any role in proposing concrete measures or schemes.[27] Thus, one of the interviewees underlined that the GAMM tried to cover different interpretations of the concept and it therefore contributed to the lack of harmonisation at the EU level.[28]

One caveat when analysing the GAMM is that it is extremely challenging to find publicly accessible information pertaining to the implementation of the Mobility Partnerships. This poses certain limitations to the assessment of this part of the EU's approach to circular migration.[29] The only scoreboard[30] currently accessible publicly is the one used to coordinate implementation of the EU-Moldova Mobility Partnership.[31] For the rest of the Mobility Partnerships, the information used in this chapter was retrieved mainly through official information requests to EU institutions as well as on the basis of interviews conducted between 2013 and 2017 with national and EU experts and representatives of NGOs implementing projects under the Mobility Partnerships. The partial information retrieved does not allow for any in-depth analysis of the measures discussed and does not claim to be exhaustive. Rather, it aims to serve illustrative purposes only, shed light on the initiatives developed, and assess them in light of the study's benchmark framework.

4.2 Entry and Re-Entry Conditions

In order to assess the entry and re-entry conditions as provided by the legal and policy instruments of the EU's circular migration approach, the following benchmarks are employed: possibility for facilitated personal travel[32]; facilitated entry for

[25] Interview # 8 with EU Council official, Belgium, February 2013, Annex I.

[26] Interview #11 with European Commission official, Belgium, May 2013, Annex I.

[27] Interview #11 with European Commission official, Belgium, May 2013, Annex I.

[28] Interview #22 with representative of international organisation, Austria, March 2017, Annex I.

[29] Also stressed in Interview # 19 with representative of international organisation, Belgium, February 2017. In this regard, see also Reslow (2017).

[30] According to the Commission, the scoreboards are internal Commission documents and constitute the 'basic monitoring tool' of the Mobility Partnerships. In European Commission (2009), p. 5.

[31] European Union-Republic of Moldova Mobility Partnership, Scoreboard: monitoring tool of the European Union – Republic of Moldova Mobility Partnership: http://scoreboard.mfa.gov.md Accessed 23 November 2017.

[32] Based on Article 1 of the European Agreement on Regulations governing the Movement of Persons between Member States of the Council of Europe, CETS No.025, 13 December 1957, entry into force 1 January 1958.

temporary visits[33]; circulation-friendly visa policies for third-country nationals, and policies to encourage circular and return migration.[34] Another benchmark pertains to the possibility of granting priority to seasonal workers who have been employed in the territory of a Member State for a significant period over other workers who seek admission to that State.[35] Multiple entry visas and a visa-free regime, as well as permits allowing periods of absence from the territory of the country of destination for long–term residents,[36] such as in the case of Sweden,[37] are all instruments that can support the implementation of these international standards.[38]

4.2.1 Legal Migration Directives

The EU approach to circular migration reflects the sectoral EU labour migration policy and therefore differs depending on the migrant category.[39] The four sectoral 'first admissions'[40] directives differentiate migrant workers on the basis of their skills and qualifications as well as on their attractiveness to Member States' labour markets, and use this as the decisive feature on which the different statuses are assigned.[41]

However, when it comes to initial entry, even the highly-skilled Blue Card holders that the EU wishes to attract face rather restrictive admission conditions.[42]

---

[33] Based on Article 1 of the European Convention on Establishment, ETS No.19, 13 December 1955, entry into force 23 February 1965.

[34] Based on Principle 15, Guideline 15.8. of ILO Multilateral Framework on Labour Migration. Non-binding principles and guidelines for a rights-based approach to labour migration, adopted by Tripartite Meeting of Experts, which convened in Geneva from 31 October to 2 November 2005.

[35] Based on Article 59 (2) of the International Convention on the Protection of the Rights of All Migrant Workers and Members of Their Families (ICRMW), G.A. res. 45/158, 18 December 1990, entered into force on 1 July 2003.

[36] Related to the right of long-term residents to return to their country of destination on the basis of Article 13 of the Universal Declaration of Human Rights (UDHR), G.A. res. 217A (III), 10 December 1948; Article 12 of the International Covenant on Civil and Political Rights (ICCPR), G.A. res. 2200A (XXI), 16 December 1966, entered into force on 23 March 1976; Article 8 of ICRMW; and Article 2 (2) and Article 3 (2) of Protocol No. 4 to the European Convention for the Protection of Human Rights and Fundamental Freedoms (ECHR), ETS No.046, 16 September 1963, entered into force on 02 May 1968. Fore more details, see Vankova (2016).

[37] See Chap. 1, Sect. 1.2.1.

[38] Vankova (2018), p. 158.

[39] Ibid., p. 160.

[40] The term 'first admissions' directives is borrowed from Barnard (2016), p. 500 and in this text refers to the Seasonal Workers' Directive, Blue Card Directive, the Intra-corporate Transferees' Directive and the Students' and Researchers' Directive.

[41] On that point, see also Eisele (2013); Wiesbrock et al.(2016), p. 968; Fridriksdottir (2017); Carrera et al. (2019).

[42] European Commission (2016a), p. 17.

The legal migration directives contain obligatory conditions for first admission related to a work contract or a binding job offer, sufficient resources, and sickness insurance. They also stipulate optional requirements: for instance, Article 5(2) of the Blue Card Directive requires the applicant to provide an address in the Member State. This condition is used by most Member States.[43] When migrants are applying from outside the EU, such optional requirements can add to the burdensome entry application procedure.[44] Another requirement, which could be widely discretionary, is that applicants should not be considered as posing a threat to public policy, public security, or public health.[45]

Furthermore, Member States retain the power to determine the volumes of admission as set out in Article 79 (5) TFEU, with regards to all 'first admissions' migration directives. They can set quotas for admission and reject applications when these quotas are reached as well as impose labour market test requirements, which could also serve as a basis for rejecting the application.[46] Another optional ground for rejecting Blue Card holders are brain drain considerations with regards to countries of origin.[47] Member States can also limit eligible applications for Blue Cards to those submitted from third countries, which can also have the effect of limiting access to the EU territory.[48] In the case of seasonal workers and ICTs, only applicants who reside outside the territory of the EU are eligible to apply for these permits.[49]

When assessing re-entry conditions and the possibility for rights-based circular migration, the Blue Card Directive stands out as the most generous of the 'first admissions' directives.[50] Blue Card holders can circulate between a Member State and a third country both before and after they qualify for an EU long-term residence status. At least on paper, they have control over their migration trajectory and can choose to circulate while also accumulating residence periods that would qualify towards a permanent residence.[51] This means that the provisions of the Directive meet the benchmark for policies encouraging circular migration and provide for permits allowing periods of absence from the host country – which is conducive to

[43] European Commission (2014a, b), p. 6.

[44] Vankova (2018), p. 159.

[45] See for instance Article 5 (1) (f) BCD, Article 7 (6) SRD, Article 5 (8) ICTD. On the wide discretion given to the Member States in this aspect, see C-544/15-Fahimian, ECLI:EU:C:2017:255.

[46] See Article 8 (3) SWD and Article 8 (2) BCD. Member States cannot apply a labour market test to Intra-corporate Transferees unless required by an Act of Accession. See Recital 21 of ICTD Preamble.

[47] Articles 8 (4) BCD. Not widely used in practice (European Commission (2014a)).

[48] Article 10 (4) BCD.

[49] Article 2 SWD and Article 2 ICTD.

[50] Vankova (2018), p. 160.

[51] See Article 16 (3) stating that for the purpose of calculating the 5-year period of legal and continuous residence in the EU required for the EU long-term residence status, periods of absence from the territory of the EU shall not interrupt this period if they are shorter than 12 consecutive months and do not exceed in total 18 months within the required 5-year period.

circular migration.[52] Nevertheless, it needs to be stressed that Member States have discretion to restrict in their national law the periods of absences to specific cases only.[53] Furthermore, despite the opportunities for a flexible migration trajectory and rights-based circulation that this Directive provides, a glance into its implementation at the national level shows that it has not been used widely due to the restrictive admission conditions and varying transposition allowing for parallel rules and procedures for admitting the same category of highly-skilled workers.[54]

In 2016, the European Commission concluded that the Blue Card Directive failed to achieve its objectives[55] and therefore proposed a recast,[56] which has so far been unsuccessful due to disagreements between Member States to maintain the parallel national schemes for admission of highly-skilled workers.[57] The changes foreseen in the Recast Proposal could ease entry conditions for such workers that are important with regards to circular migration commencement.[58] Furthermore, the impact assessment conducted by the European Commission as part of the recast proposal, spells out a 'nuanced' understanding of circular migration[59] when it comes to highly-skilled workers 'primarily considered as a spontaneous movement to achieve goals set within the migrant household'.[60] The opportunity for 'spontaneous movement' means that highly-skilled migrant workers can benefit from a migrant-led trajectory. In the Commission's view, this type of migration is 'likely to support subsistence activities in areas of origin'.[61] Therefore, the Recast Proposal aims to give migrant workers the possibility 'of longer "time-outs", enabling them to return to their country of origin without being penalised with a loss of their residence permit or expiration of the years of residence that count towards the right to long-term resident status'.[62] This means that when it comes to highly-qualified workers, the

---

52 Vankova (2018), p. 161.

53 Article 16 (5) BCD.

54 European Commission (2016b), p. 2. For data, see p. 3.

55 Ibid., p. 2.

56 Ibid., p. 3.

57 Groenendijk (2019), p. 63.

58 For instance, the required length of the work contract is shortened from at least 12 months to at least 6 months compared to the current Directive, which creates more flexibility for the employer and the employee (Article 5 (1) (a) of the Proposal). Also, the draft Article 9 (2) of the Proposal requires Member States to allow applications from both within and outside the country and the standard period of validity for the Blue Card permit is fixed at least 24 months (Article 8 (2) of the Proposal).

59 See Chaps. 1 and 3 for background information in this regard.

60 European Commission (2016c), p. 30.

61 Ibid.

62 Ibid. The Commission's Proposal reproduces the objective that is contained in Recitals 20 and 21 of the Preamble to the BCD with a view to facilitating circular migration between the EU and third countries through the same derogations that are present in the EU Long-term Residence Directive concerning periods of absence before qualifying for the status and after obtaining it. See also Recital 37 of the Preamble to the Proposal for a Directive on the conditions of entry and residence of third-country nationals for the purposes of highly-skilled employment, COM (2016) 378.

Commission's vision is to foster circular migration of continuously staying or settled third-country nationals with EU long-term residence status, who are given the opportunity to temporarily return to their countries of origin, while also enabling them to circulate before obtaining this status.

The circular migration approach with regards to seasonal workers differs from the one afforded to Blue Card holders. By way of contrast to the flexible migrant-led options provided to highly-skilled migrants, seasonal workers can benefit from short-term stays coupled with re-entry conditions.[63] According to Article 16 (1) of the Directive, Member States are obliged to 'facilitate re-entry of third-country nationals who were admitted to that Member State as seasonal workers at least once within the previous 5 years, and who fully respected the conditions applicable to seasonal workers under this Directive during each of their stays'. Even though they are required to provide rules on facilitated re-entry, the Member States are given a wide margin of discretion in doing so.[64]

Article 16 (2) of the Seasonal Workers' Directive contains possible facilitation measures such as exemption for the seasonal workers from the requirement to submit certain documents or allowing for several seasonal worker permits to be issued in a single administrative act.[65] This is a non-exhaustive list of examples and no minimum requirements are provided therein.[66] Therefore, the current provision does not demand any concrete commitments from the Member States and its effectiveness is, as a result, entirely dependent on them.[67] There still is no comprehensive information for assessing what kind of measures have been put in place in national law and whether they give priority in practice to seasonal workers already admitted or whether they contribute to circulation-friendly policies.

ICTs could also be given an option to circulate between the respective Member State and their country of origin. However, Member States are free to impose a gap period of up to 6 months (the so-called cooling off period) between the end of the maximum duration of the last transfer and another, new, application, before ICTs can return on the same grounds to the same Member State.[68] They may reject an application when fewer than 6 months have elapsed since the ICT concluded his or her previous transfer.[69] This notwithstanding, no minimum period is provided and there is no limit on the number of times ICTs can re-enter on the basis of an ICT permit.[70] This means that the 'temporary nature' of these transfer periods could span

[63] Vankova (2018), p. 161.

[64] Articles 16 (1) and (2) of the SWD.

[65] Article 16 (2) of the SWD.

[66] Fudge and Olsson (2014), p. 457. Wiesbrock et al. (2016), p. 960.

[67] Wiesbrock et al. (2016), pp. 960–961.

[68] Article 12 (2) of the ICTD.

[69] Article 7 (4) of the ICTD. For the rules stipulating refusal to renew and withdrawal of the ICT card, see Article 8 of the ICTD.

[70] Fridriksdottir (2017), p. 286.

several years and include circular elements, depending on how Member States have transposed the Intra-corporate Transferees' Directive into national law.[71]

Despite the objective of fostering the transfer of skills, as well the reference to mobility as an important element of the GAMM,[72] the Students' and Researchers' Directive does not contain any provisions on facilitating re-entry or easier access to continuous residence with the possibility for geographical mobility, as is the case of the Blue Card Directive. This means that researchers need to re-apply following the general admission procedure[73] and potentially make use of the visa facilitation instruments and visa-free regimes in the context of the Eastern Partnership.[74]

As already mentioned, the Policy Plan on Legal Migration identified the EU Long-term Residence Directive as an instrument offering '(...) the possibility for Member States to allow returning migrants to retain this status for longer than the 1 year period provided for in Article 9'.[75] According to the EU Long-term Residence Directive, this status can be withdrawn or lost in the event of an absence from the territory of the EU for a period of 12 consecutive months, unless longer periods are permitted because of specific or exceptional reasons.[76] The Commission cites the development of a project in the country of origin as an example of such a specific exceptional reason.[77] The Report on the implementation of this Directive shows, however, that only some Member States provide for the possibility to allow a longer period of absence due to exceptional circumstances: Austria, Belgium, Czech Republic, Germany, Estonia, Spain, Luxembourg, Latvia, Malta, Portugal, and Slovenia.[78] In addition, there are Member States that allow longer periods of absences as a rule, such as Finland (for a period of 2 years that can be further extended) and France (for 3 years).[79] Furthermore, in 2014, Sweden extended the permitted period of absence outside its territory to 2 years as a result of legislative changes in the country aimed at encouraging circular migration and promoting the positive impacts of migration on development.[80] The limited period of absence allowed could seriously hinder circularity because of the potential risk of loss of status, as illustrated in subsequent chapters.

---

[71] For more information on implementation at the national level, see for instance de Bie and Ghimis (2017).

[72] See Recital 6 of the Preamble to the SRD Recast Proposal and Recital 6 of the Preamble to the SRD.

[73] Articles 7 and 8 SRD.

[74] Vankova (2018), p. 161.

[75] See Chap. 3, Sect. 3.2.

[76] Article 9 (1) (c) of the LTRD.

[77] European Commission (2011a), p. 5.

[78] European Commission (2019a), p. 5. See also Bertelsmann Stiftung (2015).

[79] European Commission (2019a), p. 5.

[80] See European Migration Network (2011), p. 45. See Bertelsmann Stiftung (2015). See also the *ad hoc* query of the EMN provided on the initiative of Sweden: https://www.udi.no/globalassets/global/european-migration-network_i/ad-hoc-queries/se-emn-ncp-ad-hoc-query-on-policies-for-circular-migration_open-compilation_2014-08-12.pdf Accessed 12 April 2017.

4.2.2 *The EU Visa Policy & GAMM Instruments*

Circular migration cannot commence without first obtaining a visa, unless migrants are exempted from the visa obligation based on their nationality. The EU visa policy is a central part of the EU's Integrated Border Management Strategy,[81] which aims to 'reconcile security and freedom' by trying to both facilitate legitimate and legal access to the EU as well as guarantee security and counteract irregular migration and cross-border crime.[82] As part of this strategy visas have become one of the main EU instruments for pre-screening and the 'extra-territorialisation' of immigration control.[83] The revised EU Visa Code is another step in this direction since citizens of countries that do not cooperate with the EU on readmission of irregular migrants might be affected by higher visa fees, a requirement to present additional documents, slower processing times, and limited options for the issue of multiple-entry visas.[84] Such measures add to the already lengthy and cumbersome procedures for visas,[85] which are also set to become more expensive as of 2020 according to the latest amendments of the EU Visa Code.[86] All of this shows that it is not easy to initiate an individual circular migration project between the EU and a third country unless the applicant can use some form of facilitated access.

One of the instruments providing re-entry facilitation with regards to circular migration is the multiple-entry visa.[87] Thus far, however its use has been rather limited in practice. Due to consulates' large margin of discretion, there are wide variations in Member States' practices when it comes to its length of validity, which leads to visa shopping.[88] In general, consulates are usually reluctant to issue visas for longer than 1 year.[89] This means that frequent travellers need to go through the costly and cumbersome visa application over and over again, creating an

[81] European Council (2010), Para 5.2, p. 27.

[82] Meloni (2013), p. 155.

[83] Meloni (2013), p. 156. See also Mitsilegas (2010).

[84] See the introduced Article 25a of Regulation (EU) 2019/1155 of the European Parliament and of the Council of 20 June 2019 amending Regulation (EC) No 810/2009 establishing a Community Code on Visas, [2019] OJ L 188. For a critical assessment of the European Commission's proposal and impact assessment, see Eisele (2018).

[85] See for instance Mananashvili (2013), pp. 2–4. For a street-level policy perspective, see Infantino (2016).

[86] Visa application fees will increase from 60 EUR to 80 EUR and from 35 EUR to 40 EUR for 6–12 years in line with the amendments of Article 16 VC. Furthermore, a visa fee of 120 EUR or 160 EUR shall apply if an implementing decision is adopted by the Council under the point (b) of the new Article 25a(5) of Regulation (EU) 2019/1155 of the European Parliament and of the Council of 20 June 2019 amending Regulation (EC) No 810/2009 establishing a Community Code on Visas, [2019] OJ L 188.

[87] Vankova (2018), p. 162.

[88] European Commission (2018a), p. 3.

[89] European Commission (2018b), pp. 13–14.

administrative burden for both migrants and consulates.[90] These issues have been addressed by the revised EU Visa Code, which streamlines the current provisions on the multiple-entry visa. As of 2020, such visas will be available with three validity periods – 1, 2, or 5 years – depending on prior lawful use of a visa.[91]

The visa facilitation agreement is another instrument fostering mobility. Widely used as part of the GAMM,[92] it aims at facilitating the issuance of short-stay Schengen visas for third-country nationals.[93] The EU has signed visa facilitation agreements with 12 countries. Soon, all Eastern Partnership countries will be covered by such agreements as the Council has approved the signing of such agreement on behalf of the European Union with Belarus, which was the only country still negotiating such an instrument.[94] The visa facilitation agreement covers provisions related to, *inter alia*, a simplification of the application process, a reduction in the fees for processing visa applications, and more precise conditions for multiple-entry visas for different categories of travellers compared to the current Article 24 (2) of the EU Visa code.[95]

Usually visa facilitation agreements precede the introduction of a visa-free travel regime with a specific third country.[96] In order to establish a visa-free regime with a particular third country, the EU has introduced the so-called Visa Liberalisation Dialogues. These instruments are structured around four thematic areas in which the candidate country needs to improve its legal and policy framework and demonstrate the effective implementation thereof in relation to matters such as document security, border management, public order and security, external relations, and fundamental freedoms.[97] The Dialogues can be compared to a mini-accession process as they have a high degree of leverage.[98] One of the interviewed officials stated that the process requires '(…) a total reform of the candidate's law enforcement system, except for prisons (…)'.[99]

As a result of the successful Visa Liberalisation Dialogues, a visa-free regime for holders of biometric passports was introduced for Moldova in 2014; in 2017 Ukraine

---

[90] Ibid., p. 15.

[91] See the revised Article 24 (2) of Regulation (EU) 2019/1155 of the European Parliament and of the Council of 20 June 2019 amending Regulation (EC) No 810/2009 establishing a Community Code on Visas, [2019] OJ L 188. However, the European Commission's impact assessment and proposal did not discuss the potential links between the concept of circular migration and the harmonising procedures for issuing multi-entry visas. In Eisele (2018), p.6

[92] Van Elsuwege and Vankova (2020).

[93] Andrade et al. (2015), p. 40.

[94] Council Decision (EU) 2019/1915 of 14 October 2019 on the signing, on behalf of the Union, of the Agreement between the European Union and the Republic of Belarus on the facilitation of the issuance of visas, [2019] OJ L297/1.

[95] For more details, see Peers (2012a), p. 317–18; Trauner and Kruse (2008), p. 424.

[96] Trauner and Kruse (2008), p. 421.

[97] Hernández i Sagrera (2014), pp. 14–15; Andrade et al. (2015), p. 30.

[98] Andrade et al. (2015), p. 30. This was also mentioned in interview #11 with European Commission official, Belgium, May 2013.

[99] Interview #11 with European Commission official, Belgium, May 2013, Annex I.

and Georgia were moved to the positive list of EU Visa List Regulation 539/2001. The introduced visa-free regimes ease travel because they waive the whole visa application process and all related administrative and financial hurdles. However, it should be borne in mind that they only provide for short-term travel and thus exclude the possibility of fostering short-term labour mobility unless special authorisation is obtained. Furthermore, only those individuals who hold biometric passports can benefit from the visa-free process.

To sum up, EU multi-entry visas and visa liberalisation instruments do not directly facilitate circular migration.[100] They provide limited opportunities because they very much depend on where the migrant comes from, whether he or she has a biometric passport, or has become eligible for a multi-entry visa. These instruments could, however, foster personal travel for temporary visits, which is one of the benchmarks in this policy area. Thus, they could contribute to the initiation of individual circular migration projects and other employment possibilities because they can support job seeking for those who can access, afford, and succeed in obtaining a visa.[101]

4.2.3 Mobility Partnerships

The bilateral agreements concluded between some EU Member States and Eastern partnership countries under the auspices of the Mobility Partnerships with the aim of fostering labour mobility are identified as one of the main instruments facilitating entry and re-entry conditions for third-country nationals. One such example is the 'Bilateral Agreement on Labour Mobility Between Italy and Moldova', which was an updated version of an agreement that had been signed between the two countries in 2003.[102] The agreement aimed to establish a basis for collaboration between the two countries in order to regulate the flow of workers and develop procedures for facilitating the employment of Moldovan citizens to work in Italy in case of shortages in the local work force (Article 1). The agreement also provided for the encouragement of professional training of workers who were intending to migrate, in line with the skills that were needed in the Italian labour market (Article 3). Even though the agreement did not explicitly contain circular migration facilitation as a priority, seasonal workers were among its target groups (Article 5).

A project supporting the implementation of this agreement also aimed at testing a pilot circular migration scheme between Moldova and Italy.[103] Around 300 potential migrants from Moldova were involved in a professional language course (such

100 Vankova (2018), p. 163.

101 Ibid.

102 Interview #27 with representative of international organisation, Austria, March 2017, Annex I.

103 EU-Republic of Moldova Mobility Partnership: Information Newsletter, No 7, May 2013, p. 9. http://www.mfa.gov.md/img/docs/bulletin-information_7_en.pdf Accessed 1 May 2019.

as in construction) and included in the database of the Italian Ministry as potential candidates to be employed in Italy in case of vacant jobs corresponding to their qualification or skills.[104] The vast majority of trained workers were able to identify an employer on their own eventually.[105] Moldova has also signed another bilateral agreement with Bulgaria under the auspices of the Mobility Partnership.[106] It will be presented in detail in later chapters.

Georgia and France signed a bilateral agreement on 'Residence and Circular Migration of Skilled Professionals' in November 2013 specifically aimed at establishing an institutionalised circular migration scheme.[107] The French National Assembly did not ratify the agreement until late 2018, after a long delay due to questions pertaining to migration and security.[108] This Agreement provides for temporary residence permits for Georgian workers for up to 1 year (renewable for some categories), including young specialists between the ages of 18 and 35, as well as students graduated from French higher education institutions who want to gain professional experience in France.[109] It sets an annual quota of 500 temporary residence cards to be issued by France to workers engaged in the list of 50 professional occupations open to Georgian citizens.[110] In addition, the number of young French and Georgian qualified specialists in both countries should not exceed 150 people per year.[111] Georgia is actively pursuing additional labour mobility agreements with Germany, Bulgaria, Portugal, and Greece.[112] Finally, the International Organisation for Migration (IOM) implemented another initiative that aimed to promote circular migration under the 'Piloting Temporary Labour Migration of Georgian workers to Poland and Estonia' project.[113] It also envisaged the signing of bilateral agreements with both Member States. However, only a limited number of migrants could eventually benefit from this pilot scheme; for instance, according to the project's evaluation report, a total of 19 migrants were selected for jobs in Poland.[114]

As evident from the projects developed and on-going initiatives, Georgia has a strong interest in circular migration. According to one of the state officials interviewed, 'Circular migration is something that is developing on a daily basis in Georgia, especially with the visa liberalisation process, in terms of public awareness

[104] Mosneaga (2015), p. 26.

[105] This information was additionally provided by interviewee # 19, Belgium, February 2017, Annex I.

[106] See Chap. 5, Sect. 5.2.2.1; Chap. 6, Sect. 6.1.2; Chap. 7, Sects. 7.1.1 and 7.2.1; Chap. 8, Sect. 8.1.1.

[107] Gogolashvili (2019).

[108] Interview # 25 with state official, Georgia, March 2017, Annex I.

[109] Georgian Journal (2019).

[110] Ambassade de France à Tbilissi (2019).

[111] Ibid.

[112] Gogolashvili (2019).

[113] ICMPD (2016), p. 15. For more details, see International Organisation for Migration (2017).

[114] Verigo and Castelfranco (2019), p. 17.

and policy decision making (...)'.[115] The Georgian authorities were relying on these pilot circular migration projects because they were planning to use the experience in order to set up a state migration system.[116] The government was interested in developing outward as well as inward circular migration policies in order to foster legal economic migration of Georgians and migrants coming to Georgia and provide an alternative to the irregular migration that is undertaken by many Georgians.[117] Another interviewed expert commented that circular migration was important because the country did not want to lose any more of its citizens due to emigration.[118]

In a similar vein to Georgia, Armenia has also signed a bilateral agreement with France in October 2016 for the exchange of students, interns, young professionals between the ages of 18 and 35 years, and qualified specialists (Articles 2–4).[119] However, this agreement had not been ratified by the French side at the time of writing. The authorised duration of employment is planned to be between 6 and 12 months for young professionals, and this may be prolonged for up to a maximum of 24 months in line with their contract. The issue of work permits for Armenians who wish to work in France is conditional upon a labour market test (Article 3.4). The number of young professionals from France or Armenia admitted to work on the territory of the other contracting party must not exceed 100 per year for each party (Article 3.5). The accepted qualified specialists from Armenia are entitled to receive a 1-year residence permit with a maximum validity of 3 years, in line with the relevant French legislation (Article 4.3).

One of the interviewees commented that the elaboration of this agreement took a very long time and the draft was only finalised in late 2016.[120] It was initiated by the Armenian government and based on the good ties between the two countries, as well as Armenia's experience of concluding bilateral agreements with some of the Gulf countries that led to the circular migration of doctors and nurses. According to one interviewee, the negotiations took so long because France was 'not very keen to facilitate or encourage Armenian migration in whatever form to France because of the already existing diaspora and the pull factor that they might attract more migrants'.[121]

[115] Interview # 25 with state official, Georgia, March 2017, Annex I.

[116] Interview # 25 with state official, Georgia, March 2017, Annex I.

[117] Interview # 18 with state official, Georgia, November 2014, Annex I.

[118] Interview #27 with representative of international organisation, Austria, March 2017, Annex I.

[119] Project d'accord de partenariat migratoire entre le gouvernement de la Republique Francaise et le gouvernement de la Republique D'Armenie provided by the French Embassy in Yerevan in March 2017.

[120] Interview #20 with representative of international organisation, Armenia, February 2017, Annex I.

[121] Interview #20 with representative of international organisation, Armenia, February 2017, Annex I.

Policy measures aiming to engage settled migrants in the EU, such as diaspora members, are identified as another relevant instrument under the auspices of the GAMM focused on facilitating re-entry conditions for third-country nationals. They further demonstrate how circular migration and permanent settlement are intertwined in the context of EU's approach to circular migration. Such an example is the 'Making Migration in Moldova Work for Development' project, which aimed to test whether Moldova and Germany could establish a 'triple win' circular migration scheme by engaging diaspora members.[122] According to one interviewee, this initiative was based on an amendment of the relevant German legislation, which allowed citizens of Moldova, Georgia, Armenia, and Morocco to be absent from the territory of the country for up to 2 years without losing their permits.[123] The idea was for diaspora members to be given the possibility to go back to their countries of origin and share their knowledge, experience, and expertise, and then to return to Germany.[124] This initiative was praised as a successful circular migration policy project.[125]

Another example of diaspora engagement in the context of circular migration is the 'Temporary Return of Qualified Nationals' project funded by the Dutch Ministry of Foreign Affairs and implemented by the IOM.[126] It aims to contribute to the reconstruction and development of a number of former war countries, including Georgia and Armenia.

The added value of the Mobility Partnerships in terms of creating labour mobility opportunities for the participating third countries has been questionable so far.[127] The presented bilateral agreements and projects engaging settled migrants are seen as creating circulation-friendly policies in line with the benchmark framework of this book. However, it should be kept in mind that only a limited number of migrants can benefit from these opportunities as the measures under the Mobility Partnerships are small-scale pilot projects in most of the cases[128] or bilateral agreements envisaging caps, which can mainly be attributed to the reluctance of Member States to open new channels for legal migration.[129]

---

[122] Migration for development (2011). See also European Union – Republic of Moldova Mobility Partnership (2010).

[123] Interview #27 with representative of international organisation, Austria, March 2017, Annex I. Interview #19 with representative of international organisation, Belgium, February 2017, Annex I.

[124] Interview #19 with representative of international organisation, Belgium, February 2017, Annex I.

[125] Interview #27 with representative of international organisation, Austria, March 2017.

[126] For more details, see Leith and Rivas (2015).

[127] See for instance Reslow (2015), p. 117.

[128] Van Elsuwege and Vankova (2020).

[129] Interview #19 with representative of international organisation, Belgium, February 2017, Annex I; Interview # 20 with representative of international organisation, Armenia, February 2017, Annex I.

4.3 Work Authorisation

Work authorisation is an important policy area to consider with regards to all labour migrants, including circular migrants, because very often their initial work permits tie them to a specific employer, occupation, and locality for a specified period of time during which they cannot work for another employer.[130] For instance, in cases of job loss, migrants may be inclined to either overstay in the host country or return to their home country earlier than their work permits allow, which leads them to sustain financial losses.[131] In addition, the impossibility of changing employment and sector can increase the risk of exploitation and abuse.

The analytical framework of this study employs benchmarks pertaining to work authorisation aimed at assessing whether workers can change their employer with a maximum restriction of 2 years[132] as well as whether loss or termination of employment constitutes the sole ground for withdrawal of a migrant worker's authorisation of residence or work permit.[133] In addition, the benchmarks assess the possibility to find alternative work in case of loss or termination of employment[134] and whether seasonal workers who have already been employed on the territory of the Member States for a significant period of time are able to take up other remunerated activities.[135] The identified instrument that can support the implementation of these benchmarks is the availability of a flexible work permit that allows its holder to change both employer and occupation within the period of validity of the permit.[136]

4.3.1 Legal Migration Directives

The two legal instruments that aim to facilitate circular migration – the Blue Card Directive and the Seasonal Workers' Directive – explicitly provide for a change of employer.[137] However, it should be stressed that Member States retain discretion on

[130] Vankova (2016), p. 337.

[131] Ibid.

[132] Based on Article 14 (a) of Convention concerning Migrations in Abusive Conditions and the Promotion of Equality of Opportunity and Treatment of Migrant Workers, C143, adopted at 60th ILC session on 24 June 1975, Geneva, entry into force on 09 December 1978 (ILO Convention No. 143); and Article 52 (3a) ICRMW. For more details see Vankova (2016), pp. 342–343.

[133] Based on Article 8 (1) ILO Convention No. 143 and Article 49 (2) ICRMW.

[134] Based on Article 8 (2) ILO Convention No. 143 and Article 51 ICRMW.

[135] Based on Article 59 (2) ICRMW.

[136] Vankova (2016), p. 343.

[137] Article 15 (3) SWD; Article 12 (2) BCD. In Vankova (2018), p. 164.

how many changes to allow seasonal workers within the authorised period.[138] Furthermore, within the first 2 years, changes of employer for Blue Card holders are subject to prior authorisation of the competent authorities of the Member State of residence, in accordance with national procedures.[139] The possibility of changing employer is implicitly provided for researchers.[140] ICTs, on the other hand, are bound to their employer during the whole period of their transfer.[141]

The analysed instruments do not legislate for change of occupation – a direct result of the EU's sectoral approach to labour migration. This means that, depending on the transposition into national law, only the Blue Card Directive could fulfil the benchmark in the area of work authorisation pertaining to free access to employment in all industries and occupations with a maximum restriction of 2 years. This also means that seasonal workers cannot look for alternative employment other than seasonal work. Allowing for such a possibility is another benchmark in this policy area (see Annex V). However, Member States can provide more favourable provisions to third-country nationals who enter as seasonal workers under bilateral agreements, as stipulated in Article 4 of the Seasonal Workers' Directive.

The Blue Card Directive is the only one of the 'first admissions' directives that explicitly stipulates in Article 13 (1) that unemployment does not automatically lead to the withdrawal of the permit, unless the period of unemployment exceeds three consecutive months and occurs more than once during the validity of the permit.[142] Taking into consideration the possibility for seasonal workers to change employer within the authorised period discussed above would implicitly mean that the sole fact of unemployment could not lead to withdrawal of the permit if the worker manages to secure another job with a different employer within a reasonable time.[143] The length of the reasonable time must, according to the CJEU, be defined in national law.[144] Not allowing for such a period would take away the *effet utile* of Article 15 (3) and thus be incompatible with the EU law principle of effectiveness.[145] However, whether this is possible in practice depends on what is stipulated in national law as well as the individual case.

---

[138] See article 15 (4) SWD.

[139] Article 12 (2) BCD.

[140] See Article 21 (5) SRD.

[141] See Article 17 ICTD.

[142] Vankova (2018), p. 164.

[143] On the basis of Article 15 (3) SWD. Ibid.

[144] The 'reasonable time' criterion was introduced by the CJEU in its judgment in Tetik, C-171/95 – Tetik v Land Berlin, ECLI:EU:C:1997:31. It concerned Turkish workers with a right to continue to work under Article 6 (1) of the EEC-Turkey Association Council Decision 1/80, but on the basis of comparison with the rights of EU workers. See Paras 27, 30–32, 42 and 48. In Vankova (2018), p. 165.

[145] Ibid.

Concerning the rest of the 'first admissions' directives, unemployment of ICTs would lead to the withdrawal of their permit,[146] and the Students' and Researchers' Directive does not legislate in this regard.[147] The latter, however, does not preclude any national rules that would allow the researchers to look for another research organisation or job. This means that, in the case of researchers, whether or not unemployment leads to withdrawal of permits hinges on how the Directive is transposed in national law by the Member States.

4.3.2 *Mobility Partnerships*

The issues related to work authorisation are addressed by different project initiatives under the auspices of the Mobility Partnerships with the Eastern Partnership countries. Some of them are focused mainly on the pre-migration phase. For instance, the 'Support of Circular Migration and Re-integration Process in Armenia' project aimed to protect the rights of potential migrant workers, reintegrate labour/circular migrants, and prevent irregular migration in line with the State Action Plan for Migration (2012–2016).[148] Awareness about migrants' rights was built through individual consultations, an awareness raising media campaign, and pre-departure orientation trainings.[149]

In addition, the 'Building Institutional Capacity of the Ministry of Foreign Affairs and European Integration' project (2009–2015) established a Call Centre for Moldovan citizens within the country and Moldovan labour migrants to provide information on the rights of citizens and migrant workers.[150] Another initiative pertaining to work authorisation was organised under the Mobility Partnership with Belarus. It was implemented by ICMPD under the MIUEX Belarus Action[151] and aimed to introduce international labour migration and migration-related conventions and standards, as well as assess national legislation with a view to their possible ratification.[152]

Even though these initiatives do not directly contribute to the implementation of the identified benchmarks, they are considered as increasing awareness about migrants' rights and the standards in the field of work authorisation among third-country nationals as well as relevant stakeholders.

---

[146] Article 8 (5) (a) in connection with Article 14 ICTD.

[147] Vankova (2018), p. 165.

[148] Kalantaryan (2015), p. 32.

[149] Ibid., p. 29.

[150] Ibid., p.38.

[151] Migration EU eXpertise (MIEUX) is a joint EU-ICMPD initiative. For more information see https://www.icmpd.org/our-work/capacity-building/multi-thematic-programmes/mieux-iii/ Accessed 12 January 2018.

[152] Interview #26 with representative of international organisation, Belarus, March 2017, Annex I.

4.4 Residence Status

As outlined in Chap. 1, permanent settlement in the host country is an inherent characteristic of circular migration.[153] Furthermore, where permanent settlement occurs, it would not hinder circulation in the long term.[154] This notion of circular migration as a fluid movement, however, is generally not reflected in the legal frameworks aimed at managing circular migration.[155] In many cases, circular and temporary migrant workers who are engaged in low-skilled occupations are obliged to leave after the expiry of their work permits and are thus prevented from accessing permanent residence.[156] Therefore, one of the benchmarks of this study aims to assess whether lawful migrants have the opportunity to qualify for a prolonged or permanent residence status.[157] The possibility for migrants to access permits allowing transit from a temporary to a permanent residence status is considered a policy instrument that supports the implementation of this benchmark.[158] In addition, this section examines whether migrants have the right to mobility and choice of residence within the host country.[159]

4.4.1 *Legal Migration Directives*

The EU Long-term Residence Directive provides the general rules on access to this status, thus it is logical to discuss this directive first.[160] Third-country nationals who have resided 'legally and continuously' within the territory of the host Member State for 5 years immediately prior to the submission of the relevant application have the right to the EU long-term residence status.[161] Prior residence 'solely on temporary grounds' or where the residence permit has been 'formally limited', however, is not considered in the computation of the 5-year period.[162]

[153] Skeldon (2012), p. 53.

[154] Hugo (2013), p. 2.

[155] Vankova (2018), p. 166.

[156] Vankova (2016), p. 338

[157] On the basis of Article 2, European Convention on Establishment.

[158] Vankova (2016), p. 344.

[159] On the basis of Article 12 (1) ICCPR, Article 39 ICRMW, Article 2 (1) of the Fourth protocol to the ECHR.

[160] On the process of its adoption, see Arcarazo (2011), pp. 77–93. See also Arcarazo (2015).

[161] Recital 6 of the Preamble to the LTRD and Article 4 (1) of the LTRD.

[162] Article 4 (2) in connection with Article 3 (2) (e) of the LTRD.

Periods of absence from the host Member State shorter than six consecutive months that do not exceed a total of 10 months within the 5-year period are not counted as an interruption and therefore they should be taken into account for the purposes of calculating the necessary period.[163] Member States may allow longer periods of absence in 'cases of specific or exceptional reasons of a temporary nature and in accordance with their national law'.[164] In such cases, the relevant period of absence should not be included in the calculation of the 5-year period. Yet, Member States can also derogate from that rule and take into account absences relating to employment purposes in the calculation of the 5-year period.[165]

Furthermore, the EU Long-term Residence Directive spells out two mandatory requirements for acquiring EU long-term residence status and one optional requirement – which are all exhaustive in nature.[166] The applicant has to have 'stable and regular resources which are sufficient to maintain himself/herself and the members of his/her family, without recourse to the social assistance system of the Member State concerned', as well as sickness insurance.[167] The text of these requirements in Article 5 (1) of the EU Long-term Residence Directive are identical to the requirements in Article 7(1) of the Family Reunification Directive, as both instruments were negotiated at the same time in the Council's Working Group.[168] Therefore, these conditions should be interpreted in the light of the Court of Justice of the European Union's (CJEU) ruling in the *Chakroun* case[169]: taking into account the needs of the individual and not setting a standard amount below which an application will be refused, and also considering the income of the family members when assessing the requirement of sufficient recourses.[170] Furthermore, in the recent case *X v Belgische Staat,* the CJEU ruled that he concept of 'resources' may also cover the those made available to an applicant by a third party provided that they are considered to be stable, regular, and sufficient.[171]

In addition, Member States may impose integration conditions as an optional requirement in accordance with national law.[172] According to the Implementation Report of the European Commission, a majority of Member States require applicants for this status to have knowledge of their official language, while some also

[163] Article 4 (3) of the LTRD.

[164] Article 4 (3), second sub-paragraph, of the LTRD.

[165] Article 4 (3), third sub-paragraph, of the LTRD.

[166] Groenendijk (2012), p. 302.

[167] Article 5 (1) of the LTRD.

[168] Groenendijk (2012), p. 301.

[169] Case C-578/08 Rhimou Chakroun v. Ministervan Buitenlandse Zaken, ECLI:EU:C:2010:117.

[170] Peers (2016), p. 424 (footnote 914).

[171] Case C-302/18 X v Belgische Staat, ECLI:EU:C:2019:830, paras 41–44.

[172] Article 5 (2) of the LTRD. In this regard, see for instance Carrera (2014); Carrera (2009); Strik and Böcker (2011); Pascouau and Strik (eds) (2012).

mandate attendance of civic knowledge courses.[173] As long as these requirements are imposed after the EU long-term residence status has been obtained and do not pose risk to withdrawal of this status, they are considered to be in line with the Directive.[174]

Taking into account that the EU Long-term Residence Directive excludes third-country nationals who reside on 'temporary' or 'formally limited permits' from its scope,[175] this means that ICTs and seasonal workers need to change to another national or EU permit that would allow them to accumulate residence periods for a long-term residence status.[176] By way of contrast, Blue Card permit holders have facilitated access to permanent residence[177] and the Recast Directive is expected to further facilitate their access to this EU permit by reducing the time required from five to three years.[178] The Students' and Researchers' Directive only implicitly provides access to this status for researchers.[179] The residence permits under this Directive are renewable as long as the admission conditions for their issue are met and thus cannot be considered as falling outside the scope of the EU Long-term Residence Directive on the basis of being 'temporary' or 'limited'. Furthermore, this would run contrary to the aims of the Directive.[180] This means that only the Blue Card and the Students' and Researchers' Directives fulfil the benchmark for facilitation of prolonged or permanent residence.

The second benchmark concerning residence status aims to examine whether migrants have the right to mobility and choice of residence within the host country. All reviewed 'first admissions' directives allow for mobility within the Member States and a choice of residence.[181] It should be kept in mind, however, that seasonal workers' choice of residence could be limited to a certain extent in cases when the employer arranges it.[182] As Wiesbrock, Jöst, and Desmond point out, the Directive does not address employer-organised accommodation, which could lead to abuse and dependency.[183]

---

[173] European Commission (2019a), p. 3

[174] Case C-579/13, P and S v Commissie Sociale Zekerheid Breda and College van Burgemeester en Wethouders van de gemeente Amstelveen, ECLI:EU:C:2015:369, para 56. For more details on the integration requirements under EU law, see Carrera and Vankova (2019), pp.19–21.

[175] According to the Court ruling in the C-502/10 – Singh, ECLI:EU:C:2012:636, these are two distinct autonomous exceptions, which need to be interpreted through the prism of the integration objective of the Directive.

[176] Vankova (2018), p. 167.

[177] See Article 16 (2) and (3) BCD.

[178] Article 17 (2) BCD Recast Proposal.

[179] See the reasoning of Steve Peers on the former Researchers' Directive in Peers (2012b), p. 138.

[180] See Recital 4 of the Preamble to the SRD.

[181] Vankova (2018), p. 166.

[182] See Article 20 SWD.

[183] See Article 5 (1)(c) and Article 6 (1) (c) in connection with Article 20 SWD. See also Wiesbrock et al. (2016), p. 966.

4.4.2 *Mobility Partnerships*

As already mentioned, the understanding that circular migration entails a fluid movement that could lead to a prolonged stay in the country of destination is usually not considered by policymakers trying to facilitate this type of migration. The initiatives under the Mobility Partnerships are not an exception in this regard and therefore cannot be considered as supporting any form of access to a permanent residence status. This is not surprising given that Member States are generally reluctant to open additional legal migration channels and 'guard' access to permanent settlement by subjecting it to *inter alia* residence and language requirements.

The implementation of one of the projects under the Mobility Partnership with Georgia demonstrates the challenges associated with this. The 3-year 'Strengthening the development potential of the EU Mobility Partnership in Georgia through targeted circular migration and diaspora mobilisation' project (2013–2016)[184] piloted a circular migration scheme where workers from Georgia could work in Germany for 18 months on the basis of a work contract that they signed before their departure.[185] Twenty-eight participants were selected for placement with German employers[186] and at the end of the project, 24 Georgians (14 from the hospitality sector and 10 from the health care sector – nurses) were benefiting from employment as part of the project.[187] The project's aim was for these workers to contribute their expertise towards Georgia's development upon return, as a result of the knowledge and experience they had gained in Germany.

The project implementation partners defined circular migration as follows:

> '(...) The mobility of people between countries, including multiple temporary or *long-term* movement which may be beneficial to, and harnessing development of all involved (migrants, countries of origin and destination, including the respective societies and individuals), if occurring voluntarily and linked to the migrants' rights and competencies and their development as well as to economic opportunities of countries of origin and destination.'[188]

Some of the participants commented that the permitted period under this pilot scheme was too short and that they were just starting to develop professionally

[184] The project was part of the joint operation that was referred to as the Centre for International Migration and Development (CIM) and was managed by the Federal Employment Agency and the organisation Deutsche Gesellschaft für Internationale Zusammenarbeit (GIZ).

[185] Interview #24 with former MP project assistant, Georgia, March 2017, Annex I. Based on § 17 of the German residence act, permitting training on the job for the duration of 12 months, with a one-time possibility to prolong to a maximum of 18 months and with the obligation to return to Georgia upon completion. For more details, see Goos (2016), pp. 57–59.

[186] Interview #24 with former MP project assistant, Georgia, March 2017, Annex I.

[187] Goos (2016), p. 102.

[188] Centre for International Migration and Development, 'Circular Migration between Georgia and Germany'. https://migration-georgia.alumniportal.com/?id=1287 Accessed 26 November 2019. Emphasis added.

when the scheme was drawing to a close.[189] Therefore, it was not surprising that the vast majority of the participants opted for a prolonged duration of employment in Germany.[190] For instance, the hospitality sector professionals prolonged employment to the maximum applicable duration of stay of 18 months; most of the nurses indicated interest in prolonging employment beyond the minimum of 12 months, by between one and 3 years, stressing that they 'are interested in further deepening their knowledge and experience by prolonging their employment or in some cases continuing education in Germany, which they believe will be more useful in transposing the systems to Georgia…'.[191] According to the latest publicly available data, five hospitality professionals remained in Germany and only one nurse returned to Georgia despite the employment opportunities and financial incentives foreseen under the project as part of the reintegration measures.[192]

As one interviewee emphasised in 2014, the project could not be considered a failure just because some of the participants decided to prolong their stay in Germany, as this was an indication of a problem, thereby preventing them from returning to Georgia.[193] In line with that, the project was regarded as successful by the Georgian authorities and there was interest in continuing with this kind of scheme with other Member States on the basis of the developed Manual on Circular Migration Scheme and in line with Georgia's Migration Strategy (2016–2020) of facilitating circular migration.[194] The Manual stresses that successful voluntary return remains a challenge and that 'both binding duration of the migration cycle with mandatory return and permanent emigration jeopardises maximisation of triple win impact of labour migration'.[195] Furthermore, return cannot be enforced when the duration of legal stay under the circular migration scheme does not correspond to the opportunities offered by the legislative framework of the country of destination.[196] One of the lessons learnt from the implementation of this project is that circular migration should be seen as means for legal voluntary mobility and that therefore the duration of contracts under circular migration projects should be kept flexible.[197] Despite the broad understanding of circular migration as entailing long-term movement, the project failed to acknowledge, however, that the leakage to settlement and access to an EU long-term residence for some of the participants

[189] Interview # 24 with former MP project assistant, Georgia, March 2017, Annex I.

[190] Goos (2016), p.103.

[191] Final external evaluation report quoted in A. Goos (2016), p. 103.

[192] Mestvirishvili (2018), pp. 4–5.

[193] Interview # 18 with state official, Georgia, November 2014, Annex I.

[194] Interview # 24 with former MP project assistant, Georgia, March 2017, Annex I. Interview # 25 with state official, Georgia, March 2017, Annex I. See also, Secretariat of the State Commission on Migration Issues, Executive Summary of Assessment of Circular Migration Potential of Georgian Workforce in the EU. http://eapmigrationpanel.org/sites/default/files/study_report_on_circular_migration_in_georgia_executive_summary.pdf Accessed 19 November 2019.

[195] Goos (2016), p.103.

[196] Ibid.

[197] Mestvirishvili (2018), p. 6.

who opted to stay would not hinder circular migration in the long-term as they can continue to circulate back and forth between Georgia and Germany as EU long-term residents. Furthermore, limiting the contracts of participants or using other contractual return incentives[198] is not a guarantee that migrants will return and further engage in circular migration, and could raise human rights concerns.

4.5 Social Security Coordination

Circular migrants live transnational lives between their countries of origin and destination, and therefore contribute to the social security systems of two countries during different periods of stay as part of their movement cycle.[199] This section of Chap. 4 focuses on social security coordination and the possible solutions thereof that are available to tackle issues arising from contributions accumulated in the context of circulation. The benchmarks in this area aim to assess what kind of benefits can be exported and whether the general principles of social security coordination are covered: maintenance of the acquired rights and rights in the course of acquisition; totalisation of periods of insurance, employment or residence and of assimilated periods for the purpose of the acquisition, maintenance, or recovery of rights and for the calculation of benefits; and, equality of treatment.[200] Another benchmark focuses on whether reimbursement of social security contributions is a possible option.[201] Multilateral and bilateral agreements are considered instruments that support the implementation of these benchmarks.[202]

4.5.1 Legal Migration Directives

All 'first admissions' directives contain equal treatment clauses in regards to branches of social security,[203] as defined in Article 3 of Regulation No 883/2004 on the coordination of social security systems.[204] Except for the Blue Card Directive, however, the rest of these directives allow Member States to restrict equal treatment

[198] See Mestvirishvili (2018), p. 6; Goos (2016), p.110.

[199] Vankova (2016), p. 338.

[200] Based on, among others, Equality of Treatment (Social Security) Convention, C118, adopted at 46th ILC session on 28 June 1962, entry into force on 25 April 1964; Equality of Treatment (Accident Compensation) Convention, C019, adopted at seventh ILC session on 05 June 1925, entry into force on 8 September 1926. See Annex V for details.

[201] In line with Article 27 (2) of ICRMW and Article 9 (1) ILO Convention No 143.

[202] Vankova (2016), p. 346.

[203] Article 23 (1) (d) SWD; Article 14 (1) (e) BCD; Article 18 (2) (c) ICTD; Article 12 (1) (e) SPD; Article 22 (1) SRD.

[204] Regulation (EC) No 883/2004 of the European Parliament and of the Council of 29 April 2004 on the coordination of social security systems [2004] OJ L 166.

in the field of social security, mostly with regards to family benefits.[205] Member States do not have unlimited freedom to restrict these equal treatment provisions, as stressed by the Court in the case *Kamberaj*.[206] Restrictions to the equal treatment of social rights of third-country nationals also need to be considered 'in the light of a human rights based approach', and Member States should not be allowed to breach their human rights obligations.[207]

Concerning export of benefits, the Blue Card Directive allows for export of statutory old age pensions, 'at the rate applied by virtue of the law of the debtor Member State(s) when moving to a third country'.[208] In line with this provision, Member States are required to pay pensions to the former Blue Card holders when they move to a third country, even when there is no bilateral social security agreement in force between the two respective countries.[209] The only requirement is that the Member State provides this type of social security export for its own nationals, as this is an equal treatment clause.[210] Nevertheless, this provision applies without prejudice to existing bilateral agreements.[211] In addition, it must be stressed that this provision covers only old-age pensions. Since Blue Card holders are not excluded from the scope of the Single Permit Directive, they can also benefit from the rights under Article 12 (4) of the Single Permit Directive, which allows for a wider range of invalidity and death pensions to be exported.[212]

According to the last paragraph of Article 23 (1) of the Seasonal Workers' Directive, seasonal workers or the survivors of such workers residing in a third country are entitled to statutory pensions based on the seasonal worker's previous employment under the same export conditions available for nationals when they move to a third country.[213] Most importantly, this equal treatment clause does not depend on the existence of an agreement with the respective third country. Still, seasonal workers cannot benefit from invalidity and death pensions because neither the Seasonal Workers' Directive nor the Single Permit Directive, which excludes seasonal workers from its scope, provide for an entitlement in this regard for these migrant workers.[214] One needs to examine the implementation of the Seasonal Workers' Directive in national law in order to establish whether these workers could

[205] Article 18 (3) ICTD; Article 22 (2) (b) and (c) SRD; Article 23 (2) (i) SWD; Article 12 (2) (b) SPD. Concerning the latter, see also C-449/16 – Martinez Silva, ECLI:EU:C:2017:485 In Vankova (2018), p. 168.

[206] Case C-571/10 Kamberaj, ECLI:EU:C:2014:233, points 78, 86, 79–81 and 92. See Groenendijk (2015), p. 557.

[207] Beduschi (2015), pp. 224–225.

[208] Article 14 (1) (f) of the BCD.

[209] Verschueren (2016), p. 383.

[210] Ibid.

[211] In other words, if there is a bilateral agreement between the two respective countries, it will apply.

[212] Groenendijk (2015), p. 558.

[213] Vankova (2018), p. 169.

[214] Ibid., p. 390.

benefit from such pensions since in some cases statutory pensions could also cover invalidity and survivors' benefits.

ICTs or their survivors are entitled to the export of old-age, invalidity, and death statutory pensions under the same conditions and at the same rates as the nationals of the Member State concerned when they move to a third country.[215] In addition, Article 12 (4) of the Single Permit Directive provides that third-country nationals and their survivors are entitled to payments of old age, invalidity, and death pensions under the same conditions and rates as for nationals of the Member State when they move abroad. This is also valid in situations where there is no bilateral social security agreement in force between the Member State and the third country.[216] Yet, the Implementation report on the Single Permit Directive identified problems with the implementation of this provision in several Member States.[217] Finally, researchers are guaranteed equal treatment with nationals of the Member State and can benefit from export of benefits on the basis of Article 12 (4) of the Single Permit Directive if the nationals are entitled to such rights.[218]

The 'first admissions' directives do not have the same approach to the different categories of migrants and allow Member States room to provide exceptions to the equal treatment provisions. Furthermore, they are relevant for the social security rights of third-country nationals, but are not instruments that coordinate social security systems. For instance, these directives do not contain any provisions on aggregation of periods of insurance, employment, and residence. For migrant workers this could mean that even in cases where they have fulfilled such periods in their home country, they might not be able to bring these into account in order to obtain the right to social security benefits that, according to the national legislation of the host Member State, depend on having fulfilled such waiting periods.[219] Therefore, the social security coordination between the Member States and third countries remains subject to the conclusion of bilateral agreements between the individual states.[220] In order to understand the actual rights of third-country nationals in practice and examine whether they fulfil benchmarks in the current study, one needs to examine these agreements in detail.[221] Finally, none of the instruments discussed above provide for the reimbursement of social security contributions, which constitutes another benchmark in this policy area (see Annex V).

---

[215] Article 18 (2) (d) ICTD.

[216] Verschueren (2016), p. 387.

[217] European Commission (2019b), pp. 10–11.

[218] Article 22 (1) SRD. In Vankova (2018), p. 169.

[219] On that, see also Verschueren (2016).

[220] Vankova (2018), p. 169.

[221] Ibid.

4.5.2 *Mobility Partnerships*

Ensuring portability of social security rights is part of the operational priorities of the GAMM's first pillar, 'Organising and facilitating legal migration and mobility',[222] and therefore is reflected in all Mobility Partnership with the Eastern Partnership countries. For instance, the Annexes of the Mobility Partnership with Moldova, Armenia, and Georgia contain a proposal by Bulgaria to negotiate bilateral agreements in the area of social security. In line with the objective of ensuring portability of social security rights, Moldova, for instance, has already concluded social security agreements with several Member States including Romania, Portugal, Bulgaria, Luxembourg, Austria, Estonia, Czech Republic, Poland, Hungary, Belgium, Lithuania, and Germany.[223] Furthermore, this priority has also been implemented through projects aiming to increase the capacity of the administration of some of the Eastern partnership countries[224] or through information sessions for potential migrants, such as the one described above in the 'Work authorisation' section. Therefore, it can be concluded that the Mobility Partnerships are used by some countries as platforms to foster bilateral cooperation in the field of social security coordination, which supports the implementation of the benchmarks in this field. This is good practice, but its implementation is yet rather uneven and limited.

4.6 Entry and Residence Conditions for Family Members

Even though the right to family reunification for circular migrants is not a policy area that can influence the ability of migrants to circulate,[225] it still needs to be considered as it can help determine what kind of policies the EU is putting in place: a rights-based approach or a revival of the guest-working model?[226] Guest workers were prevented from bringing their families as a guarantee that they would return to their countries of origin, and this is still the case with many of the current time-bound migration schemes. It is, therefore, important to assess whether circular migrants, including seasonal workers and other temporary migrants,[227] can reunite

[222] European Commission (2011b), p.13.

[223] ICMPD (2016), p. 39. Eastern Partnership Panel on Migration, Mobility and Integrated Border Management (2017), 'Moldovans to receive social security benefits after signing an agreement with Germany', 2 November 2017. http://eapmigrationpanel.org/en/news/moldovans-receive-social-security-benefits-after-signing-agreement-germany Accessed 26 November 2019.

[224] See for instance the project 'Strengthening the administrative capacity of the authorities of the Republic of Moldova in relation to transfers of social security benefits': http://scoreboard.mfa.gov.md/ Accessed 26 November 2019.

[225] Vankova (2016), p. 338.

[226] Vankova (2018), p. 170.

[227] Benchmark developed on the basis of ILO Guidelines on Special Protective Measures for Migrant Workers in Time-bound Activities (Doc. MEIM/1997/D.4) adopted by the Tripartite Meeting of Experts on Future ILO Activities in the Field of Migration, ILO, Geneva, Annex 1, Para. 6.1. For the rest of the benchmarks, see Annex V.

with their family members under EU law instruments in the field of legal migration. In addition to this benchmark, this section also considers the following two policy measures that can facilitate family reunion: waiting periods, which should not exceed 12 months, and housing conditions, which are not as restrictive as to prevent family reunification.[228]

4.6.1 Legal Migration Directives

The Family Reunification Directive goes further than the universal human rights instruments and the case law of the ECtHR, stipulating a right to entry and residence for nuclear family members.[229] The most important feature of this Directive is that it sets out a general rule for the authorisation of family reunification, according to the ruling of the CJEU in *Chakroun,*[230] and all of the conditions for family reunification, as well as the exceptions and derogations to this general rule, should be regarded as exhaustive and interpreted strictly.[231]

Article 3 of the Family Reunification Directive stipulates that in order to reunite with family members, sponsors need to hold 'a residence permit issued by a Member State for a period of validity of 1 year' or have 'reasonable prospects of obtaining the right of permanent residence, if the members of his or her family are third country nationals of whatever status'. The Report on the implementation of the Directive showed that most Member States allow migrants who are holders of temporary residence permits to reunite with their family members, but they subject this to a minimum period of residence that varies among the Member States.[232]

In its Guidelines for the application of Directive 2003/86/EC on the right to family reunification, the Commission stresses that Member States have discretion on how to consider 'reasonable probability of obtaining the right of permanent residence'.[233] However, it also underlines that 'holders of residence permits issued for a specific purpose with a limited validity and that are not renewable cannot, in principle, be considered to have a reasonable prospect of obtaining the right to permanent residence'. Thus, the scope of the Family Reunification Directive excludes forms of temporary stay, such as those of temporary or seasonal workers, and residence permits that are valid for less than 1 year.[234] This means that the benchmark

---

[228] For more details, see Vankova (2016), p. 347.

[229] Peers (2012c), p. 248; Groenendijk (2014), p. 331.

[230] Case C-578/08, Chakroun, ECLI:EU:C:2010:117, Para. 43.

[231] Peers (2012c), p. 249.

[232] European Commission (2008). The latest edition of MIPEX reconfirms this conclusion. See Huddleston et al. (2015).

[233] European Commission (2014b), p. 4.

[234] See Article 3 (1) of the FRD.

on family reunion of seasonal workers and 'special purpose workers' in this policy area is not met.[235]

In addition, before authorising the entry of family members, Member States have the discretion to impose additional requirements, among which are the normal accommodation requirement[236] and a waiting period.[237] With regards to the requirement for the sponsor to have normal accommodation for a comparable family in the same region and which meets the general health and safety standards in force in the Member State, the Guidelines state that its evaluation is left to the discretion of the Member States. However, 'the criteria adopted may not be discriminatory and this provision defines the upper limit of what may be required'.[238] It also emphasises that the adopted criteria need to be transparent and clearly specified in the national legislation, and that 'the fulfilment of this requirement may be judged on either the situation of the sponsor at the moment of the application, or on a reasonable prognosis of the accommodation that can be expected to be available when the sponsor will be joined by his/her family member(s)'.[239] Therefore this requirement is considered as being in line with the benchmark on accommodation mentioned above.

Member States are also free to impose a waiting period of up to 2 years of lawful stay before the migrant's family members can join.[240] This provision is subject to a derogation, allowing Member States to retain a 3-year waiting period between the submission of an application for family reunification and the issue of a resident permit to a family member on the basis of national law, if the relevant legislation in force on the date of the adoption of the Family Reunification Directive took account of Member State's reception capacity.[241] This provision was among the three clauses challenged by the European Parliament in the case of *European Parliament v. Council of the European Union*. The CJEU affirmed the validity of this provision and stressed that it did not violate the right to family life.[242] The provision allows Member States to delay reunification in accordance with their margin of appreciation in order to provide for the better integration of family members.[243] If Member States decide to use this option, then they should make an individual case-by-case assessment and should not impose a general blanket waiting period, as per Article 17 of the Family Reunification Directive.[244] Member States should also take into consideration the best interests of any minor children.[245]

---

235 Vankova (2018), p. 170.

236 Article 7 (1) (a) FRD.

237 Article 8 FRD.

238 European Commission (2014b), p. 11.

239 Ibid.

240 Article 8 of the FRD.

241 Article 8 of the FRD. Only Austria applied this derogation in the past. See Peers (2012c).

242 Case C-540/03 Parliament v Council, ECLI:EU:C:2006:429, paras 97–103.

243 Ibid, paras 97–98.

244 European Commission (2014b), p. 17.

245 Article 5 (5) of the FRD.

In its Guidelines, the Commission recommends to Member States to keep those waiting periods as short as possible so as to avoid affecting the right to family life in a disproportionate way.[246] Member States should take into account any 'lawful stays' under national law from day one, authorised through residence permits or other status allowing such a legal stay.[247] They can require that the lawful stay is continuous but certain interruptions, such as temporary absences for business trips or visits to the country of origin, should be allowed.[248] Even if circular migrants have a permit that is not excluded from the scope of the Directive, this requirement can seriously hinder migrants' family life if it is too lengthy.[249] Therefore the benchmark on waiting periods in this policy area sets the limit at up to 12 months.

A comparison between the 'first admissions' directives shows that all highly-skilled categories – Blue Card holders, ICTs, and Researchers – have facilitated access to family reunification on the basis of derogations to the Family Reunification Directive.[250] They are exempted from the requirement to have reasonable prospects of obtaining the right to permanent residence and from the waiting period requirement, which means that they could enter and stay on temporary permits and still reunite with their family members. In contrast, seasonal workers who also stay temporarily are the only category of migrant workers excluded from the scope of the Family Reunification Directive and the right to family reunion, along with other temporary permits under national law.[251] This means that the EU's family reunification policy confirms the trend of providing rights-based circular migration solutions only with regard to highly-skilled migrants. Therefore, the benchmark on obligations to facilitate family reunion can be considered as only partially fulfilled.[252]

4.6.2 Mobility Partnerships

The focus on families as part of the priorities of the Mobility Partnerships with the Eastern partnership countries concerns mainly social protection[253] and children left behind.[254] However, some of the bilateral agreements aimed at promoting circular

[246] European Commission (2014b), p. 17.

[247] Ibid.

[248] Ibid.

[249] Vankova (2018), p. 171.

[250] Article 15 BCD; Article 26 SRD; Article 19 ICTD.

[251] Vankova (2018), p. 171.

[252] Ibid.

[253] See for instance EU-Moldova Mobility Partnership, p. 13; EU-Belarus Mobility Partnership, p.7; EU-Azerbaijan-Mobility Partnership p.5.

[254] See the following projects under the EU-Moldova Mobility Partnership: 'Strengthen the development dimension of migration', 'Addressing the Negative Effects of Migration on Minors and Families Left Behind', 'Study on Children and Elderly left behind in Moldova and Georgia', 'Bridge the gap' – End of life-care for terminally ill and elderly people left behind in Moldova and

migration discussed in Sect. 4.2.3 contain family reunification facilitation. For instance, the bilateral agreement between Armenia and France for the exchange of students, interns, and young professionals between the ages of 18 and 35 years as well as qualified specialists – which is pending ratification by the French side[255] – contains family reunification provisions (Article 4.4.) allowing spouses and minor children to benefit from a residence permit of the same validity as that of the qualified specialist who serves as a sponsor. The agreements concluded by Bulgaria under the auspices of the Mobility Partnerships with Armenia (Article 3.2)[256] and Moldova (Article 3.2)[257] also provide for family reunification in line with the applicable national legislation of the receiving state.[258] However, this is not the case in the bilateral agreement between Georgia and France on 'Residence and Circular Migration of Skilled Professionals'. Therefore, the initiatives under the Mobility Partnerships with Eastern Partnership countries are considered to only partially fulfil the benchmarks on family reunification facilitation.

4.7 Recognition of Qualifications

Another policy area that can influence the willingness of circular migrants to engage in this type of migration is the recognition of qualifications. As already mentioned in Chap. 1, the 'triple win' proponents claim that it enables skills transfer back to the countries of origin, which in turn supports development and counteracts 'brain drain'. Nonetheless, some evaluations of circular migration schemes stress that there are cases when migrants return home and their new skills cannot be recognised or are not needed.[259] Therefore, this policy area is considered important for the

Ukraine: http://scoreboard.mfa.gov.md/ Accessed 26 November 2019; the project 'Mitigating social consequences of labour migration and maximizing migrants involvement in local development' under the EU-Armenia Mobility Partnership assessed by UNICEF: https://www.unicef.org/evaldatabase/index_101734.html Accessed 26 November 2019.

[255] Project d'accord de partenariat migratoire entre le gouvernement de la Republique Francaise et le gouvernement de la Republique D'Armenie provided by the French Embassy in Yerevan in March 2017.

[256] Agreement between the Republic of Bulgaria and the Republic of Armenia on labour migration management, approved by Council of Ministers Decision No. 176 of 22 March 2018, entered into force 9 October 2018/ Спогодба между Република България и Република Армения за регулиране на трудовата миграция, утвърдена с Решение No. 176 от 22 март 2018 г. на Министерския съвет. В сила от 9 октомври 2018 г.

[257] Agreement between the Government of the Republic of Bulgaria and the Government of the Republic of Moldova on labour migration management, approved by Council of Ministers Decision No 492 of 13 July 2018, entered into force on 11 September 2018/ Спогодба между правителството на Република България и правителството на Република Молдова за регулиране на трудовата миграция, утвърдена с Решение No. 492 на Министерския съвет от 13 юли 2018 г. В сила от 11 септември 2018 г.

[258] For more details, see Chap. 8, Sect. 8.1.1.

[259] Hooper and Sumption (2016), pp. 20–21.

purposes of this research, especially when it comes to recognition of regulated professions, such as in the health sector.

The benchmarks in this policy area focus on the availability of provisions on the recognition of occupational qualifications that have been acquired outside the EU, including certificates and diplomas, and other means for the recognition of professional qualifications.[260] Among the instruments that can implement these benchmarks are international cooperation instruments and active information policy in relation to the recognition of academic qualifications that would make circular migration beneficial for the circular migrant and support both skill and knowledge transfer.[261]

In addition, the Convention on the Recognition of Qualifications concerning Higher Education in the European Region (the Lisbon Recognition Convention) stipulates that holders of qualifications shall have adequate access to an assessment of these qualifications in another country, and provides other basic principles that can be used as benchmarks in the field of the academic recognition of qualifications (see Annex V).[262] The establishment of national information centres, offering advice on the recognition of foreign qualifications to parties or persons, is among the instruments that support the implementation of the recognition of academic qualifications.[263]

4.7.1 Legal Migration Directives

The Professional Qualifications Directive 2005/36/EC[264] provides for the system of recognition of professional qualifications in the EU. These EU standards in the field of the recognition of professional qualifications can also apply to certain third-country nationals on the basis of equal treatment provisions, such as to EU long-term residents, Blue card holders, ICTs, single permit holders, and researchers. Even though these standards can be applied in order to make circular migration more beneficial, hitherto, they do not necessarily bind Member States to apply them to third-country nationals. Nonetheless, all these categories of third-country nation-

[260] Based on Article 14 (b) of ILO Convention No. 143; Paragraph 6 of Migrant Workers Recommendation, 1975 (No. 151), adopted at 60th ILC session on 24 June 1975; Principle 12, Guideline 12.6 of the ILO's Multilateral Framework on Labour Migration.

[261] Vankova (2016), p. 348.

[262] Article III.1 of Convention on the Recognition of Qualifications concerning Higher Education in the European Region, CETS No.165, adopted 11 April 1997, entry in force 1 February 1999. See also Sections 4, 5 and 6 of the LRC.

[263] Article IX.2 of the LRC. For more details see Vankova (2016), p. 349.

[264] Directive 2005/36/EC of the European Parliament and of the Council of 7 September 2005 on the recognition of professional qualifications [2005] OJ L 255. Amended by Directive 2013/55/EU of the European Parliament and of the Council of 20 November 2013 amending Directive 2005/36/EC on the recognition of professional qualifications and Regulation (EU) No 1024/2012 on administrative cooperation through the Internal Market Information System [2013] OJ L 354.

als can also benefit from rights in this respect under the specific EU labour migration directives providing for equal treatment provisions concerning recognition of qualifications.

Article 14 (1) (d) of the Blue Card Directive provides for the equal treatment of Blue Card holders in relation to the recognition of diplomas, certificates, and other professional qualifications in accordance with the relevant national procedures. Recital 19 of the Preamble to the Blue Card Directives provides further details: professional qualifications acquired by a third-country national in another Member State should be recognised in the same way as those of EU citizens. Furthermore, qualifications that have been acquired in a third country should be taken into account in conformity with the Professional Qualifications Directive 2005/36/EC.[265] In addition, the deadline for examining the application for an EU Blue Card should not include the time for the recognition of professional qualifications, if so required.[266] The Recast Proposal envisages a requirement for the Member States 'to recognise professional experience as an alternative to education qualifications'.[267] There is also a new draft requirement for Member States to facilitate the validation and recognition of documents which attest the relevant higher professional qualifications for unregulated professions.[268]

The equal treatment clause that is contained in the Single Permit Directive also covers the recognition of diplomas, certificates, and other professional qualifications in accordance with the relevant national procedures in Article 12 (1) (d) thereof. Recital 23 of the Preamble to the Single Permit Directive provides more information in this regard, stating that Member States should recognise professional qualifications that have been acquired by third-country nationals in another Member State in the same way as those of EU citizens, and should take into account any qualifications that have been acquired in a third country in accordance with the Professional Qualifications Directive 2005/36/EC. Therefore, if migrant workers are admitted for the first time to the EU and possess a qualification from a third country, they can have this recognised on the basis of national law. If migrant workers have a prior EU qualification, they can benefit from the recognition procedures that are contained in the Professional Qualifications Directive 2005/36/EC. Researchers also benefit from the equal treatment clause that is contained in the Single Permit Directive.[269]

The Seasonal Workers' Directive also contains an equal treatment clause on the recognition of qualifications in Article 23 (1)(h) thereof. Finally, ICTs enjoy equal treatment with nationals of the Member State where the work is carried out in relation to the recognition of diplomas, certificates, and other professional qualifications in accordance with relevant national procedures.[270]

[265] Recital 19 of the Preamble to the BCD.

[266] Recital 12 of the Preamble to the BCD.

[267] See BCD Recast Proposal, p. 13 and Article 2 (g) thereof.

[268] Article 5 (6) of the BCD Recast Proposal.

[269] Article 12 (1) (d) of the SRD.

[270] Article 18 (2) (b) of the ICTD.

Despite the existence of EU instruments in this field, research shows that the qualification recognition systems continue to differ depending on which country is in charge of the recognition procedure.[271] There are a variety of definitions of regulated and non-regulated professions as well as various types of recognition procedures and methods of assessment that are applicable in each case. According to the Migrant Immigration Policy Index (MIPEX), procedures for recognising skills and foreign qualifications are, in general, recent and only facilitated by some countries.[272] This plethora of different instruments can hinder access to the labour market and the use of migrants' qualifications obtained in their home countries, and can thus impede circularity.[273] Furthermore, equal treatment under the Directives applies only after migrants have received their authorisation to enter the respective Member State.[274] In order to assess what these equal treatment clauses mean in practice and what the provisions and measures available to circular migrants are, one must analyse the national law provisions to gain further insight.

4.7.2 *Mobility Partnerships*

All Mobility Partnerships with the Eastern Partnership countries contain the facilitation of recognition of skills and qualifications amongst their priorities.[275] Accordingly, some of the Mobility Partnerships envisaged initiatives under their Annexes, such as the conclusion of bilateral agreements.[276]

The scoreboard of the Mobility Partnership between the EU and Moldova lists several projects related to the recognition of qualifications in cooperation with the European Training Foundation (ETF) aimed at building capacity to manage labour and return migration.[277] The 'Better managing the mobility of health professionals in the Republic of Moldova' project (2011–2014) aimed to promote legal and circular migration, diminish the negative effects of brain drain and brain waste, and facilitate the reintegration of health professionals returning to the Moldovan health system.[278] It also envisaged the promotion of bilateral agreements regulating migration for health personnel between Moldova and the EU Member States in line with

[271] Schuster et al. (2013), p. 22.

[272] Nine countries assessed with MIPEX: AU, CA, CY, DE, EE, IS, NL, SE, UK. See more at: http://www.mipex.eu/labour-market-mobility

[273] See Schuster et al. (2013).

[274] S. Carrera et al. (2019), p. 63.

[275] See EU-Moldova Mobility Partnership, p.4; EU-Belarus Mobility Partnership, p. 6; EU-Azerbaijan Mobility Partnership, p.6; EU-Georgia Mobility Partnership, pp. 3 and 4; EU-Armenia Mobility Partnership, pp. 4–5.

[276] See EU-Moldova Mobility Partnership, p. 15; EU-Georgia Mobility Partnership, p. 8; EU-Armenia Mobility Partnership, pp. 10–12.

[277] For more details see Mosneaga (2015).

[278] World Health Organisation (2015).

the WHO Global Code of Practice for International Recruitment of Health Personnel. As mentioned in Chap. 3, countering the 'brain drain' from third countries' health sectors is among the reasons for introducing the circular migration concept in EU migration policy.[279]

Similarly, ETF has also implemented a project focused on the recognition of qualifications and piloting the validation of non-formal and informal learning in Armenia in the field of tourism and the hospitality sector.[280] In addition, the MISMES report on Armenia gives another example of an initiative in this field, namely the National Information Centre for Academic Recognition and Mobility, which is part of the international network of ENIC-NARIC organisations and plays an important role in facilitating the recognition of foreign qualifications in Armenia.[281]

Another initiative that contained a qualifications recognition component was the 3-year 'Strengthening the development potential of the EU Mobility Partnership in Georgia through targeted circular migration and diaspora mobilisation' project discussed above.[282] It piloted a circular migration scheme through which nurses from Georgia could work in Germany.[283] Since this is a regulated profession, Georgian citizens had to go through the recognition of qualifications procedure in line with the relevant German legislation because there were 'major differences in the qualification "nurse" between Georgia and Germany'.[284] Applicants undertook German language courses whilst they were still in Georgia, and while working in Germany received professional training, which took up to 1 year and concluded with an exam.[285] The project's aim was for these workers to contribute their expertise towards Georgia's development as a result of the knowledge and experience they gained in Germany. According to one interviewed state official, the Georgian nurses would not need to repeat the recognition process upon their return.[286] However, in the future an additional recognition system for returning migrants would have to be elaborated,[287] as facilitation of these procedures among other factors was considered one of the incentives for circular migrants to return.[288] This is also part of the lessons learned from the implementation of the circular migration pilot projects in Georgia.[289]

---

[279] See Chap. 3, Sect. 3.2.

[280] Kalantaryan (2015), p. 21.

[281] Ibid, p. 22.

[282] The project was part of the joint operation that was referred to as the Centre for International Migration and Development (CIM) and was managed by the Federal Employment Agency and the German organisation GIZ.

[283] Interview #24 with former MP project assistant, Georgia, March 2017, Annex I.

[284] Interview # 18 with state official, Georgia, November 2014, Annex I. For more details, see Goos, (2016), p.48

[285] Interview # 24 with former MP project assistant, Georgia, March 2017, Annex I.

[286] Interview # 18 with state official, Georgia, November 2014, Annex I.

[287] Interview # 18 with state official, Georgia, November 2014, Annex I.

[288] Goos (2016), p. 56.

[289] Mestvirishvili (2018), pp. 5–6.

The review of the latest available Mobility Partnership scoreboards and reports shows that some of the Eastern Partnership countries are investing targeted efforts in establishing workable recognition of qualifications systems in order to ensure mobility of skills and qualifications and prevent skill underutilisation.[290] The initiatives and projects identified vary from a country to country, depending on the stage of finalisation of important educational reforms, but are nonetheless considered to support the implementation of the benchmarks in this field.

4.8 Conclusions

The analysis in this chapter has demonstrated that circular migration has a rather marginal role in the EU labour migration *acquis*. This is in contrast with the EU policy documents adopted by the European Commission, which serve as the basis for the development of the EU's approach to circular migration. Nevertheless, two main concepts of circular migration can be outlined in the EU's labour migration policy. On the one hand, a spontaneous pattern of circularity that can be facilitated through a legislative framework, such as in the context of the Blue Card Directive; and on the other, a temporary migration scheme with a re-entry component is regulated through the Seasonal Workers' Directive.[291] These two policy approaches underline the unequal treatment between the different categories of migrant workers.

This chapter shows that rights-based circular migration is reserved for highly-skilled migrants, which the EU wants to attract. The most desirable category of migrants – the Blue Card holders – benefit from migrant-led trajectories and are given the opportunity to settle permanently according to the EU labour migration *acquis*.[292] Furthermore, Blue Card holders are the only category of migrants with the explicitly-provided right to switch employer and remain unemployed for up to 3 months without the risk of losing their permits. Nevertheless, even this desired category of migrants does not have facilitated entry to the EU.

Seasonal workers and ICTs, on the other hand, are admitted for a limited period of time and provided with the possibility to re-enter on the basis of different conditions.[293] They do not benefit from migrant-led trajectories and the high protection of rights that is guaranteed to Blue Card holders. For instance, seasonal workers are not entitled to family reunification, and neither category has access to obtaining an EU long-term residence permit – which is not in line with the benchmarks of the study based on international and European standards (see Annex V). Furthermore, the Seasonal Workers' Directive does not address employer-organised accommodation,

[290] See for instance Kalantaryan (2015), p. 39; Badurashvili (2015), p. 39; Mosneaga (2015), p. 5, pp. 31–32.

[291] Vankova (2018), p. 172.

[292] Ibid.

[293] Ibid.

which could lead to abuse and dependency of this category of workers. The analysis also demonstrates that researchers, who are considered a mobile group, do not benefit from any special provisions in relation to the facilitation of circular migration.

Despite the equal treatment clauses concerning social security coordination, the possibility of exporting social benefits depends, to a large extent, on national provisions and the existing bilateral agreements signed between Member States and third countries.[294] The same applies to the equal treatment provisions regarding recognition of qualifications, which need to be examined through analysis of national law.

This chapter also focused on the implementation of circular migration-related initiatives under the framework of the GAMM and the Eastern Partnership. Despite the myriad initiatives, dialogues, and agreements, in practice circular migration is hardly facilitated as part of the GAMM. There are several bilateral agreements that could lead to small-scale facilitation if the respective Member States are able to overcome their reluctance to open new channels for legal migration. Apart from that, most of the initiatives are based on pilot projects with uncertain futures. The projects analysed through the prism of the book's benchmark framework vary in terms of content and implementation from country to country. However, initiatives such as the bilateral agreements providing opportunities for migrant workers to reunite with their families and focusing on social security coordination, as well as projects fostering recognition of qualification, can support rights-based circular migration in the long-run.

As the European Commission stresses in its Legal Migration Fitness Check Conclusions, the limited opportunities for circular migration permitted by the legal migration directives 'means that it is up to Member States to develop initiatives in this area'.[295] Given that only a few Member States have done so, 'there could be scope to strengthening the legal framework in this area and to further use funding possibilities for initiatives and projects promoting circular migration'.[296]

References

Ambassade de France à Tbilissi (2019) Accord franco-géorgien sur le séjour et la migration circulaire de professionnels, 1 February 2019. https://ge.ambafrance.org/Accord-franco-georgien-sur-le-sejour-et-la-migration-circulaire-de Accessed 19 November 2019.

Andrade, P.G., I. Martin, and S. Mananshvili (2015) *EU cooperation with third countries in the field of migration*. Study for the LIBE Committee. European Parliament, Brussels.

Arcarazo, D. A. (2011) *The Long-Term Residence Status as a Subsidiary Form of EU Citizenship. An Analysis of Directive 2003/109*. Leiden: Brill/Nijhoff.

Arcarazo, D. A. (2015) Civic Citizenship Reintroduced? The Long-Term Residence Directive as a Post-National Form of Membership. *European Law Journal*, 21 (2): 200–219.

[294] Vankova (2018), p. 172.

[295] European Commission (2019c), p. 152.

[296] Ibid.

Badurashvili, I. (2015) Inventory of Migrant Support Measures from an Employment and Skills Perspective. Georgia. European Training Foundation.
Barnard, C. (2016) The Substantive Law of the EU: The Four Freedoms. Fifth Edition. Oxford: Oxford University Press.
Bertelsmann Stiftung (2015) A Fair Deal on Talent – Fostering Just Migration Governance. Lessons from Around the Globe.
Beduschi, A. (2015) An Empty Shell? The Protection of Social Rights of Third-Country Workers in the EU after the Single Permit Directive. *European Journal of Migration and Law* 17 (2–3): 210–238.
Carrera, S. (2009) *In Search of the Perfect Citizen?: The Intersection Between Integration, Immigration and Nationality in the EU*. Immigration and Asylum Law and Policy in Europe. Leiden: Martinus Nijhoff Publishers.
Carrera, S. (2014) Integration of Immigrants in EU Law and Policy: Challenges to Rule of Law, Exceptions to Inclusion. In L. Azoulai and K. D. Vries (eds), *EU Migration Law: Legal Complexities and Political Rationales*. Oxford: Oxford University Press, DOI:https://doi.org/10.1093/acprof:oso/9780198708537.003.0006.
Carrera, S. and Z. Vankova (2019) Human Rights Aspects of Immigrant and Refugee Integration Policies. A comparative assessment in selected Council of Europe member states at the request of the Special Representative of the Secretary General on Migration and Refugees of the Council of Europe, Strasbourg.
Carrera, S., L. Vosyliute, Z. Vankova, N. Laurentsyeva, M. Fernandes, J. Dennison, and J. Guerin (2019) The Cost of Non-Europe in the Area of Legal Migration: at the request of the European Added Value Unit of the Directorate for Impact Assessment and European Added Value within the Directorate General for Parliamentary Research Services. *CEPS Paper in Liberty and Security in Europe*, No 2019-01, Brussels.
Council of the European Union (2008) Joint Declaration on a Mobility Partnerships between the European Union and the Republic of Moldova, 9460/08, ADD 1, 21 May 2008, Brussels.
Council of the European Union (2009) Joint Declaration on a Mobility Partnership between the European Union and Georgia, 16396/09, 20 November 2009, Brussels.
Council of the European Union (2011a) Outcome of Proceedings of Working Party on Integration, Migration and Expulsion, 8485/11, 5 April 2011, Brussels.
Council of the European Union (2011b) Joint Declaration on a Mobility Partnership between the European Union and Armenia, 14963/11, ADD 1, 6 October 2011, Brussels.
Council of the European Union (2013) Joint Declaration on Mobility Partnership between the Republic of Azerbaijan and the European Union and its participating Member States, 16399/13, 26 November 2013, Brussels.
de Bie, A. and A. Ghimis (2017) The Intra-corporate Transferees Directive: a revolutionary scheme or a burden for multi-nationalcompanies? *ERA forum, Academy of European Law* (18): 199–211.
Eisele, K. (2013) Why come here if I can go there? Assessing the 'Attractiveness' of the EU's Blue Card Directive for 'Highly Qualified' Immigrants. *CEPS Paper on Liberty and Security in Europe No. 60*, Brussels.
Eisele, K. (2018) Briefing. Initial Appraisal of European Commission Impact Assessment. Revision of the visa code: Impact Assessment (SWD(2018) 77, SWD(2018) 78 (summary)) of a Commission proposal for a regulation amending Regulation (EC) No 810/2009 establishing a Community Code on Visas (Visa Code) (COM(2018) 252). European Parliamentray Research Service. April 2018, Brussels.
European Commission (2007) Proposal for a Council Directive on the conditions of entry and residence of third-country nationals for the purposes of highly qualified employment, COM (2007) 637, 23 October 2007, Brussels.
European Commission (2008) Report on the application of Directive 2003/86/EC on the right to family reunification, COM (2008) 610 final, 8 October 2008, Brussels.
European Commission (2009) Mobility Partnerships as a tool of the Global Approach to Migration, SEC (2009) 1240, 18 September 2009, Brussels.

European Commission (2011a) Report on the application of Directive 2003/109/EC concerning the status of third-country nationals who are long-term residents, COM (2011) 585 final, 28 September 2011, Brussels.

European Commission (2011b) Communication 'The Global Approach to Migration and Mobility', 5. Operational priorities, COM (2011) 743 final, 18 November 2011, Brussels.

European Commission (2013) Proposal for a Directive of the European Parliament and of the Council on the conditions of entry and residence of third-country nationals for the purposes of research, studies, pupil exchange, remunerated and unremunerated training, voluntary service and au pairing, COM (2013) 0151 final, Brussels.

European Commission (2014a) Communication on the implementation of Directive 2009/50/EC on the conditions of entry and residence of third-country nationals for the purpose of highly qualified employment, COM (2014) 287 final, 22 May 2014, Brussels.

European Commission (2014b) Communication on guidance for application of Directive 2003/86/EC on the right to family reunification, COM (2014) 210 final, 3 April 2014, Brussels.

European Commission (2015) Commission decision on Joint Declaration establishing a Mobility Partnership between the Republic of Belarus and the European Union and its participating Member States, C (2015) 3955 final 15 June 2015, Brussels.

European Commission (2016a) Communication 'Towards a reform of the common European asylum system and enhancing legal avenues to Europe', COM (2016) 197 final, 6 April 2016, Brussels.

European Commission (2016b) Proposal for a Directive on the conditions of entry and residence of third-country nationals for the purposes of highly skilled employment, COM (2016) 378 final, 7 June 2016, Strasbourg.

European Commission (2016c) Commission Staff Working Document, Impact assessment accompanying the document Proposal for a Directive of the European Parliament and the Council on the conditions of entry and residence of third-country nationals for the purposes of highly skilled employment and repealing Directive 2009/50/EC, SWD (2016) 193 final, Part 6/6, 7 June 2016, Strasbourg.

European Commission (2018a) Communication 'Adapting the common visa policy to new challenges', COM (2018) 251 final, 14 March 2018, Brussels.

European Commission (2018b) Impact assessment accompanying the document 'Proposal for a Regulation of the European Parliament and of the Council amending Regulation (EC) No 810/2009 establishing a Community Code on Visas (Visa Code)', SWD (2018) 77 final, 14 March 2018, Brussels.

European Commission (2019a) Report on the implementation of Directive 2003/109/EC concerning the status of third-country nationals who are long-term residents, COM (2019) 161 final, 29 March 2019, Brussels.

European Commission (2019b) Report on Directive 2011/98/EU of the European Parliament and of the Council of 13 December 2011 on a single application procedure for a single permit for third-country nationals to reside and work in the territory of a Member State and on a common set of rights for third-country workers legally residing in a Member State, COM (2019) 160 final, 29 March 2019, Brussels.

European Commission (2019c) Fitness Check on EU Legislation on legal migration, SWD (2019) 1055 final, 29 March 2019, part 2/2, Brussels.

European Council (2010) Stockholm Programme: An open and secure Europe serving and protecting the citizens, OJ C 115/1, 4 May 2010.

European Migration Network (2011) 'Temporary and Circular Migration: empirical evidence, current policy practice and future options in EU Member States. Report on Sweden', prepared by Migrationsverkert.

European Union – Republic of Moldova Mobility Partnership (2010) Scoreboard – monitoring tool of the European Union – Republic of Moldova Mobility Partnership, http://scoreboard.mfa.gov.md/project/view/262 Accessed 23 November 2017.

Fridriksdottir, B. (2017) *What Happened to Equality? The Construction of the Right to Equal Treatment of Third-Country Nationals in Europan Union Law on Labour Migration.* Immigration and Asylum Law and Policy in Europe, Volume 43. Leiden: Brill.

Fudge, J. and P. Olsson. (2014). The EU Seasonal Workers Directive: When Immigration Controls Meet Labour Rights. *European Journal of Migration and Law* (16): 439–466.

Georgian Journal (2019) Georgian citizens are granted the right to work in France, 24 January 2019. https://www.georgianjournal.ge/business/35484-georgian-citizens-are-granted-the-right-to-work-in-france.html Accessed 19 November 2019.

Gogolashvili, K. (2019) A New Chance for Circular Labor Migration between Georgia and the EU. *Rondeli Blog*. 15 January 2019. https://www.gfsis.org/blog/view/918 Accessed 27 November 2019.

Goos, A. (2016) *Manual on Circular Migration Scheme*. Centre for International Migration and Development. http://migration.commission.ge/files/pcms_en_final.pdf Accessed 27 November 2019.

Groenendijk, K. (2012) Long-Term Residents. In S. Peers, E. Guild, D. A. Arcarazo, K. Groenendijk and V. Moreno-Lax (eds), *EU Immigration and Asylum Law (Text and Commentary): Second Revised Edition. Volume 2: EU Immigration Law*. Leiden: Martinus Nijhoff Publishers, pp. 287–353.

Groenendijk, K. (2014) Recent Developments in EU Law on Migration: The Legislative Patchwork and the Court's Approach. *European Journal of Migration and Law* (16): 313–335.

Groenendijk, K. (2015) Equal treatment of workers from third countries: the added value of the Single Permit Directive. *ERA Forum* (16): 547–561.

Groenendijk, K. (2019) Legal migration. In M. De Somer, P. De Bruycker and J.-L. De Brouwer (eds) *From Tampere 20 to Tampere 2.0: Towards a New European Consensus on Migration*. European Policy Centre, Brussels.

Hernández i Sagrera, R. (2014) The Impacts of Visa Liberalisation in Eastern Partnership Countries, Russia and Turkey on Trans-border Mobility. Study for the LIBE Committee, Brussels.

Hooper, K. and M. Sumption (2016) *Reaching a 'Fair' Deal on Talent. Emigration, Circulation and Human Capital in Countries of Origin*. Migration Policy Institute, Washington DC.

Hugo, G. (2013) *What We Know About Circular Migration and Enhanced Mobility*. Migration Policy Institute, Washington DC.

Huddleston, T., Ö. Bilgili, A. Joki and Z. Vankova (2015) Migrant Integration policy Index (MIPEX). CIDOB and MPG, Brussels/Barcelona.

Infantino, F. (2016) State-bound Visa Policies and Europeanized Practices. Comparing EU Visa Policy Implementation in Morocco. *Journal of Borderlands Studies* 31 (2): 171–186.

ICMPD (2016) Overview of the EU Mobility Partnership Instrument. Summary of the Initiatives in Georgia and Moldova, MOBILAZE project, May 2016. https://www.pragueprocess.eu/en/migration-observatory/publications/document?id=161 Accessed 26 November 2019.

International Organisation for Migration, Mission to Georgia (2017) Temporary Labour Migration of Georgian Workers to Poland and Estonia Pilot Project Kicks Off. http://iom.ge/1/temporary-labour-migration-georgian-workers-poland-and-estonia-pilot-project-kicks Accessed 12 December 2017.

Kalantaryan, S. (2015) Inventory of Migrant Support Measures from an Employment and Skills Perspective. Armenia. European Training Foundation.

Leith, J. and A. Rivas (2015) The Temporary Return of Qualified Nationals Project (TRQN). III. Evaluation Report for the International Organization for Migration. https://iom-nederland.nl/images/Publications/Evaluation_TRQN__Response_IOM.pdf Accessed 23 November 2019.

Mananashvili, S. (2013) Access to Europe in a Globalised World: Assessing the EU's Common Visa Policy in the light of the Stockholm Guidelines. *EUI Working Papers, RSCAS* 74, Fiesole.

Meloni, A. (2013) EU Visa and Border Control Policies: What Roles for Security and Recirpocity?. In D. Bigo, E. Guild, and S. Carrera (eds), *Foreigners, Refugees or Minorities? Rethinking People in the Context of Border Controls and Visas*. Farnham: Ashgate, pp. 151–169.

Mestvirishvili, N. (2018) Circular Migration Schemes in Georgia: Lessons Learned and Ways Forward. *Prague Process Policy Brief*. https://www.pragueprocess.eu/documents/repo/91/Policy_Brief_GEO_Circular%20Migration_EN.pdf Accessed 23 November 2018.

Migration for development (2011) Making Migration in Moldova Work for Development. http://www.migration4development.org/en/projects/making-migration-moldova-work-development Accessed 23 November 2017.

Mitsilegas, V. (2010) Extraterritorial Immigration Control in the 21st Century: The Individual and the State Transformed. In B. Ryan and V. Mitsilegas (eds), *Extraterritorial Immigration Control. Legal Challenges*. Leiden: Martinus Nijhoff Publishers, pp. 39–68.

Mosneaga, V. (2015) Inventory of Migrant Support Measures from an Employment and Skills Perspective. Republic of Moldova. European Training Foundation.

Pascouau, Y. and T. Strik (eds) (2012) *Which Integration Policies for Migrants? Interaction between the EU and its Member States*. Oisterwijk: Wolf Legal Publishers.

Peers, S. (2012a) Visa Facilitation. In S. Peers, E. Guild, and J. Tomkin (eds), *EU Immigration and Asylum Law (Text and Commentary): Second Revised Edition. Volume 1: Visas and Border Controls*. Leiden: Martinus Nijhoff Publishers, pp. 315–336.

Peers, S. (2012b) Admission of Researchers. In S. Peers, E. Guild, D. A. Arcarazo, K. Groenendijk and V. Moreno-Lax (eds), *EU Immigration and Asylum Law (Text and Commentary): Second Revised Edition. Volume 2: EU Immigration Law*. Leiden: Martinus Nijhoff Publishers, pp. 129–164.

Peers, S. (2012c) Family Reunion. In S. Peers, E. Guild, D. A. Arcarazo, K. Groenendijk and V. Moreno-Lax (eds), *EU Immigration and Asylum Law (Text and Commentary): Second Revised Edition. Volume 2: EU Immigration Law*. Leiden: Martinus Nijhoff Publishers, pp. 247–286.

Peers, S. (2016) *EU Justice and Home Affairs Law. Volume I: EU Immigration and Asylum Law.* Fourth Edition. Oxford: Oxford University Press.

Reslow, N. (2013) *Partnering for mobility? Three-level Games in EU External Migration Policy'*. Datawyse/Universitaire Pers Maastricht.

Reslow, N. (2015) EU 'Mobility' Partnerships: An Initial Assessment of Implementation Dynamics. *Politics and Governance* 3 (2): 117–128.

Reslow, N. (2017) Not everything that counts can be counted: Assessing 'success' of EU external migration policy, *International Migration* 55 (6): 156–169.

Schuster, A., M. V. Desiderio, and G. Urso (2013) Recognition of Qualifications and Competences of Migrants. International Organization for Migration, Brussels.

Skeldon, R. (2012) Going Round in Circles: Circular Migration, Poverty Alleviation and Marginality. *International Migration* 50 (3): 43–60.

Strik, T. and A. Böcker (2011) Language and Knowledge tests for Permanent Residence Rights. Integration tests, help or hindrance? *European Journal of Migration and Law,* 13 (2): 157–184.

Trauner, F. and I. Kruse (2008) EC Visa Facilitation and Readmission Agreements: A New Standard EU Foreign Policy Tool?, *European Journal of Migration and Law* 10 (4): 411–438.

Van Elsuwege, P. and Z. Vankova (2020) Migration and Mobility in the EU's Eastern Neighbourhood: Mapping out the Legal and Political Framework. In F. Ippolito, G. Borzoni, and F. Casolari (eds), *Bilateral relations in the Mediterranean: Prospects for migration issues.* Cheltenham: Edward Elgar.

Vankova, Z. (2016). EU Circular Migration Policies: Dead or Alive? Developing a Rights-based Benchmark Framework for Policy Assessment. *Journal of Immigration, Asylum and Nationality Law* 30 (4): 332–352.

Vankova, Z. (2018) The EU's Approach to Circular Migration in the Context of the Eastern Partnership Neighbourhood. In S. Carrera, L. den Hertog, D. Kostakopoulou, and M. Panizzon (eds), *EU external migration policies in an era of global mobilities: Intersecting policy universes*. Immigration and Asylum Law and Policy in Europe, Vol. 44. Leiden: Brill/Nijhoff, pp. 155–176.

Verigo, T. and A. Castelfranco (2019) Evaluation report: Ex-post internal evaluation of the project 'Piloting Temporary Labour Migration of Georgian Workers to Poland and Estonia'. Commissioned by IOM Country Office Georgia.

Verschueren, H. (2016) Employment and Social Security Rights of Third-Country Labour Migrants under EU Law: An Incomplete Patchwork of Legal Protection. *European Journal of Migration and Law* 18 (4): 373–408.

World Health Organisation (2015) Better managing the mobility of health professionals in the Republic of Moldova – context, challenges, and lessons learned. http://www.euro.who.int/en/countries/republic-of-moldova/news/news/2015/03/better-managing-the-mobility-of-health-professionals-in-the-republic-of-moldova-context,-challenges,-and-lessons-learned Accessed 23 November 2019.

Wiesbrock, A., T. Jöst and A. Desmond (2016) Seasonal Workers Directive 2014/36/EU. In K. Hailbronner and D. Thym (eds), *EU Immigration and Asylum Law. A Commentary*. (2nd edition). Munchen/Oxford/Baden-Baden: C. H. Beck/Hart/Nomos, pp. 928–973.

Chapter 5
National Approaches to Circular Migration in Bulgaria and Poland

This chapter aims to illustrate national variances in instruments conducive to circular migration as part of the implementation of the EU's approach. It serves as an introduction to the two different national approaches to circular migration facilitation developed in Bulgaria and Poland, which were chosen as case studies. First, the chapter examines the respective strategies of these two countries in their pre-accession periods that influenced the instruments developed in Bulgaria and Poland. Second, it zooms in on the national contexts in order to put some flesh on the different instruments seen as fostering circular migration at the national level.

5.1 Bulgaria and Poland: Two Different Pre-accession Conditionality Strategies[1]

The migration policies of both Poland and Bulgaria were, to a great extent, a product of the Europeanisation and conditionality pre-accession processes. The reason for this was the limited development of these types of policies during communist rule, which mainly aimed at preventing unwanted emigration of their own citizens.[2] The Europeanisation process that had already started after 1989 filled the 'institutional lacunae' of the communist legacy.[3] Many institutions as well as legal and policy measures designed in Western Europe were transferred to Poland and Bulgaria as part of the Europeanisation process on the basis of policy transfer and

[1] Parts of this chapter were previously published in Vankova (2017).

[2] Grabbe (2006), p. 167; Kicinger et al. (2007), p. 181; Bobeva (1994), p. 222; Markova and Vankova (2014), p. 41.

[3] Grabbe (2006), p. 167.

Z. Vankova, *Circular Migration and the Rights of Migrant Workers in Central and Eastern Europe*, IMISCOE Research Series,
https://doi.org/10.1007/978-3-030-52689-4_5

policy learning through various channels[4] and led to the establishment of restrictive migration policies.[5] This created a paradoxical situation, where the foundations of their migration policies were being laid in a top-down manner as a result of the conditionality pressure rather than being based on an actual necessity created by immigration processes.[6]

Even though both countries had to adapt to the EU conditionality, their approaches differed. In the field of Justice and Home Affairs, the Schengen *acquis* transferring rules on visa and borders was the EU policy that caused the most controversy at the national level because it had the potential to affect political and economic relations with neighbours and partners in Central and Eastern Europe (CEE).[7] First of all, CEE countries were very sensitive to the possibility of creating a new 'Iron Curtain'. Secondly, both countries had close historical, cultural, and economic ties, albeit of a differing character, with countries like Ukraine and Russia. Nevertheless, both countries knew that failure to comply with Justice and Home Affairs conditionality could veto their EU accession and that they needed to respond to EU demands.[8] As a result, the implementation of the EU visa policy at the national level is illustrative of the different strategies adopted by Poland and Bulgaria.

Poland's strategy was 'combative, involving tough negotiating stances and slow implementation of the policies that caused most domestic controversy'.[9] Poland had a clear national interest when it came to its eastern borders. Cross-border trade was the main economic activity for many Polish citizens living at the eastern borders, as well as for many Ukrainians. With time, the difference between cross-border mobility and labour migration became vague[10] and Ukrainians gradually began dominating certain types of jobs and occupations such as horticulture, housekeeping, and construction that were the least attractive employment niches for the Poles.[11]

Furthermore, the visa-free regime with Ukraine, Belarus and Russia, which was preserved after 1989, was based on the former 'socialist brotherhood' ties and reflected Poland's foreign policy interest in maintaining close contacts in the Eastern neighbourhood in order to achieve a stable and predictable geopolitical order in the region.[12] Therefore, Poland's strategy was to postpone the introduction of the visa requirements for as long as it could and introduce them at the last possible moment.[13] In addition, shortly before the pending changes, the Foreign Ministry launched an

---

[4] Kicinger and Koryś (2011), p. 371. See also Pawlak (2015).

[5] Lesińska et al. (2010), p. 67; Iglicka and Gmaj, (2013), p. 170; Jileva (2002), p. 81.

[6] Kicinger and Koryś (2011), p. 371.

[7] Grabbe (2006), p. 168. Another was the refugee protection system and the requirements of the Dublin II Regulation. In Kicinger and Koryś (2011), p. 372. See also Jileva (2002), p. 81.

[8] Ibid., p. 183.

[9] Ibid., p. 111.

[10] Stola (1998), p. 88.

[11] Okólski (2001), p. 102.

[12] Kicinger et al.(2007), pp. 188–189.

[13] Iglicka (2007), p. 265.

information campaign encouraging people in neighbouring countries to apply for multi-entry visas to soften the impact of the visa requirements and the burden the new rules were expected to pose on administrative capacity.[14]

On the other hand, Bulgaria's approach was 'catch-up and imitation'.[15] The driving force behind this strategy was entrenched in the desire to leave the EU's negative visa list and show its EU partners that it could be trusted to implement the *acquis*.[16] After the decision to lift EU visa requirements for Bulgarians was reached, the country introduced a visa regime for Ukraine faster than Poland, despite being further from an accession date.[17] With regards to Russia, Bulgaria delayed the introduction of visas due to the special relations between the two countries.[18] It sent a readmission agreement to Russia in an attempt to circumvent the visa problem, but since Russia did not have an agreement of this kind with any other country, this attempt failed and Bulgaria introduced visas for Russian citizens – again, years ahead of Poland. The economic effects of this decision were significant, especially with regards to trade with Russia, as well as tourism, which was very popular among Russians and Ukrainians.[19]

5.2 National Instruments Fostering Circular Migration Developed After Accession to the EU

5.2.1 Poland

The policy measures and regulations fostering circular migration that Poland developed after its accession to the EU were to a great extent a response to the barriers erected by the Schengen *acquis*.[20] Poland continued its pre-accession strategy as a participant in the EU decision-making process. It transferred its national foreign policy to the EU agenda and started advocating for a visa-free regime between the EU and the neighbouring countries from the Commonwealth of Independent States (CIS), incorporating them in the prevention of undesired migration to the EU and establishing exterritorial means of migration control.[21] Along with Sweden, it initiated the European Union's Eastern Partnership, which, among others, provides for the gradual opening of the EU borders for citizens of these CIS countries on the

[14] Grabbe (2006), p. 174.

[15] Ibid., p. 111.

[16] Gros-Tchorbadjiyska (2010), p. 421; Grabbe (2006), p. 175; Jileva (2002), p. 81.

[17] Grabbe (2006), p. 175.

[18] Jileva (2002), p. 82; Gros-Tchorbadjiyska (2010), p. 255.

[19] Jileva (2002), p. 83.

[20] Iglicka and Gmaj (2013), p. 170.

[21] Müller (2014), p. 135.

basis of agreements on mobility and security.[22] In addition, with a view to Poland's accession to the Schengen area, the country supported the introduction of a separate local border traffic regime at the external borders of the EU,[23] and signed a local border traffic agreement with Ukraine in 2008.[24] The existing border traffic agreement between the two countries had to be renounced before the EU accession date.[25]

On a national level, after the visa introduction, Poland managed to secure its interests by liberalising the issuance of regular visas and the provision of no-fee visas for Ukrainian nationals and Russians residing in Kaliningrad until 2007, when the whole Schengen *acquis* was implemented.[26] After that, in line with Poland's policy towards its eastern neighbours, citizens of Ukraine, Moldova, and Belarus were exempted from paying the consular fees for processing visa applications when applying for a Polish national visa (visa type D).

Furthermore, Poland's EU accession and the introduction of visas for Ukrainians had a strong impact on the availability of a flexible labour force provided in Poland based on the existing pattern of circularity of Ukrainian migrants; most had a legal stay in Poland because they did not need any visas to enter, but were in irregular employment due to the restrictive labour market legislation that made it very hard to obtain a work permit.[27] The adoption of the EU *acquis* on migration control and irregular migration 'rendered the Polish "tacit tolerance" more difficult'.[28]

Greater economic growth combined with a diminished labour supply due to the massive exodus of Polish workers after the EU accession caused gaps in labour market sectors such as construction, agriculture, and horticulture.[29] This strengthened the employers' organisations position to lobby for opening the labour market to foreign workers – which was echoed by politicians and quickly entered the public discourse.[30] All these factors, combined with the pressure exerted by farmers and fruit growers in need of foreign workers for the upcoming harvest, along with the participation of the farmers' party *Samoobrona* in the government coalition, led to the prompt introduction of a simplified procedure for hiring foreign workers referred in the text as the *Oświadczenie* procedure.[31]

Beyond the impact on the border regions, as a result of the implementation of the Schengen *acquis* the new visa regime was also perceived in the public debate as endangering relations with the Polish diaspora in the East.[32] Therefore, after

---

[22] European Migration Network (2012), p. 32.

[23] Refer to Regulation (EC) No. 1931/2006 of the European Parliament and of the Council of 20 December 2006 laying down rules on local border traffic at the external land borders of the Member States and amending the provisions of the Schengen Convention [2006] OJ L 405.

[24] Kicinger et al. (2007), p. 189; Lesińska et al. (2010), p. 66. See Chap. 6, Sect. 6.3.1.

[25] Gros-Tchorbadjiyska (2010), p. 267.

[26] Kicinger et al. (2007), p. 189; Lesińska et al. (2010), p. 64.

[27] Müller (2014), p. 138.

[28] Ibid.

[29] Lesińska et al. (2010), p. 69.

[30] Ibid.

[31] Ibid, pp. 69–70; SOPEMI (2011), p. 310.

[32] Müller (2014), p. 148.

Poland's accession to the EU, the country continued with its policy to strengthen ties with foreigners of Polish descent living abroad.[33] It was no coincidence that the Pole's Card Act,[34] which was in the process of preparation since the late 1990s,[35] came into force on 29 March 2008, 1 day before the final step for full Schengen integration.[36] The Act simplified the procedure for obtaining a multiple-entry national visa and exempted the holders thereof from paying Schengen visa fees.[37]

5.2.1.1 The Simplified *Oświadczenie* Procedure

The concept of circular migration did not officially become a national policy term until after the policy document 'Migration Policy of Poland: state of play and proposed actions' was adopted in 2012.[38] The document stated:

> 'Conditions for legal circular migration should be created (for instance on the basis of visa facilitation), while at the same time ensuring the possibility of transposition, into Polish law, of certain solutions that are contained in the forthcoming Directive concerning admission of foreigners for seasonal work. It is worth emphasizing that circular migration, on the one hand, contributes to a decrease in the number of hired foreigners who do not possess work permit and reduces the phenomenon of illegal immigration. On the other hand, however, depending on the particular needs and through the use of appropriate actions and instruments, circular migration is a potential source of a verifiable permanent migration. It is also worth noting that circular migration does not undermine the human potential of sending countries and creates a favourable system of mutual economic and interpersonal relations'.[39]

This strategic document demonstrated the broad understanding of the concept by the Polish government as encompassing different instruments and not excluding permanent migration. It recommended the creation of proper conditions for circular migration, for instance through the further development of the *Oświadczenie* procedure,[40] and thus linked the established simplified procedure to circular migration. The 2019 draft strategic document 'Migration Policy of Poland', which was

---

[33] Ibid.

[34] Act on the Pole's Card / Ustawa z dnia 7 września 2007 r. o Karcie Polaka (Dziennik Ustaw z 2019 r., poz. 1598). See Kozak et al. (2014), p. 188. This is a different procedure from the one determining Polish origin. For the differences, see ibid., pp. 185–189.

[35] Zogata-Kusz (2013), p. 160.

[36] The final integration step of Poland into the Schengen area occurred on 30 March 2008, when the border controls on internal EU flights were lifted. In Gros-Tchorbadjiyska (2010), p. 265.

[37] Ibid.

[38] Polityka migracyjna Polski – stan obecny i postulowane działania (2012), p. 9, p. 113, p. 122. This policy document was repealed in October 2016 and at the time of writing the Polish government had not yet adopted a new migration policy strategy. Council of Ministers (2012).

[39] See Chapter I: Legal migrations, Section 3-Labour migrations, 34 Subsection b: Recommendations (author's translation).

[40] Zogata-Kusz (2013), p. 216; European Migration Network (2011a), p. 13; Unterschütz (2016), p. 162.

distributed only unofficially, also stressed that as a result of the introduction of the *Oświadczenie* procedure, an important channel for the inflow of economic migrants as part of legal circular migration had been launched.[41] The draft document gave an indication of Poland's future migration policy, according to which circular migration that complements Poland's economic needs remains the preferred type of immigration.[42] However, since immigration can contribute to Poland's economic development, it is advisable to treat circular migration as potentially leading to settlement, which forces reconciliation of immigration policy with integration policy.[43]

The report on circular migration, which was prepared by the Polish National Contact Point at the request of the European Migration Network in 2011, also highlighted the *Oświadczenie* procedure as one of the national instruments conducive to circular migration.[44] It emphasised that the *Oświadczenie* procedure was not a 'circular migration instrument *per se*' or one of the typical circular migration programmes based on international bilateral agreements aiming to attract 'guest workers'.[45] Nonetheless, it had features that promoted this form of migration on the basis of national law, which amongst other things, facilitated access for foreigners to the Polish labour market.[46]

When asked how they understood the term circular migration, several interviewees representing different stakeholders directly referred to the *Oświadczenie* procedure that is operational between Poland and Ukraine as an example of such migration.[47] Even though it is still 'not present *expresis verbis* in existing national legal acts'[48] or developed as part of an official policy on circular migration,[49] this is the primary national instrument that is considered to both 'represent' and promote circular migration, and thus is one of the national instruments that is assessed in this book.

The procedure was stipulated in a Regulation issued by the Minister of Labour and Social Policy,[50] who was also a member of the *Samoobrona* party. The adoption

[41] Polityka Migracyjna Polski, Zespół do Spraw Migracji, Redakcja: Departament Analiz i Polityki Migracyjnej MSWiA, Draft of 10 June 2019, p. 6. Ministry of Interior and Administration (2019).

[42] Ibid., p. 55.

[43] Ibid.

[44] European Migration Network (2011a), p. 15, p. 19.

[45] Ibid., p. 5, p. 20. Also in Interview # 1 with officials, Poland, November 2016, Annex II.

[46] Ibid., p. 18. Interview # 15 with academic, Poland, November 2016, Annex II.

[47] Interview #1 with officials, Poland, November 2016, Annex II; Interview #2 with academic, Poland, November 2016, Annex II; Interview #2 with official, Poland, November 2016, Annex II; Interview #5 with civil society representative, Poland, November 2016, Annex II.

[48] European Migration Network (2011a), p. 8.

[49] Iglicka and Gmaj (2013), p. 170.

[50] Regulation of the Minister of Labour and Social Policy of 30 August 2006 regarding the performance of work by foreigners without the need to obtain a work permit / Rozporządzenie Ministra Pracy i Polityki Społecznej z dnia 30 sierpnia 2006 r. w sprawie wykonywania pracy przez cudzoziemców bez konieczności uzyskania zezwolenia na pracę (Dziennik Ustaw z 2006 r nr 156, poz. 1116).

of a bylaw was the preferred approach by the Ministry because it was both expeditious enough to respond to farmers' pressing need for a flexible option that allowed for amendment in case the procedure needed to be changed[51] and it also exempted workers from the general rule requiring work permits. As stressed by one of the interviewees, it did not require the approval of the whole cabinet or parliamentary scrutiny.[52] Furthermore, it reflected the idea that Poland did not need an official labour migration policy and that the *Oświadczenie* procedure introduced was merely a temporary solution.[53]

The procedure allowed workers from the neighbouring third countries to work in the agricultural and horticulture sectors without a work permit for a period of 3 months in every 6 months on the basis of a declaration of intent to entrust work to a foreigner.[54] This programme expanded to all economic sectors in 2007 and the permissible work duration period was changed from 3 to 6 months in 12-month periods in 2008.[55] Furthermore, 2009 saw Moldova and Georgia being added to the list of countries that could benefit from this scheme as part of the initiatives under the Mobility Partnerships signed with them and different Member States, including Poland.[56]

In 2010, the Ministry of Labour and Social Policy took a decision to indefinitely extend the *Oświadczenie* procedure, which had initially been introduced as a pilot programme.[57] In 2014, Armenia joined the group of countries covered by this process. Currently the full list of countries whose citizens can benefit from *Oświadczenie* is provided in the Regulation of the Minister of Labour and Social Policy on the citizens of countries that are subject to certain provisions concerning a seasonal work permit, as well as provisions concerning the declaration on entrusting work to a foreigner[58] on the basis of Article 90 (10) 2 of the Act on Employment Promotion and Labour Market Institutions.[59]

Registered abuses of the system led to changes in the procedure in 2018 in line with the new governmental Strategy for Responsible Development until 2020 (with

---

[51] Interview #1 with officials, Poland, November 2016, Annex II.

[52] Interview #15 with academic, Poland, November 2016, Annex II.

[53] Interview #15 with academic, Poland, November 2016, Annex II.

[54] Kepinska and Kindler (2014), p. 274. Interview #1 with officials, Poland, November 2016, Annex II.

[55] Ibid.

[56] Interview #1 with officials, Poland, November 2016. See also 'Migration Policy of Poland: state of play and proposed actions', 2012, pp. 9, 122.

[57] SOPEMI (2011), p. 310.

[58] Rozporządzenie Ministra Rodziny, Pracy i Polityki Społecznej z dnia 8 grudnia 2017 r. w sprawie państw, do których obywateli stosuje się niektóre przepisy dotyczące zezwolenia na pracę sezonową oraz przepisy dotyczące oświadczenia o powierzeniu wykonywania pracy cudzoziemcowi (Dziennik Ustaw z 2017 r., poz. 2349)

[59] Act of 20 April 2004 on Employment Promotion and Labour Market Institutions / Ustawa o promocji zatrudnienia i instytucjach rynku pracy (Dziennik Ustaw z 2019 r., poz. 1482, consolidated text).

a perspective to 2030)[60] as well as Poland's obligation to transpose the Seasonal Workers Directive, which was significantly delayed. The procedure, previously based on a Regulation issued by the Minister of Labour and Social Policy,[61] was explicitly stipulated in the Act on Employment Promotion and Labour Market Institutions. In addition, the sectors of agriculture, forestry, fishery, services, gastronomy, and hospitality were excluded from the scope of the *Oświadczenie* procedure on the basis of the Regulation of the Minister of Family, Labour and Social Policy of 8 December 2017, which reserved these sectors for the holders of EU seasonal work permits.[62] Another novelty introduced concerns the employer, who is currently obliged to inform the respective authorities whether the foreigner, for whom a declaration has been registered, has started the work in question or not (Article 88z (13) of the Act on Employment Promotion and Labour Market Institutions).

5.2.1.2 Pole's Card (*Karta Polaka*)

According to academic literature and the indications of the Polish focus group participants, another instrument that allows for the circulation of a special category of migrants is the Pole's Card (*Karta Polaka*).[63] This became another 'gate' that migrants of Polish origin could use to gain access to their 'fatherland', along with the repatriation provisions regulated in the 2000 Act on Repatriation – the access to permanent settlement for foreigners who can prove that they are of Polish origin (in line with Article 52 (5) of the Polish Constitution) – and the scholarship opportunities for Polish students abroad.[64] The scope of the Act on the Pole's Card was broader than the Act on Repatriation because it could be used by all foreigners of Polish origin from the former USSR, and not only those from the Asian parts of the

[60] Strategia na rzecz Odpowiedzialnego Rozwoju do roku 2020 (z perspektywą do 2030 r.), p.152. Council of Ministers (2017).

[61] In para 1 (20) of the Regulation of the Minister of Labour and Social Policy of 21 April 2015 concerning cases when commission of work to foreigners in the territory of the Republic of Poland is permitted without the requirement of obtaining a work permit/ Rozporządzenie Ministra Pracy i Polityki Społecznej w sprawie przypadków, w których powierzenie wykonywania pracy cudzoziemcowi na terytorium Rzeczypospolitej Polskiej jest dopuszczalne bez konieczności uzyskania zezwolenia na pracę (Dziennik Ustaw z 2018 r., poz. 2273, consolidated text).

[62] Regulation of the Minister of Family, Labor and Social Policy of 8 December 2017 regarding activity subclasses according to the Polish Classification of Activities (PKD), in which seasonal work permits for a foreigner are issued/Rozporządzenie Ministra Rodziny, Pracy i Polityki Społecznej z dnia 8 grudnia 2017 r. w sprawie podklas działalności według Polskiej Klasyfikacji Działalności (PKD), w których wydawane są zezwolenia na pracę sezonową cudzoziemca (Dziennik Ustaw z 2019 r., poz. 1845 consolidated text).

[63] Kindler et al. (2016), p. 10. Focus groups with Russian and Ukrainian migrants, Warsaw, November 2016. Interview #10 with staff member in private recruitment agency, Poland, October 2016.

[64] Lesińska et al. (2010), pp. 85–86

country.[65] The Pole's Card affirmed affinity to the Polish nation and applied to those who could not be granted Polish nationality because their countries of residence do not allow dual citizenship.[66]

Applicants for the Pole's Card have to demonstrate their links with Polish ancestry by possessing at least a basic knowledge of the Polish language, which they see as their mother tongue, and a knowledge and cultivation of Polish traditions and customs.[67] They need to submit a written declaration of belonging to the Polish nation before a Polish consul, the *Podlaski voivode*, or designated proxy.[68] In addition, the applicants need to prove that at least one parent or grandparent, or two great-grandparents, are or were of Polish nationality, or were Polish citizens. The requirement for proving Polish nationality can be replaced by an attestation from a Polish organisation or a Polish diaspora organisation that is active in one of the countries whose nationals are eligible to apply for the *Karta Polaka*, stating that the applicant had been actively involved in Polish cultural and linguistic activities within the Polish community in their region for a period of at least 3 years prior.[69]

Finally, applicants have to declare that neither they nor their ascendants have repatriated themselves or have been repatriated from the territory of the Republic of Poland or the Polish People's Republic on the basis of repatriation agreements concluded in the years 1944–1957 by the Republic of Poland or the Polish People's Republic with the Belarusian Soviet Socialist Republic, Ukrainian Soviet Socialist Republic, the Lithuanian Soviet Socialist Republic, or the Union of Soviet Socialist Republics onto the territory of one of the countries that were party to these agreements.

Until 2019, the countries whose citizens or non-citizen residents were eligible to apply for the *Karta Polaka* were Armenia, Azerbaijan, Belarus, Estonia, Georgia, Kazakhstan, Kyrgyzstan, Lithuania, Latvia, Moldova, Russia, Tajikistan, Turkmenistan, Ukraine, and Uzbekistan.[70] Also eligible for the *Karta Polaka* were foreigners from one of the above-mentioned countries whose Polish origin has been established according to the procedure stipulated in the Act on Repatriation. Since July 2019, citizens of all countries can apply for *Karta Polaka* if they can

[65] Ibid., p. 87.

[66] Kepinska and Kindler (2014), p. 276.

[67] Kozak et al. (2014), p. 188.

[68] Since the amendments in 2017, the Council of Ministers may also designate a governor as the competent authority to receive applications for issuing a Pole's Card and indicating the states whose citizens will be eligible to submit their applications via this route. For more details, see Regulation of the Council of Ministers of 5 October 2017 regarding the appointment of a governor competent to conduct proceedings for granting or extending the validity of a Pole's Card, pursuant to Article 12 (4) of the Pole's Card Act/ Rozporządzenie Rady Ministrów z dnia 5 października 2017 r. w sprawie wyznaczenia wojewody właściwego w zakresie prowadzenia postępowań o przyznanie lub przedłużenie ważności Karty Polaka (Dziennik Ustaw z 2017 r., poz. 1900).

[69] Ibid. A list of such organisations can be accessed here: http://www.migrant.info.pl/organizacje-polskie-i-polonijne.html

[70] Article 2, second paragraph, Act on the Pole's Card.

demonstrate their Polish origin and speak Polish.[71] Foreigners who are granted a *Karta Polaka* are also entitled to apply free of charge for a permit to settle in Poland[72] and since the amendments in 2017 can claim a 9-month financial allowance after submitting an application for a permanent residence permit (Article 8a.1 of the Act on the Pole's Card).

5.2.2 Bulgaria

Bulgaria did not respond as actively as Poland to the barriers introduced by the adoption of the Schengen *acquis*, postponing the introduction of visas only with regards to its closest neighbours until the country's EU accession.[73] Bulgaria concluded bilateral intergovernmental agreements with North Macedonia and Serbia regarding mutual travel of their citizens, which allowed visas to be issued for the citizens of those countries at the border for a stay of up to 10 days.[74] These agreements also provided for rules that can facilitate short-stay visas as well as the travel of citizens of North Macedonia and Serbia to Bulgaria. In addition, two more consulates were opened in Bitola (North Macedonia) and Nis (Serbia) respectively.

The 'catch-up and imitation' model continued after Bulgaria's accession to the EU, this time with regards to the accession to the Schengen Area. The process of Europeanisation advanced through a rushed 'copy and paste' transposition of EU law, which lead to restrictive and unpractical provisions.[75] As one of the interviewed lawyers said: 'Bulgarian lawmakers create rules artificially and impose them on Bulgarian society even if they are not adapted to our reality'.[76] A representative of the administration confirmed that the current state of the legal framework on legal and irregular migration showed that the integration of international and European legislation into Bulgarian law had led to fragmentation.[77] The Act on Normative Acts[78] imposed a requirement to perform an analysis of the existing legal framework on the relevant topic and, if necessary, amendments were proposed in order to avoid contradictory regulatory decisions. Despite that, the interviewee stressed that there were cases where the amendments had not passed through Parliament or legal acts

[71] Act of 16 May 2019 amending Act on the Pole's Card / Ustawa z dnia 16 maja 2019 r. o zmianie ustawy o Karcie Polaka (Dziennik Ustaw z 2019 r., poz. 1095).

[72] Article 195 (1) 3 and 9 AF.

[73] Gros-Tchorbadjiyska (2010), p. 269.

[74] Ibid., p. 267.

[75] Interview #9 with lawyer, Bulgaria, July 2016, Annex III.

[76] Interview #9 with lawyer, Bulgaria, July 2016, Annex III.

[77] Interview #13 with state official, Bulgaria, September 2016, Annex III.

[78] Закон за нормативните актове (SG No. 27/3 April 1973, last amendment SG No. 34/3 May 2016).

had to be amended again due to an inaccurate or incomplete implementation of the requirements of European legislation.[79]

Unlike Poland, Bulgaria had not been an active Member State and acted primarily as 'a policy taker', meaning that it had generally followed the mainstream views in relation to the EU's migration policy.[80] This was also confirmed by one of the interviewees who was previously engaged in the EU policymaking process: 'Every time we go to a meeting on integration and asylum, Bulgaria is absent. They don't really participate'.[81] This could be attributed mainly to the weak administrative capacity in migration policy management and the pre-accession inertia as a 'policy taker',[82] as well as the imposed Cooperation and Verification Mechanism[83] and the efforts of Bulgaria to prove that it was a loyal EU Member State in line with the country's ambition to join the Schengen Area.[84] The latter also explains why the country's approach to migration has been primarily 'securitised', e.g., based on anticipation of soft security challenges in this area.[85]

Even though the foundations for its development were laid because of the conditionality pressure, migration policy was established as a national public policy on a strategic level just after Bulgaria's accession to the EU.[86] Bulgaria developed four national migration strategies after 2007. The first, the National Strategy on Migration and Integration (2008–2015), claimed to set the grounds for the development of a consistent national policy on managing migration and integration. In fact, one of the main reasons for creating a strategy in this field could be found in the establishment of the EU general program on 'Solidarity and Management of Migration Flows' and the available funds for all EU Member States.[87] The second strategy adopted was the National Strategy in the Field of Migration, Asylum and Integration (2011–2020) developed as part of the Bulgarian government's efforts to meet the requirements of accession to the Schengen Area.[88] In 2014 and 2015 two other strategies were developed, mainly in response to the refugee crisis resulting from the conflict in Syria.

Three of the national strategies on migration highlighted the fact that circular migration needed to be both encouraged and promoted.[89] The following extract

---

[79] Interview #13 with state official, Bulgaria, September 2016, Annex III.

[80] Lessenski (2009), p. 8, p. 46; Vankova (2011), p. 82.

[81] Interview #12 with academic, Italy, May 2013, Annex III.

[82] Vankova (2011), p. 82.

[83] European Commission (2006), Decision of 13 December 2006 establishing a mechanism for cooperation and verification of progress in Bulgaria to address specific benchmarks in the areas of judicial reform and the fight against corruption and organised crime, 2006/929/EC, OJ L 354.

[84] Lessenski (2009), p. 46.

[85] Ibid.

[86] Krasteva et al. (2011).

[87] Angelov et al. (2011), p. 173.

[88] Markova and Vankova (2014), p. 42.

[89] Council of Ministers (2008), p. 26, p. 35; Council of Ministers (2011), p. 40; Council of Ministers (2015), p. 49.

from the 2008 Strategy shows how Bulgarian policymakers had understood the concept of circular migration:

> 'The large-scale acceptance of third- country nationals is not a good solution and has to be avoided, as the experience of other countries shows. An organised and balanced reception of third country nationals is undertaken. Their return to the country of origin is regulated after the expiry of their contract. In this way the European initiative at the Community level for the promotion of the so-called „circular" migration is implemented in practice'.[90]

According to the report on circular migration prepared by the Bulgarian National Contact Point at the request of the European Migration Network (EMN), the notion of temporary and circular migration persisted in its strategic documents even though circular migration was not a priority of Bulgaria's migration policy.[91] In most cases, 'circular migration' is referred to as an EU term that is derived from EU migration policy to which Bulgaria adheres to as part of its national strategy on migration.[92]

Furthermore, one of the interviewed officials stressed that this term had lately been forgotten at the EU level and that he had to remind his colleagues at high-level meetings that this concept existed – especially in the context of the GAMM.[93] The interviewee stated that the EU had adopted this term, but that it was up to the national authorities to provide the concept with further substance: 'We need to rather see what is behind this term in practice; or more precisely, which existing processes can be likened to or classified as a section, subsection, or type of circularity'.[94] He understood this concept as primarily trying to avoid the negative effects of 'brain drain'[95] as it signified a process whereby persons moved between their country of origin and an EU Member State. According to this state official, the circular migration process would lead to a win-win situation where, on the one hand, the persons would be able to enhance their qualifications, acquire know-how, experience, knowledge, and skills while working in the destination country and assisting its economy; on the other hand, by returning from their country of destination, they could pass on this experience to others who would benefit from 'the enrichment' on the basis of the 'training of trainers' principle.[96] This concept was described in a similar way by an interviewee working for an international organisation in Bulgaria.[97]

Unlike the interviewees in Poland, the interviewed stakeholders in Bulgaria did not immediately link this concept to a specific legal instrument or national policy.

---

[90] Council of Ministers (2008), p. 23.

[91] European Migration Network (2011b), p. 10.

[92] Council of Ministers (2008), p. 23; Council of Ministers (2011), p. 40; Council of Ministers (2015), p. 49.

[93] Interview #1 with official, Bulgaria, July 2016, Annex III.

[94] Interview #1 with official, Bulgaria, July 2016, Annex III.

[95] Interview #1 with official, Bulgaria, July 2016, Annex III.

[96] Interview #1 with official, Bulgaria, July 2016, Annex III.

[97] Interview #12 with expert from international organisation, Bulgaria, June 2016, Annex III.

Rather, they described it as something that was not taking place at present. One of the interviewed lawyers emphasised that she did not have any clients who had experienced problems as a result of their circulation because voluntary circulation was 'mission impossible'.[98] Asked how the term circular migration was to be understood, another interviewee said that in order for the country to facilitate circular migration, politicians had to first gain an understanding that labour migration was necessary and that people who would come to work for this purpose had to be provided with a circle of rights.[99] The interviewee stressed that in practice, however, this approach was not mentioned in the public, political, or legislative discourses. She added that, in the current climate, none of the politicians in the country were prepared or willing to clearly state that the country needed immigrants.[100]

The representative of one employers' organisation gave a similar response, stating that current Bulgarian politicians 'did not want to hear about migration'.[101] This interviewee added that there was a large discrepancy between the actions of politicians and the administration because the latter was obliged to develop migration policies as a result of Bulgaria's legal obligations to the EU. Another interviewee, who represented a different employers' organisation, also stressed that the country 'unfortunately' needed foreign workers in specific sectors;[102] the interviewee remarked that for business organisations it was better if this migration was of a temporary and circular character because this would better match the labour market's changing needs. Unlike other interviewees, this one opined that the current legal framework was facilitating circular migration.[103] The trade union representatives, on the other hand, noted that the current legislation was 'fairly reasonable' and that they did not see the need for reforms in relation to the migration of third-country nationals.[104]

Moreover, neither the Migration Strategies nor the EMN report on circular migration explicitly pointed to any national instruments that would support the implementation of the EU's circular migration approach in Bulgaria. The EMN report stressed that 'the understanding of optimal migration' was based on the notion of temporary and circular migration: 'the economic situation is very dynamic, the labour market is flexible; migration, which can quickly and precisely respond to its changing requirements, is seen as the best option'.[105] Therefore, the EMN study concluded that some of the instruments that could be identified in the first 2008 Strategy – such as the determination of quotas or the identification of labour deficits in certain professions with the participation of social partners – pertained to the

[98] Interview #8 with lawyer, Bulgaria, July 2016, Annex III.

[99] Interview #9 with lawyer, Bulgaria, July 2016, Annex III.

[100] Interview #9 with lawyer, Bulgaria, July 2016, Annex III.

[101] Interview #6 with representative of employers' organisation, Bulgaria, August 2016, Annex III.

[102] Interview #16 with representative of employers' organisation, October 2016, Annex III.

[103] They referred to Chapter II 'Labour Migration from third countries' of the ALMLM.

[104] Interview #5 with trade union representative, July 2016, Annex III.

[105] European Migration Network (2011b), p. 12.

notion of temporary migration.[106] Previous research on this topic, as well as the interviews that were conducted as part of this study, however, highlighted the bilateral labour agreements as being among the main national instruments with the potential to encourage circular migration.[107]

Unlike Poland, which introduced the Pole's Card as a quasi-citizenship mechanism, Bulgaria provided a fast-track procedure for acquiring Bulgarian citizenship on the basis of Bulgarian ethnic origin[108] and, over the years, developed numerous policies to attract and retain foreign citizens of Bulgarian origin, ranging from facilitated access to the labour market and permanent residence to scholarships to study in Bulgaria.[109] For instance, persons of Bulgarian ancestry are exempted from the majority of obligations that one needs to meet under the general naturalisation regime: they only need to have reached the age of majority and not have been sentenced by a Bulgarian court for a premeditated crime of a general nature or subject to criminal proceedings for such a crime unless the person concerned has been rehabilitated (Article 15 (1) 1 of the Act on Bulgarian Citizenship[110]). Ethnic Bulgarian origin is certified through the State Agency for Bulgarians Abroad, inter alia, based on the birth certificates of applicants' parents and grandparents, their mother tongue, membership of a Bulgarian Church, school, or the former Bulgarian citizenship of their parents.[111]

Such communities of foreign citizens of Bulgarian origin are considered to reside inter alia in countries like the Czech Republic, Slovakia, Romania, Hungary, Serbia, Moldova, Ukraine, and Croatia, where they gained the status of national minorities.[112] The biggest Bulgarian community in the Balkan region was considered to be located in North Macedonia.[113] The 2018 Strategy stressed that these were foreigners 'who would have fitted without any difficulties in the Bulgarian society due to their knowledge of the Bulgarian language, customs, and culture'.[114]

5.2.2.1 Bilateral Agreements

With the 2008 Strategy, the government sought to pursue a 'balanced approach' based on the EU's circular migration concept, i.e., the return of the immigrants to their country of origin after the expiry of their employment contract had to be

---

106 Ibid.

107 Vankova (2009), p. 57. Interview #1 with official, Bulgaria, July 2016, Annex III. See also Lessenski (2011), p. 10.

108 For more details, see Smilov and Jileva (2009).

109 For more details see Vankova (forthcoming).

110 Закон за българското гражданство (SG No. 136/18 November 1998, last amendment SG No. 77/18 September 2018).

111 Smilov and Jileva (2009), p. 225.

112 Council of Ministers (2008), p. 6. For more details see Vankova (forthcoming).

113 Ibid., p. 5.

114 Ibid., p. 17.

regulated in advance.[115] Apart from individual admission and sector quotas, bilateral labour migration agreements – where possible paired with a bilateral agreement for social security coordination – were envisaged as suitable mechanisms that were considered in line with EU trends.[116]

The bilateral agreements for exchange of labour were the main policy instrument for managing labour migration before, as well as after, the fall of the communist regime. The signing of such agreements was considered the preferred way to resolve labour market problems, regulate labour migration, and at the same time limit discrimination against Bulgarian workers in foreign labour markets.[117] Therefore, it was not a surprise that this instrument was among the main policy mechanisms concerning admission of third-country nationals envisaged by the Strategy.

One interviewee, representing the state administration, stated that the Migration Strategies aimed to create a balance between the policies on immigration, emigration and those that were targeted at ‘Bulgarians abroad’.[118] Therefore, the 2008 Strategy envisaged the conclusion of ‘bilateral agreements for labour migration regulation’ with third countries, which were important in relation to the latter policy.[119] This means that in the case of Bulgaria, the implementation of the EU’s circular migration approach through bilateral agreements is aligned with the country’s policy of attracting ethnic Bulgarians who, as already discussed above, are also the target group for facilitated permanent settlement policies and citizenship.

In 2008, the draft bilateral labour migration agreements developed on the basis of a template approved by the National Council on Labour Migration were sent to four countries – Moldova, Macedonia, Ukraine, and Armenia – and a process of consultation was initiated.[120] According to public officials interviewed in 2010 as part of a study for the Open Society Institute in Sofia, and within the parameters of the current study, no agreements were ever signed due to the financial crisis.[121]

The last National Strategy on Migration, Asylum and Integration (2015–2020) stressed that Bulgaria had to temporarily apply stricter rules for access to the Bulgarian labour market in order to reduce the number of work permits issued and tackle rising unemployment in the country as a result of the economic crisis.[122] It acknowledged, nevertheless, that when the country’s economy recovers, bilateral labour migration agreements with specific third countries would be used as the main instrument for foreign labour recruitment. The Eastern Partnership countries were

[115] Vankova (2011), p. 73–4.

[116] Council of Ministers (2008), p. 28.

[117] SOPEMI (1993), p. 117.

[118] This term is usually used to refer to foreigners of Bulgarian origin rather than Bulgarian citizens who emigrated. For more details see Vankova (forthcoming).

[119] Interview #1 with official, Bulgaria, July 2016, Annex III.

[120] Vankova (2011), p. 78.

[121] Ibid.

[122] Council of Ministers (2015), p. 37.

identified once again as the countries suitable for concluding such agreements.[123] The Strategy emphasised that EU instruments in the field of legal migration, such as the Blue Card and the Single Permit Directives, that were transposed in the country's legislation were not considered sufficient for enabling Bulgaria to compete economically and socially with other EU Member States as well as with global competitors such as the USA, Canada, and Australia in the race to attract the best specialists.[124]

In 2016, the adopted Act on Labour Migration and Labour Mobility (ALMLM)[125] provided a section on the possibility to conclude these types of agreements, which were meant to create options for circular migration.[126] As envisaged by the 2008 Strategy, priority was to be given to those countries with which there was an ongoing negotiation for the conclusion of bilateral social security agreements or had already signed such agreements (Article 62 (3) of the ALMLM).

In late June 2017, the government approved a framework labour migration agreement with Armenia, Moldova, and Ukraine.[127] It was remarked that 'signing agreements with Armenia, Moldova and Ukraine will be an opportunity to provide workers for economic sectors where there is a shortage of labour'.[128] So far, two bilateral labour agreements have been concluded with Armenia and Moldova.[129] Negotiations of a draft agreement with Ukraine have commenced and the Council of Ministers has approved two draft agreements with Belarus and Georgia. In addition, two consultations on draft bilateral social security agreements with South Caucasus countries have been carried out.[130]

5.3 Conclusions

This chapter demonstrated that the EU's circular migration concept entered the migration policy agendas of Bulgaria and Poland as part of the policy transfer driven by the Europeanisation process. Depending on the national specifics, this concept was interpreted differently at the national level. In Poland it was understood broadly as encompassing various instruments and not excluding permanent settlement. In Bulgaria, on the other hand, this notion was interpreted as close to the guest-workers model, where migrants were expected to leave after the end of their contracts.

---

[123] Ibid., p. 37.

[124] Ibid., p. 39.

[125] Закон за трудовата миграция и трудовата мобилност (SG No. 33/ 26 April 2016, last amendment No. 34/ 23 April 2019).

[126] See Chapter 5 of the ALMLM. Interview #1 with official, Bulgaria, July 2016, Annex III.

[127] Mediapool.bg (2017).

[128] Ibid.

[129] National Council on Migration and Integration (2018), p. 3.

[130] Ibid.

Bulgaria's and Poland's pre-accession models provide a background for the instruments developed to foster circular migration. Poland's 'combative' strategy, characterised by a strong national stance on maintaining existing close contacts in the Eastern neighbourhood, led to the development of instruments at both EU and national levels allowing for circular migration of migrants coming from the CIS region to a great extent as a reaction to the Schengen barriers created. In contrast, Bulgaria has mainly been a 'policy taker' at the EU level and resorted to bilateral agreements as the main instrument to facilitate circular migration, which has been a preferred model of labour migration management in Bulgaria since the communist regime. These agreements are targeting mainly states with communities of ethnic Bulgarians, including the Eastern partnership countries, in line with Bulgaria's national strategy to attract these foreigners for settlement as well as for circular migration purposes.

This book has so far illustrated that circular migration, as a concept, is an empty shell filled by a combination of EU instruments introduced in Chap. 4, as well as existing national instruments labelled as part of the circular migration umbrella, including ethnic origin-based policy solutions. These national approaches to circular migration, however, do not exist in a vacuum and in order to assess whether they promote rights-based circular migration, they need to be examined in more detail against the background of general admission legal frameworks, including the transposed EU labour migration directives as well as other policies pertinent to circular migration.

References

Angelov, G., Z. Vankova and I. Ivanova (2011) Conclusions and Recommendations. In I. Ivanova (ed.) *Current Trends in Cross-border Workforce Migration and the Free Movement of People – Effects for Bulgaria*. Open Society Institute, Sofia.

Bobeva, D. (1994) Emigration from and Immigration to Bulgaria. In H. Fassmann and R. Münz (eds.), *European migration in the late twentieth century. Historical patterns, actual trends, and social implications*. Hants: Edward Elgar Publishing, pp. 221–238.

Council of Ministers (2008) National Strategy of Republic of Bulgaria on Migration and Integration (2008–2015), adopted with Council of Ministers Protocol No. 22.21/5 June 2008, Sofia.

Council of Ministers (2011) National Strategy in the field of Migration, Asylum and Integration (2011–2020), adopted with Council of Ministers Decision No. 8.17/23 February 2011, Sofia.

Council of Ministers (2012) Migration Policy of Poland: state of play and proposed actions, adopted with Council of Ministers resolution on 31 July 2012, Warsaw. Repealed in October 2016.

Council of Ministers (2015) National Strategy in the field of Migration, Asylum and Integration (2015–2020), adopted with Council of Ministers Decision No. 437/12 June 2015, Sofia.

Council of Ministers (2017) Strategy for Responsible Development until 2020 (with a perspective to 2030), adopted by resolution of the Council of Ministers on 14 February 2017.

European Commission (2006) Decision of 13 December 2006 establishing a mechanism for cooperation and verification of progress in Bulgaria to address specific benchmarks in the areas of judicial reform and the fight against corruption and organised crime. 2006/929/EC, OJ L 354.

European Migration Network (2011a) Temporary and Circular Migration in Poland: empirical evidence, current policy practice and future options in EU Member States. Prepared by the National Contact Point to the European Migration Network, Warsaw.

European Migration Network (2011b) Temporary and Circular Migration in Bulgaria: empirical evidence, current policy practice and future options in EU Member States. Research Study within the framework of the European Migration Network. Ministry of Interior of Republic of Bulgaria, Sofia.

European Migration Network (2012) Visa policy as migration channel in Poland. National Report. Prepared by the National Contact Point to the European Migration Network, Warsaw.

Grabbe, H. (2006) *The EU's Transformative Power. Europeanization Through Conditionality in Central and Eastern Europe*. Palgrave Studies in European Union Politics. London: Palgrave Macmillan UK.

Gros-Tchorbadjiyska, A. (2010) *The Europeanization of Visa Policy: A Transfer of Sovereignty Shaped by enlargement*. Leuven: Katholieke Universiteit Leuven.

Iglicka, K. (2007) Poland. In A. Triandafyllidou and R. Gropas (eds), *European Immigration: A Sourcebook*. 1st edition. Farnham, UK: Ashgate, pp. 263–277.

Iglicka, K. and K. Gmaj (2013). Circular Migration Patterns between Ukarine and Poland. In A. Triandafyllidou (ed.), *Circular migration between Europe and its neighbourhood : choice or necessity?* Oxford: Oxford University Press, pp. 166–186.

Jileva, E. (2002) Larger Than the European Union: The Emerging EU Migration Regime and Enlargement. In S. Lavenex and E. M. Uçarer (eds), *Migration and the Externalities of European Integration*. Lanham, MD: Lexington Books, pp. 75–89.

Kepinska, E. and M. Kindler (2014) Poland. In A. Tryandafillidou and R. Gropas (eds), *European Immigration: A Sourcebook*. 2nd edition. Farnham, UK: Ashgate, pp. 273–286.

Kicinger A. and I. Koryś (2011) The case of Poland. In M. Borkert, R. Penninx, and G. Zincone (eds), *Migration Policymaking in Europe: The Dynamics of Actors and Contexts in Past and Present*. IMISCOE Research Series. Amsterdam University Press, pp. 347–376.

Kicinger, A., A. Weinar and A. Gorny (2007) Advanced yet Uneven: the Europeanization of Polish Immigration Policy. In T. Faist and A. Ette (eds), *The Europeanization of National Policies and Politics of Immigration: Between Autonomy and the European* Union. London: Palgrave Macmillan, pp. 181–200.

Kindler, M., A. Kordasiewicz and M. Szulecka (2016) *Care needs and migration for domestic work: Ukraine-Poland*. Global Action Programme on Migrant Domestic Workers and their Families An ILO/UE project. Geneva: International Labour Office.

Kozak, I., M. Tota and B. Wojcik (2014) *Long-term Residence of Foreigners in Poland in the Light of Act on Foreigners of 12 December 2013*. In T. Sieniow (ed.), Prawo I Polityka Migracyjna II. Lublin: Rule of Law Institute Foundation.

Krasteva, A., I. Otova and E. Staykova (2011) *Labor Immigration to Bulgaria*. European Migration Network report. http://annakrasteva.wordpress.com/2011/02/09/labor-migration-in-bg/#_ftn2 Accessed 1 December 2017.

Lesińska, M. et al. (2010) Migration Policy in Poland and its impact on the inflows and settlement of immigrants. In A. Górny, I. Grabowska-Lusińska, M. Lesińska and M. Okólski (eds), *Immigration to Poland: Policy, Employment, Integration*. Wydawnictwo Naukowe Scholar, Warszawa.

Lessenski, M. (2009) The EU New Member States as Agenda Setters in the Enlarged European Union. Country Report: Bulgaria. European Policies Initiative. Open Society Institute, Sofia.

Lessenski, M. (2011) Executive summary. In I. Ivanova (ed.) *Current Trends in Cross-border Workforce Migration and the Free Movement of People – Effects for Bulgaria*. Open Society Institute, Sofia.

Markova, E. and Z. Vankova. (2014) Bulgaria. In A. Tryandafillidou and R. Gropas (eds), *European Immigration: A Sourcebook*. 2nd edition. Farnham, UK: Ashgate, pp. 41–54.

Mediapool.bg (2017) Valeri Simeonov: We accelerate the granting of Bulgarian citizenship, 56 000 people have been waiting for a long time. http://www.mediapool.bg/valeri-simeonov-

uskoryavame-davaneto-na-balgarsko-grazhdanstvo-56-000-dushi-chakat-otdavna-news265934.html Accessed 26 November 2017.
Ministry of Interior and Administration (2019) Poland's Migration Policy. Unofficial draft of 10 June 2019, Warsaw.
Müller, A. (2014) *Governing mobility beyond the state: centre, periphery and the EU's external borders. Palgrave studies in European political sociology*. London: Palgrave Macmillan.
National Council on Migration and Integration (2018) Report on the 2018 Action Plan for the implementation of the National Strategy in the field of Migration, Asylum and Integration 2015–2020.
Okólski, M. (2001) The transformation of spatial mobility and new forms of international population movements: Incomplete migration in Central and Eastern Europe. In J. W. Dacyl (ed.), *Challenges of Cultural Diversity in Europe*. Stockholm: CEIFO, pp. 57–109.
Pawlak, M. (2015) Research-Policy Dialogues in Poland. In P. Scholten, H. Entzinger, R. Penninx, S. Verbeek (eds), *Integrating Immigrants in Europe: Research-Policy Dialogues*. IMISCOE Research Series. Cham: Springer, pp. 253–273.
Smilov, D. and E. Jileva (2009) The politics of Bulgarian citizenship: National identity, democracy and other uses. In R. Bauböck, B. Perchinig and W. Sievers (eds), *Citizenship Policies in the New Europe*. Amsterdam University Press, pp. 211–246.
SOPEMI (1993) Trends in International Migration. SOPEMI Annual Report 1993. Paris: OECD.
SOPEMI (2011) *International Migration Outlook 2011*. Paris: OECD.
Stola, D. (1998) Poland as a Migration Middle Zone at the EU Eastern Border. In K. Hakola (ed.), *Migration and Refugee Policy on the Eastern Border of the European Union*. University of Jyväskylä, pp. 84–91.
Unterschütz, J. (2016) National Report on Implementation of EU Migration Directives in Poland. In R. Blanpain, F. Hendrickx and P. Olsson (ed.), *National Effects of the Implementation of EU Directives on Labour Migration from Third Countries*. Alphen aan den Rijn: Wolters Kluwer, pp. 161–192.
Vankova, Z. (2009) Subsection B.7: Migration. In A. Kavrakova and E. Markova (eds). *The Unfinished Business of the Fifth Enlargement Countries. Country Report: Bulgaria*. European Policies Initiative, Open Society Institute, Sofia.
Vankova, Z. (2011) Specifics of Migration Management Policies. In I. Ivanova (ed.) In I. Ivanova (ed.) *Trends in Cross-border Workforce Migration and the Free Movement of People – Effects for Bulgaria*. Open Society Institute, Sofia.
Vankova, Z. (2017) *Implementing the EU's Circular Migration Approach: Legal and Migrant Perspectives on Entry and Re-Entry Conditions in Bulgaria and Poland.* Robert Schuman Centre for Advanced Studies Research Paper No. RSCAS 2017/34. Florence: European University Institute.
Vankova, Z. (forthcoming) diaspora policies, consular services and social protection for Bulgarian citizens abroad. In J.-M. Lafleur & D. Vintila (Eds.), *Migration and social protection in Europe and beyond (volume 2). Comparing consular services and diaspora policies*. Cham: Springer
Zogata-Kusz, A. (2013) *Labour immigration policy in a country known for emigration: Poland's policy towards economic immigration after EU accession.* Olomouc: Univerzita Palackého v Olomouci.

Chapter 6
Implementation of National Approaches to Circular Migration: Entry and Re-entry Conditions for Migrant Workers in Bulgaria and Poland

This chapter brings together the EU and national instruments conducive to circular migration developed in Bulgaria and Poland and assesses their implementation against the backdrop of the study's benchmarks concerning entry and re-entry conditions for migrant workers.[1] In order to do that, the chapter first presents the national general admission frameworks and the specific instruments identified as favourable to circular migration; as a second step, it focuses on the implementation of EU instruments in the national laws of Bulgaria and Poland. This analysis is complemented by insights into the implementation dynamics of the EU and national instruments on the basis of data collected through focus groups with migrants from Ukraine and Russia as well as interviews conducted with stakeholders and data on permits retrieved from the national administrations of both countries. The chapter ends with an assessment of whether the instruments developed provide options for facilitated entry for migrants from the Eastern partnership countries and Russia, as well as for circulation-friendly policies – for instance, the possibility to grant priority to seasonal workers who have been employed in the territory of a Member State for a significant period over other workers who seek admission to that State.

[1] In line with the benchmark framework of the study. See Chaps. 1 and. 4, Sect. 4.2 and Annex V for more details. Parts of this chapter were previously published in Vankova (2017).

Z. Vankova, *Circular Migration and the Rights of Migrant Workers in Central and Eastern Europe*, IMISCOE Research Series,
https://doi.org/10.1007/978-3-030-52689-4_6

6.1 Entry and Re-entry Conditions According to National Instruments in Poland and Bulgaria

6.1.1 General Admission Frameworks

The Polish Act on Foreigners of 2013[2] regulates the granting, extension, and revocation or cancellation of visas,[3] including Schengen visas (type C), that permit its holder to remain in the territory of Poland for a maximum of 90 days during a 180-day period,[4] and national long-stay visas (type D). A national visa authorises foreigners[5] to enter Poland and stay in its territory uninterruptedly or within several consecutive stays lasting, in total, more than 90 days within the period of the visa's validity.[6] The validity of the visa is not longer than 1 year[7] and the periods of stay are determined according to the purpose of the stay as specified by the applicant.[8] Article 60 (1) of the Polish Act on Foreigners lists 30 purposes for which a Schengen or a national visa may be issued, including 'performing work, for a period not exceeding 6 months in the next 12 months, based on declaration entered in the register on commissioning work to a foreigner' referred in the text as the *Oświadczenie* procedure,[9] 'carrying out scientific research or development work',[10] or 'enjoying the rights of a holder of Pole's Card'.[11] Furthermore, in a flexible manner, the Act allows a foreigner to provide another reason for arriving in Poland that is not indicated in the Act.[12]

In line with Poland's policy towards its eastern neighbours, citizens of three countries are exempt from paying the consular fees for processing visa applications when applying for a Polish national visa. The citizens of Belarus have been exempted from visa application fees since January 2011 after a unilateral decision taken by the Polish Minister of Foreign Affairs.[13] Before that, the citizens of Belarus had to pay

[2] Polish Act on Foreigners of 2013/ Ustawa o cudzoziemcach z dnia 12 grudnia 2013 (Dziennik Ustaw z 2018r., poz. 2094, consolidated text).

[3] See Section IV. For more details, see Kozak et al. (2014), p. 132.

[4] See Article 3 (22) AF.

[5] In this book 'foreigner' and 'third-country national' are used as interchangeable terms. In Polish and Bulgarian legislation 'foreigner' refers to a person who does not have the country's citizenship and excludes EU citizens and members of their families. For more details, see Articles 2 and 3 (2) of the Polish AF and Article 2 of the Bulgarian AFRB.

[6] Article 59 (1) AF.

[7] Article 59 (3) AF.

[8] Article 59 (2) AF.

[9] Article 60 (1) 5 AF.

[10] Article 60 (1) 13 AF.

[11] Article 60 (1) 20 AF.

[12] Article 60 (1) 25 AF. For more details, see Kozak et al. (2014), p. 132.

[13] European Migration Network (2012), p. 17. In line with the Act of the Minister of Foreign Affairs of the Republic of Poland from 22 December 2015, concerning the reduction of consular fees for

a fee of 20 EUR. Following the adoption of a bilateral intergovernmental agreement, citizens of Ukraine have also been exempted since August 2012 on the basis of the principle of reciprocity.[14] A similar decision was also taken regarding the citizens of Moldova, who as of 1 May 2013 are also exempt from paying the visa processing fees. In addition, foreigners who have obtained the Pole's Card are entitled to receive, free of charge, a special long-stay visa for multiple crossings of the Polish border, issued under Article 60 (1) 20 of the Act on Foreigners.[15]

The Polish Act on Foreigners stipulates that a foreigner is obliged to either leave the territory of Poland or apply for a residence permit within the period of validity of the national visa.[16] However, Article 82 also lists certain conditions that may lead to the extension of the visa's validity. The most relevant condition for circular migration purposes is that the extension is justified by 'vital personal or professional interests of the foreigner' or when 'the foreigner is unable to leave the territory before the expiry of the national visa or before the end of the authorised period of stay'.[17] The other conditions are related to unforeseeable events that occur independently of the will of the foreigner at the time of the visa application, when there are circumstances that do not indicate that the purpose of the foreigner's stay will be different than the one declared, and, finally, when there are no grounds on which the foreigner should be refused a visa.[18]

When foreigners apply for a work visa based on Article 60 (1) 4–6 of the Act on Foreigners, they need to either obtain a work permit or have a written declaration of commissioning work to a foreigner in line with the *Oświadczenie* procedure.[19] The application and issue of a work permit is regulated by the Act of 20 April 2004 on Employment Promotion and Labour Market Institutions and its implementing regulations.[20] Article 88 (1) of that Act lists all the cases when an employer needs to

---

citizens of the Republic of Belarus who apply for a national visa in Belarus, the processing fee for a national visa application has been waived through 31 December 2020.

[14] National Contact Point to the European Migration Network (2013).

[15] For more details see Chap. 5, Sect. 5.2.1.2.

[16] It is beyond this study's scope to discuss refusal to issue visa in details. For more information on this see Kozak et al. (2014), pp. 135–137.

[17] Article 82 (1) 1 AF.

[18] Article 82 (1) 2-4 AF.

[19] Unterschütz (2016), p. 163. It must be stressed that Article 60 (1) 4 focuses on entrepreneurs who require different documents than a work permit or declaration in line with the *Oświadczenie* procedure as they register and run companies; thus, they would need a work permit only in limited cases.

[20] Act of 20 April 2004 on Employment Promotion and Labour Market Institutions/ Ustawa o promocji zatrudnienia i instytucjach rynku pracy (Dziennik Ustaw z 2019 r., poz. 1482, consolidated text). In addition, as of September 2019 there are five main Regulations that need to be considered: Regulation of the Minister of Family, Labor and Social Policy of 8 December 2018 regarding activity subclasses according to the Polish Classification of Activities (PKD), in which seasonal work permits for a foreigner are issued/Rozporządzenie Ministra Rodziny, Pracy i Polityki Społecznej z dnia 7 września 2018 r. zmieniające rozporządzenie w sprawie podklas działalności według Polskiej Klasyfikacji Działalności (PKD), w których wydawane są zezwolenia na pracę

apply for a work permit.[21] When a foreigner carries out work in Poland under an employment or civil law contract with an employer whose seat or residence is in the territory of Poland, a work permit is issued by the governor *(wojewoda)* of the province *(województwo)*, if the remuneration offered to the migrant worker is not lower than the remuneration of Polish employees performing work of a comparable type or comparable work, and if the results of a labour market test are negative.[22]

According to Polish law, a labour market test is performed by the Local Labour Office (*Powiatowy Urząd Pracy*) on behalf of the local governor *(starosta)* upon a notification of a vacancy by the employer.[23] The competent Local Labour Office, depending on the place where the work will be performed, issues a decision about either the lack of possibilities to meet the workforce needs of the employer on the basis of a review of the register of the unemployed and job-seekers (within 14 days from the date of submitting the vacancy to the Labour Office) or about possible recruitment among those registered at the Labour Office (within 21 days from the date of submitting the vacancy to the Labour Office). The decision issued by the governor is then attached to the work or single permit application. The labour market test also takes into account the priority of access to the labour market for Polish and EU citizens, as well as other categories of foreigners stipulated in Article 87 (1)

sezonową cudzoziemca (Dziennik Ustaw z 2018 r., poz. 1749); Regulation of the Minister of Labor and Social Policy of 29 January 2009 on determining cases in which a work permit for a foreigner is issued regardless of the detailed conditions for issuing work permit for foreigners/ Rozporządzenie Ministra Pracy i Polityki Społecznej z dnia 29 stycznia 2009 r. w sprawie określenia przypadków, w których zezwolenie na pracę cudzoziemca jest wydawane bez względu na szczegółowe warunki wydawania zezwoleń na pracę cudzoziemców (Dziennik Ustaw z 2019 r., poz. 154, consolidated text); Regulation of the Minister of Family, Labor and Social Policy of 8 December 2017 regarding the fees for filing an application for a work permit or seasonal work permit and a declaration of commissioning work to a foreigner/Rozporządzenie Ministra Rodziny, Pracy i Polityki Społecznej z dnia 8 grudnia 2017 r. w sprawie wysokości wpłat dokonywanych w związku ze złożeniem wniosku o wydanie zezwolenia na pracę lub zezwolenia na pracę sezonową oraz złożeniem oświadczenia o powierzeniu wykonywania pracy cudzoziemcowi (Dziennik Ustaw z 2017 r., poz. 2350); Regulation of the Minister of Labour and Social Policy of 8 December 2017 regarding the citizens of countries that are subject to certain provisions concerning a seasonal work permit, as well as provisions concerning the declaration on commissioning work to a foreigner/ Rozporządzenie Ministra Rodziny, Pracy i Polityki Społecznej z dnia 8 grudnia 2017 r. w sprawie państw, do których obywateli stosuje się niektóre przepisy dotyczące zezwolenia na pracę sezonową oraz przepisy dotyczące oświadczenia o powierzeniu wykonywania pracy cudzoziemcowi (Dziennik Ustaw z 2017 r., poz. 2349); Regulation of the Minister of Family, Labor and Social Policy of 7 December 2017 regarding the issue of a work permit for a foreigner and the entry of a declaration on commissioning work to a foreigner in the register of declarations/ Rozporządzenie Ministra Rodziny, Pracy i Polityki Społecznej z dnia 7 grudnia 2017 r. w sprawie wydawania zezwolenia na pracę cudzoziemca oraz wpisu oświadczenia o powierzeniu wykonywania pracy cudzoziemcowi do ewidencji oświadczeń (Dziennik Ustaw z 2017 r., poz. 2345).

[21] For a detailed description of the different work permits, see Unterschütz (2016), 167–168.

[22] A labour market test is performed as part of the application procedure for type A work permits. See Article 88c (1) AEPLMI.

[23] See Article 88c (1) 2 AEPLMI.

of the Act on Employment Promotion and Labour Market Institutions and having the right to work in Poland.

Nonetheless, there are several cases in which specific categories of migrant workers can be exempted from the obligation to perform a labour market test, such as when the job is covered by the list of occupations which are in great demand, as provided for on the national level in a Regulation of the Minister of Labour and Social Policy[24] or listed in an Ordinance issued by a governor of the province.[25] In case of a successful application, a work permit is issued for a specified period of time, which cannot be longer than 3 years but which can be extended.[26] In addition, the Act on Employment Promotion and Labour Market Institutions allows for many exemptions from the requirement to obtain a work permit listed under Article 87 (2). This Article is given further enumeration by a Regulation issued by the Minister of Labour and Social Policy.[27]

In comparison, the Bulgarian Act on Foreigners in the Republic of Bulgaria (AFRB)[28] establishes the terms and procedures under which foreigners may enter, reside, and leave the Republic of Bulgaria (Article 1 of the AFRB).[29] As mentioned above, the visas that are relevant to this research are short-stay visas for the purposes of a planned stay on the territory of the Republic of Bulgaria (visa type C) and long-stay visas (visa type D).[30] Short-stay visas with the purpose of a planned stay must be issued for a period not exceeding 90 days, within any 6-month period, to be counted from the date of first entry into the Republic of Bulgaria (Article 14 (1) of the AFRB).

---

[24] For the most recent list of occupations, see Regulation of the Minister of Family, Labour and Social Policy of 28 June 2018 amending the Regulation on determining cases in which a work permit for a foreigner is issued irrespective of the detailed conditions for issuing work permits for foreigners/ Rozporządzenie Ministra rodziny, pracy i polityki społecznej z dnia 28 czerwca 2018 r. zmieniające rozporządzenie w sprawie określenia przypadków, w których zezwolenie na pracę cudzoziemca jest wydawane bez względu na szczegółowe warunki wydawania zezwoleń na pracę cudzoziemców (Dziennik Ustaw z 2018 r., poz. 1264).

[25] Article 10 (4) 1 AEPMLI.

[26] Article 88e (1) AEPLMI.

[27] Regulation of the Minister of Labour and Social Policy of 21 April 2015, concerning cases when commissioning work to foreigners in the territory of the Republic of Poland is permitted without the requirement of obtaining a work permit/Rozporządzenie Ministra Pracy i Polityki Społecznej w sprawie przypadków, w których powierzenie wykonywania pracy cudzoziemcowi na terytorium Rzeczypospolitej Polskiej jest dopuszczalne bez konieczności uzyskania zezwolenia na pracę. (Dziennik Ustaw z 2018 r., poz. 2273, consolidated text).

[28] Закон за чужденците в Република България (SG No. 153/23 December 1998, last amendment SG No. 58/23 July 2019).

[29] Visas are not required in line with Council Regulation (EC) No. 539/2001 of 15 March 2001, other acts of the European Union, international agreements or acts of the Council of Ministers, as well as in cases where the foreigner has a valid permit for continuous, long-term, or permanent residence in Bulgaria (Article 8 (2) and (3) AFRB).

[30] For more details see Article 9a in conjunction with Article 14 AFRB. Transit visas regulating these types of stays fall outside the scope of this study and therefore are not presented in detail.

National long-term visas with a validity of up to 6 months that grant the right to stay in Bulgaria for up to 180 days can be issued when the foreigner intends to apply for continuous, long-term, or permanent residence[31] on one of the grounds stipulated by the Act (Article 15 (1) of the AFRB). Long-stay visas with a validity of up to 1 year and with right of stay for up to 360 days may be issued to foreigners only in specific cases that are listed in the Regulation on the terms and procedure for issuing visas and for determining the visa regime.[32] Once a residence permit has been issued, the visa is annulled (Article 15 (4) of the AFRB).[33]

Unlike Poland, Bulgaria's visa policy did not provide any facilitation for Eastern Partnership countries or for Russia until 2018, when Bulgaria signed bilateral labour migration agreements with Armenia and Moldova (discussed in the next section of this chapter). Applicants are still required to pay all the appropriate fees, unless they fall under one of the exemptions contained in the EU visa facilitation agreements. The only other facilitation for nationals of Ukraine and Russia is provided through the outsourced Visa Application Centres operated by the VFS Global Company.[34] In 2015, several members of the Bulgarian Parliament introduced a draft resolution seeking to abolish visas for Russian citizens – a measure which, in itself, was contrary to EU Visa List Regulation 539/2001.[35] The Ministries of Interior and Foreign Affairs intimated their positions to the Parliamentary Commission on Economic Policy and Tourism, stressing that facilitations already existed, namely the outsourced Visa Application Centres and the visa-free regime for holders of Schengen visas issued by other Member States.

---

[31] According to Article 23 (2) AFRB, continuous residence means an authorized stay for up to 1 year, except in the cases provided for in the law. Article 24 (1) enumerates the cases when a foreigner can apply for such permit. To obtain a continuous permit, for the period of residence foreigners should have insured accommodation, compulsory health and social security insurance, sufficient means of subsistence without resorting to the social assistance system not less than the minimum monthly salary, minimum stipend, or minimum pension according to the legislation of Bulgaria (Article 24 (2) AFRB).

[32] Наредба за условията и реда за издаване на визи и определяне на визовия режим (adopted by Council of Ministers Decree No 198/ 11 July 2011, SG No. 55/19 July 2011, last amendment SG No. 27/2 April 2019). According to Article 9 (2), these are foreigners who conduct scientific research or who are students on 1-year educational programmes, post-graduate students or trainees, foreigners sent on a business trip by a foreign employer in order to perform specific tasks, related to the control and coordination of fulfilment of a contract for tourist services, as well as to foreigners sent on a business trip by a foreign employer for making and maintaining investments, certified following the procedure laid down in the Investment Promotion Act.

[33] It is beyond the study's scope to discuss refusal to issue visa in details. For more information on this see Article 10 AFRB for full list of the grounds on which visas may be refused.

[34] Visa application centres operate in the following Russian cities: Moscow, Sankt Petersburg, Ekaterininburg, Novosibirsk, Kazan, Samara, Nijni Novgorod, Krasnodar, Krasnoyarsk, Rostov on Don, Vladivostok, Khabarovsk, Irkutsk, Ufa, Sochi and Kaliningrad; and in Kiev, Odessa, Lvov, Lutsk, Ivano-Frankovsk, Donetsk, Kharkov and Simferopol in Ukraine. Source: Ministry of Foreign Affairs (2018).

[35] National Assembly of the Republic of Bulgaria/Народно събрание на Република България/ (2015). For more details on the EU visa policy, see Chap. 4, Sect. 4.2.2.

According to the Act on Labour Migration and Labour Mobility (ALMLM),[36] the initial application for access to the Bulgarian labour market is only possible when the applicant is outside the territory of Bulgaria (Article 5 (2) of the ALMLM). The authorisation for access to the labour market is subject to the negative results of a labour market test that is to be performed by the employer and proof that in the previous 12 months, the total number of third-country nationals working for the local employer did not exceed 20% of the average number of workers employed (35% in the case of medium and small enterprises).[37] In addition, foreigners also need to be offered working conditions that are no less favourable than the conditions offered Bulgarian citizens in the respective labour category. Finally, the foreign worker must possess specialised knowledge, skills, and professional experience that is required for the post in question (Article 7 (1) of the ALMLM).

The employer carries out the labour market test by publishing a job advert containing information about the requirements for filling the specific position, remuneration, and other social benefits. The job vacancy must be published in media with national coverage, as well as in the Local Labour Office (*Бюро по труда*) at the prospective place of work of the applicant (Article 4 (1) of the Implementing Regulation of ALMLM[38]). The duration of this labour market test ranges between 15 days and 3 months. The results of the test are used by the Employment Agency in order to establish the 'objective impossibility' of the employer to employ a Bulgarian citizen, a citizen of another EU Member State or country that is party to the Agreement on the European Economic Area or the Swiss Confederation, or any other foreigner legally residing in Bulgaria who has the right to fill the vacant position.[39]

There are several exceptions to the work authorisation procedure that are predominantly based on EU law and applicable to the following types of workers who do not need a work permit in order to work in Bulgaria: ship crew members; workers with long-term and permanent residence permits and their family members; recognised refugees and their family members in line with the Act on Asylum and Refugees; foreigners whose employment is a result of the implementation of an international treaty; family members of Bulgarian and EU citizens; diplomatic mission members and accredited journalists from foreign media outlets; and foreigners pending expulsion after 1 year of the issue of the expulsion order (Article 9 (1) of the ALMLM).

In line with the priorities of the national migration policy, migrants of Bulgarian origin are also entitled to facilitated labour market access without a work permit

[36] Закон за трудовата миграция и трудовата мобилност (SG No. 33/ 26 April 2016, last amendment No. 34/ 23 April 2019).

[37] According to Article 14 (1) ALMLM, the Minister of Labour and Social Policy, where proven appropriate, may authorise labour market access on a case-by-case basis outside these limitations.

[38] Правилник за прилагане на Закона за трудовата трудовата миграция и трудовата мобилност (SG No. 79/ 7 October 2016, last amendment SG No. 27/ 2 April 2019).

[39] The categories are listed under Article 9 (1) 2-6 ALMLM.

under Article 8 (2) 1 and (3) of the ALMLM. Currently they only need to be registered by the employer in the Employment Agency in line with Article 30a of the Implementing Regulation of the ALMLM. The provisions concerning labour market access of migrants of Bulgarian origin were amended in May 2018 and replaced the former, rather ambiguous, formulation of Article 8 (2) 1 and Article 15 (4) 1 of the ALMLM.[40] As mentioned in Chap. 5, these foreigners of Bulgarian origin often come from the Eastern Partnership countries and are entitled to facilitated access to naturalisation.

According to Article 7 (4) of the ALMLM, the work permits are issued with a period of validity of 1 year only and the overall duration of the work authorisation can be extended for up to 3 years if circumstances for its issue have not changed (Article 7 (5) of the ALMLM). After this 3-year period has elapsed, foreigners can submit a follow-up application from outside the territory of Bulgaria. Furthermore, according to Article 7 (1) of the Implementing Regulation of the ALMLM, upon reaching the maximum work authorisation period of 3 years under Article 7 (5) of the ALMLM, a new work authorisation application can only be submitted after a 3-month interruption between the expiration of the third-country national's permit and the request for a new starting period of employment.

6.1.2 National Instruments Conducive to Circular Migration

One of the categories of foreigners in Poland exempt from the requirement to obtain a work permit are workers admitted on the basis of the *Oświadczenie* procedure, which has been identified as one of the instruments facilitating circular migration.[41] This procedure allows employers to recruit foreigners to work in Poland for a period not exceeding 6 months within a 12-month period on the basis of a declaration of commissioning work to a foreigner. Nationals of Armenia, Belarus, Moldova, Russia, Georgia, and Ukraine – in line with the Regulation of the Minister of Labour and Social Policy on the citizens of countries subject to certain provisions concerning a seasonal work permit as well as provisions concerning the declaration on commissioning work to a foreigner on the basis of Article 90 (10) 2 of the Act on Employment Promotion and Labour Market Institutions – are eligible candidates for the *Oświadczenie* procedure.[42]

The respective Local Labour Office, depending on the employer's office location or place of permanent residence, is required to register declarations of commissioning work to a foreigner only in cases when the following conditions are fulfilled (in

[40] See SG No. 24/ 16 March 2018. A new Para 3 was introduced.

[41] On the basis of Article 87 (3) AEPLMI. See Chap. 5, Sect. 5.2.1.1 for more details.

[42] Rozporządzenie Ministra Rodziny, Pracy i Polityki Społecznej z dnia 8 grudnia 2017 r. w sprawie państw, do których obywateli stosuje się niektóre przepisy dotyczące zezwolenia na pracę sezonową oraz przepisy dotyczące oświadczenia o powierzeniu wykonywania pracy cudzoziemcowi (Dziennik Ustaw z 2017 r., poz. 2349).

line with Article 88z (2) of the AEPLMI): applicants are citizens of the abovementioned countries; they are not engaged in the sectors of agriculture or hospitality and catering, which are reserved for holders of EU seasonal work permits; and, the period of work does not exceed 6 months in total within the following 12 months, independent of the number of entities (employers) commissioning a work to foreigners on the basis of such a declaration or declarations already entered in the registry at a respective Local Labour Office.

The Local Labour Office shall enter declarations of commissioning work to a foreigner in the register of declarations or refuse to do so on behalf of the local governor by way of a decision within 7 working days from the date of receipt of the declaration in cases not requiring additional verification.[43] If such verification is required, this should be done no later than in 30 consecutive days from the date of receipt of the declaration. The employer is obliged to inform the respective authorities whether the foreigner, for whom a declaration has been registered, has started the job in question or not (Article 88z (13) of the AEPLMI). The legal basis for refusing to register a declaration, the possibility to conduct additional verification, and the obligation of the employers to register presence and work of the foreigner were three important amendments introduced with the reform of the *Oświadczenie* procedure in 2018.

The declaration that the employer needs to submit for registration shall indicate: information concerning the entity commissioning the foreigner to perform a given job; information concerning the worker in question; and, data concerning the job offered to the foreigner.[44] It also states that the employer is acquainted with the legal regulations on entrusting a job to a foreigner in the territory of the Republic of Poland.[45] Another requirement is that the declaration needs to indicate the type of contract serving as the basis for performing the said job[46] as performing work on the basis of the declaration is only legal when, after their entry into Poland, foreigners conclude an employment or service (referred also as 'civil') contract with the employer.[47] A signed and registered declaration can serve as the basis for the work visa application.[48]

The other instrument in Poland identified as circulation-friendly is the Pole's Card.[49] Foreigners who have obtained a *Karta Polaka* are also exempted from the requirement to apply for a work permit in line with Article 87 (2) 6 of the Act on Employment Promotion and Labour Market Institutions. This facilitates their access to the Polish labour market.

---

[43] Article 88z (4) AEPLMI. For the grounds of refusal, see Article 88z (5) and (6) AEPLMI.

[44] See Article 88z (1) AEPLMI for more details.

[45] Article 88z (1) 1 (h) AEPLMI.

[46] Article 88z (1) 3 (e) AEPLMI.

[47] For more details on the possible financial penalty and Criminal Code regulation penalising untrue statements see Kozak et al. (2014), p. 142.

[48] Kozak et al. (2014), p. 142.

[49] See Chap. 5, Sect. 5.2.1.2 for more details.

As already mentioned in Chap. 5, bilateral labour migration agreements were identified as the main Bulgarian instrument aimed at facilitating circular migration. The agreements concluded between Bulgaria and Armenia[50] and Bulgaria and Moldova,[51] respectively, cover citizens of these countries that have concluded labour contracts under these agreements and obtained the necessary residence authorisations for the particular host-country (Article 2 (1) of both Agreements). Migrant workers from Armenia and Moldova can be employed in Bulgaria without a work permit for an initial period of up to 1 year with the possibility of subsequent prolongation for a total of 3 years. Seasonal workers from these countries can work in Bulgaria for the maximum period of 9 months in line with the Seasonal Workers' Directive (Article 2 (1) б of both Agreements).

Migrant workers coming to Bulgaria on the basis of these agreements are required to sign a declaration obliging them to return to the territory of their sending state upon expiration of their legal residence and to present their passport personally to its consulate within 1 month of their return (Article 23 of the Agreement of Moldova/ Article 13 of the Agreement with Armenia). Non-fulfilment of this obligation might impact a future application for a residence permit submitted to the Bulgarian authorities. All this shows that the agreements provide for a circular migration model closer to a temporary migration scheme.

In addition, as already mentioned, these agreements provide for visa facilitation measures: 'Visa applications under this agreement shall be processed with priority by the competent consular service of the receiving party on the territory of the sending party, in accordance with the laws of the parties and in accordance with the relevant provisions of the international treaties binding to both parties' (Article 8 (1) of both Agreements). This is a positive development compared, for instance, to the labour migration agreement concluded between Bulgaria and Israel that does not provide for any explicit visa facilitation measures (see Article 7 (2)).[52]

---

[50] Agreement between the Republic of Bulgaria and the Republic of Armenia on labour migration management, approved by Council of Ministers Decision No. 176 of 22 March 2018, entered into force 9 October 2018/Спогодба между Република България и Република Армения за регулиране на трудовата миграция, утвърдена с Решение № 176 от 22 март 2018 г. на Министерския съвет, в сила от 9 октомври 2018 г.

[51] Agreement between the Government of the Republic of Bulgaria and the Government of the Republic of Moldova on labour migration management, approved by Council of Ministers Decision No 492 of 13 July 2018, entered into force on 11 September 2018/Спогодба между правителството на Република България и правителството на Република Молдова за регулиране на трудовата миграция, утвърдена с Решение № 492 на Министерския съвет от 13 юли 2018 г., в сила от 11 септември 2018 г.

[52] Agreement between the Government of the Republic of Bulgaria and the Government of the State of Israel, regarding recruitment and temporary employment of citizens of the two countries, adopted with Council of Ministers Decree No 852/24 November 2011, in force from 20 December 2011/Спогодба между правителството на Република България и правителството на Държавата Израел за посредничество и временна заетост на граждани на двете държави.

6.2 Entry and Re-entry Conditions Provided by National Instruments – Implementation Dynamics

Table 6.1 presents the number of declarations of commissioning work to a foreigner by employers since 2009 as part of the *Oświadczenie* procedure in Poland. It needs to be stressed, however, that this number does not reflect the real number of people who came to Poland with visas issued under these declarations – which was smaller.[53] The data demonstrate a steady rise in the declarations since 2014, leading to a peak of 1,824,464 declarations in 2017. The decline since 2018 can be attributed mainly to legal amendments of the *Oświadczenie* procedure, the transposition of the Seasonal Workers' Directive that came in force in the same year, and the introduction of the visa-free regime with Ukraine (only for biometric passport holders), which is the country of origin of more than 90% of the declarations.[54]

Even though a larger share of foreigners work on the basis of a registered declaration,[55] the picture of the migration dynamic in Poland will not be complete without data on the number of work permits issued. This has increased gradually

Table 6.1 Registered declarations under the *Oświadczenie* procedure in the period 2009 – first half of 2019

	Citizenship						
Year	Belarus	Russia	Ukraine	Moldova	Georgia	Armenia	Total per year
2009	4860	674	180,133	2747	–	–	**188,414**
2010	3623	595	169,490	5912	453	–	**180,073**
2011	4370	963	239,646	13,024	1774	-	**259,777**
2012	7636	1624	223,671	9421	1384	-	**243,736**
2013	5194	1260	217,571	9248	2343	–	**235,616**
2014	4017	1227	372,946	6331	2103	774	**387,398**
2015	5599	1939	762,700	9575	1366	1043	**782,222**
2016	23,400	3937	1,262,845	20,650	1698	1597	**1,314,127**
2017	58,046	6150	1,714,891	31,465	11,126	2786	**1,824,464**
2018	62,805	6718	1,446,304	36,742	28,008	1648	**1,582,225**
2019 (first half)	32,237	4915	764,759	20,426	22,295	1297	**845,929**

Source: Ministry of Family, Labour and Social Policy, Poland (Employment of foreigners in Poland/Zatrudnianie cudzoziemców w Polsce. https://psz.praca.gov.pl/rynek-pracy/statystyki-i-analizy/zatrudnianie-cudzoziemcow-w-polsce Accessed 13 December 2019)

[53] For more details see Górny et al. (2018), pp. 26–27.

[54] See ibid. See also Pawlak and Lashchuk (forthcoming).

[55] Ibid.

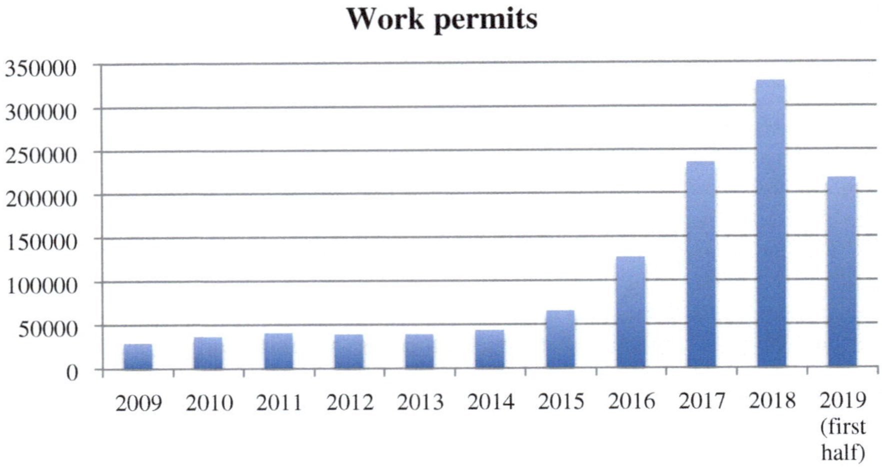

Fig. 6.1 Work permits in Poland in the period 2009 – first half of 2019. Source: Author's own elaboration on the basis of data of the Ministry of Family, Labour and Social Policy, Poland (Work permits for foreigners/Zezwolenia na pracę cudzoziemców. https://psz.praca.gov.pl/web/urzad-pracy/-/8180075-zezwolenia-na-prace-cudzoziemcow Accessed 13 December 2019)

more than ten times over the last decade to reach 328,768 in 2018 (see Fig. 6.1). Here again, it should be noted that the data cover both new and extended work permits with a validity of up to 3 years. Ukrainians are the group with the most work permits, reaching 238,334 in 2018. Another Eastern partnership country that stands out is Belarus; the number of work permits issued to citizens of this country increased in 2018 to reach 19,233.

The Ukrainian and Russian participants in the focus group conducted in late 2016 in Poland shared very diverse experiences with regards to circular migration, which in most of the cases presented a picture of circularity consisting of a chain of different trajectories supported by various instruments. Almost all of the Ukrainians and a few of the participants in the Russian focus group experienced, at some point in time, the *Oświadczenie* procedure, which they used both as an entry mechanism to Poland and as means of circulation.[56] Several of them had experience of obtaining a 'fake'[57] declaration that enabled them to enter Poland and look for a job, then afterwards obtain a new *Oświadczenie* (declaration) stating the real employment or use other ways to 'legalise their stay', e.g., by starting a course of study. The fake declarations were usually obtained through friends, acquaintances, or colleagues

[56] Focus groups with Russian and Ukrainian migrants, Poland, November 2016, Annex IV. Interview #6 with lawyers, Poland, November 2016, Annex II.

[57] Meaning that the employer who issued the declaration had no intention of offering a job to the foreigner in question and its sole purpose was to enable the foreigner to obtain a visa and legally enter Poland to seek a job.

who registered a declaration, stating that they needed the person in question. They could also be bought online through different social media channels for 100 USD[58] or through intermediaries in Ukraine for 200–250 USD[59] in cases when the migrants did not know anyone in Poland who could register a declaration on their behalf.[60]

The analysis of the data gathered through the focus groups and interviews shows that the migrants using the *Oświadczenie* procedure for the purposes of circular migration did not face any problems with regards to entry and re-entry conditions when the registered declaration was filled in properly.[61] One of the participants in the Ukrainian focus group remarked that he was deported due to a discrepancy between the information about his work place submitted in the declaration by the employer and the actual work place[62]; he appealed this decision successfully, stating that this was a mistake by the employer, and was eventually allowed to return to Poland.

The participants using the *Oświadczenie* procedure as an entry mechanism shared that the majority of them had trouble finding a job in Poland that would have allowed them to come on a work visa.[63] Using the declaration system gave them time to come to Poland, look for a job, and convince an employer to hire them so that they could begin the procedure for obtaining the single permit.[64] In the case of the recently arrived Ukrainians, this was the fastest way of leaving Ukraine on account of the political situation (instead of applying for asylum) or actually entering Poland as soon as possible before 'some new regulations were introduced in January 2017' (the then-pending changes due to the transposition of the Seasonal Workers' Directive).[65]

The interviews conducted with representatives of the administration brought another perspective to the use of this procedure. According to the officials who were interviewed in late 2016, due to its liberal character, the procedure was often abused by both migrant workers and employers.[66] One of the most frequent problems was caused by employers who issued more declarations than the actual number of

58 Focus group with Russian migrants, Poland, November 2016, Annex IV.

59 Interview #1 with officials, Poland, November 2016, Annex II. Interview #6 with lawyers, Poland, November 2016, Annex II. Interview #5 with civil society representative, Poland, November 2016, Annex II. Interview #10 with staff member of a private recruitment agency, Poland, October 2016, Annex II.

60 Focus group with Ukrainian migrants, Poland, November 2016, Annex IV.

61 Focus groups with Russian and Ukrainian migrants, Poland, November 2016, Annex IV. Interview #9 with employer, Poland, December 2016, Annex II.

62 Focus group with Ukrainian migrants, Poland, November 2016, Annex IV.

63 Focus groups with Russian and Ukrainian migrants, Poland, November 2016, Annex IV.

64 Focus groups with Russian and Ukrainian migrants, Poland, November 2016, Annex IV. Interview #6 with lawyers, Poland, November 2016, Annex II.

65 Focus group with Ukrainian migrants, Poland, November 2016, Annex IV. Interview #7 with official, Poland, November 2016, Annex II. Interview #6 with lawyers, Poland, November 2016, Annex II. Amendments were introduced as of January 2018.

66 Interview #1 with officials, Poland, November 2016, Annex II. See also Górny et al. (2018).

seasonal workers required as a way of ensuring that some migrant workers would eventually come and work for them. Another issue stemmed from migrants who used the *Oświadczenie* procedure to access the Schengen zone and did not begin working for the employer who registered the declaration. The officials shared that they received information from their colleagues in other Member States that there were irregularly-working migrants who entered on a visa issued on the basis of the *Oświadczenie* procedure.[67] Very often migrant workers would use this procedure when they wished their relatives to visit them because it was cheaper and faster than applying for a tourist visa.[68] These problems contributed to the overburdening of Ukrainian consulates, which had to handle the increased volume of visa applications. The large queues for submitting visa applications caused delays and was conducive to corruption that was facilitated by intermediaries, who were in fact 'selling' places in the queues.[69]

Another instrument related to the circular migration experience of the focus group participants was the *Karta Polaka*.[70] The scarce public information available on the number of cards issued demonstrates that mostly Belarusians (111,932) and Ukrainians (101,934) have benefited from this instrument in the period 2008–2017.[71] In the same period, 5586 Russians, 1629 Moldovans, 156 Georgians, 125 Azerbaijanis, and 89 Armenians also obtained such cards.

Both Russians from Kaliningrad and Ukrainians used this instrument and found it very useful in aiding their circularity between Poland and their countries of origin. Furthermore, they shared that it saved them from having to go through many bureaucratic hurdles ('Queues, losses of documents, prolonging issue time, and all those horrible things'[72]) that most of their friends faced when applying for residence permits.[73] One of the participants shared:

> 'I obtained Karta Polaka and since then I had no problems with visas, because the consulate has a very good attitude to Polish people. You come with Karta Polaka and all will be done fast and without any problems. There are some special regulations, which simplified the procedure and allowed to obtain visa in easier way. So it is not because they like us but there is a legislation which allows them to do it in this way.'[74]

---

[67] Interview #1 with officials, Poland, November 2016, Annex II.

[68] Also in Interview #3 with civil society actor, Poland, November 2016, Annex II and Interview #2 with academic, Poland, November 2016, Annex II.

[69] Interview #1 with officials, Poland, November 2016, Annex II; also in interview #3 with civil society actor, Poland, November 2016, Annex II and Interview #2 with academic, Poland, November 2016, Annex II.

[70] See Chap. 5, Sect. 5.2.1.2 for background information.

[71] Statistics Poland (2018), p. 454.

[72] Focus group with Russian migrants, Poland, November 2016, Annex IV.

[73] Focus group with Russian migrants, Poland, November 2016, Annex IV.

[74] Focus group with Russian migrants, Poland, November 2016, Annex IV. The latest amendments to the preamble of the Act on the Pole's Card in 2019, however, clearly show that the Polish government perceives these cardholders as Poles and therefore they are treated differently than regular visa applicants. See the Act of 16 May 2019 amending the Act on the Pole's Card/Ustawa z dnia

Another entry mechanism used by some of the focus group participants was an education visa, which led to a circular migration trajectory at a certain point in time. Some of them studied in Poland or obtained educational grants from Polish institutions; they returned to their countries of origin several times and then migrated back to Poland for employment purposes, where they eventually changed their status by applying for the EU long-term residence in one of the cases or for the *Karta Polaka*.[75]

Unlike Poland, Bulgaria does not regularly publish public data on the employment of foreigners. The available data are scarce and dispersed among ad hoc reports and other governmental documents, which all have the Bulgarian Employment Agency as a source. The only way to obtain data for a particular period is through official requests for access to public information. Yet, when the Employment Agency's data obtained in different years are compared, one can see discrepancies in the number of permits reported in a given year. Furthermore, according to the different publicly available reports, the number of foreigners who gained access to the Bulgarian labour market varies in that same year despite all quoting the Employment Agency data as the original source.[76] Finally, there are discrepancies between the same type of data provided by different institutions, such as the Ministry of Interior and the Employment Agency. These discrepancies could be attributed to different definitions used by institutional actors as well as data corrections that are not being communicated publicly. Keeping in mind the above caveats, the data obtained through official requests for information are presented in this section in order to provide a comparison with the Polish case and an indication of labour migration trends in Bulgaria.

According to the Employment Agency's data, close to 10,000 work permits – both new and extended – have been issued to foreigners in the period between 2009 and the first half of 2019.[77] This number indicates all foreigners who have obtained authorisation for access to the Bulgarian labour market, such as single permit holders, Blue Card holders, ICTs, seasonal workers residing up to 9 months on the basis of a work permit, and posted workers from third countries. It excludes, however, seasonal workers residing in Bulgaria for a period of up to 90 days. The data show that after stabilising at around an average 600 work permits annually for a period of 7 years after the economic crisis, there was a gradual increase in the number of work

16 maja 2019 r. o zmianie ustawy o Karcie Polaka (Dziennik Ustaw z 2019 r., poz. 1095). The author wishes to thank prof. Witold Klaus for this comment.

[75] Focus groups with Russian and Ukrainian migrants, Poland, November 2016, Annex IV.

[76] For instance, according to different sources, which all refer to the Employment Agency's data, the number of work authorisations issued to foreigners in 2018 ranges from 1904 to 7056. The first figure is published in the report of the National Council on Migration and Integration (2019); the second is from the report on Bulgaria of the European Migration Network (2018) (again produced by the administration). The data obtained through an official request for information indicate that 1642 work authorisations were issued in 2018.

[77] Information obtained via an official request for information under Decision No. РД-08-2156 from 12.08.2019 of the general secretary of the Employment Agency.

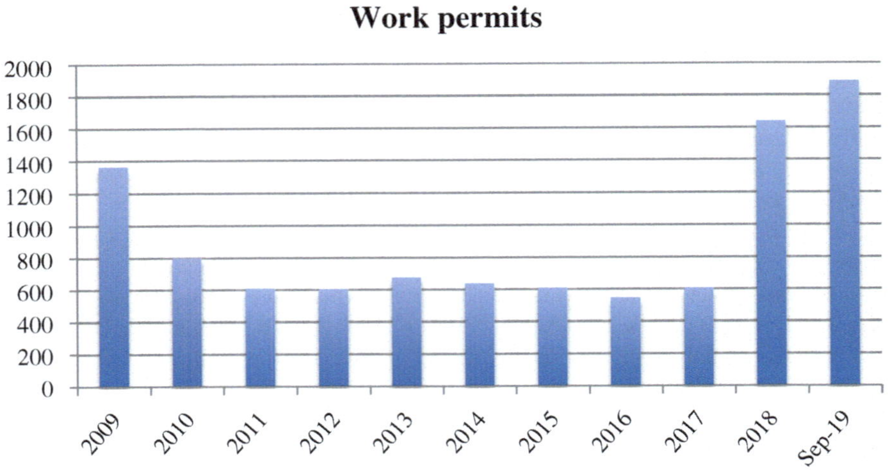

Fig. 6.2 New and extended work permits in Bulgaria in the period 2009 – first half of 2019. Source: Author's own elaboration on the basis of data of the Employment Agency, Bulgaria (Information obtained via an official request for information under decision No. РД-08-2156 from 12.08.2019 of the general secretary of the Employment Agency)

permits issued in the last 2 years (see Fig. 6.2).[78] Most of the work permits in 2018 were granted to Turkish workers (640), followed by Ukrainians (294) and Serbians (118). In the first half of 2019, the number of work permits issued to Ukrainians (404) surpassed the number of permits granted to Turkish workers, and there was an increase in the number of work permits issued to Moldovans (95) and Kyrgyzstanis (247).[79]

In addition, the number of seasonal workers residing in Bulgaria for a period of up to 90 days has increased exponentially (see section 6.3.3.4 for more details).[80] This increase in labour migration can be attributed to the country's economic growth and rising demand for foreign labour, as well as the procedures established for the recruitment of third-country nationals as seasonal workers as a result of the implementation of the Seasonal Workers Directive and the liberalised access to the Bulgarian market for certain categories of workers.

During the recruitment process for the focus groups in Bulgaria, it was a challenge to find migrant workers employed on labour contracts. This was due to the restrictive entry conditions that were dependent on the performance of a labour market test as a requirement for obtaining a work permit. These impediments forced

[78] Ibid.

[79] Ibid.

[80] Ibid.

most of the foreigners to resort to different channels to circumvent the entry procedure.[81]

According to one of the lawyers interviewed, the labour market test led to migrants being dependent on the employer and thus created a breeding ground for corruption. Furthermore, it was remarked that:

> '[t]hey need to find someone to do this test for them. This test is a formality, does not prove anything and it leaves discretion in the hands of the administration that can refuse to issue a permit. Who is going to fight for you? You need to be exceptional for the employer to be willing to go through all this.'[82]

One of the employers interviewed mentioned that he wished to employ Russian-speaking foreigners, however, she stressed that this was very difficult and that there were many hurdles impeding migrants' ability to obtain the necessary visa type D.[83] The interviewee explained that she would like to hire a cook but that it would be very difficult to obtain a visa for such person just because it is 'an ordinary job'. One of the lawyers interviewed also commented on this procedure, saying that it was very hard to prove that a migrant was 'unique', e.g., the cook needed to be hired to work in an 'exotic' restaurant in order to be able to pass the labour market test.[84] Therefore, the employer interviewed preferred to hire migrants who were already in the country: 'I cannot imagine how much money, time and nerves I would need to waste in order to obtain a visa D for a foreigner I want to hire!'.[85]

The state officials interviewed, on the other hand, claimed that the test was a requirement that aimed to protect the labour market and 'keep Bulgarians with high qualifications here'.[86] In line with this policy, the fee that the employer was required to pay for a work permit was deliberately high compared to the fees of other administrative documents for Bulgarians until 2017 when it was reduced to 100 BGN (50 EUR) as part of the amendment to the ALMLM.[87] Foreigners were supposed to come on a temporary basis in order to train Bulgarian workers, and the aim of the 3-year period of validity of the work permit was to allow Bulgarians to qualify for the position in question.[88]

According to the migrants interviewed, restrictive access to the Bulgarian labour market and the labour market test requirement meant that many foreigners entered Bulgaria via registration of a trade representation and sought

[81] Interviews #8 and #9 with lawyers, Bulgaria, July 2016, Annex III; focus groups with Russian and Ukrainian migrants, Bulgaria, September 2016, Annex IV.

[82] Interview # 9 with lawyer, Bulgaria, July 2016, Annex III; also in focus group with Russian migrants, Bulgaria, September 2016, Annex IV.

[83] Interview #19 with foreign employer, Bulgaria, September 2016, Annex III.

[84] Interview #8 with lawyer, Bulgaria, July 2016, Annex III.

[85] Interview #19 with foreign employer, Bulgaria, September 2016, Annex III.

[86] Interview #10 with officials, Bulgaria, July 2016, Annex III.

[87] See SG No. 97/ 5 December 2017.

[88] Interview #10 with officials, Bulgaria, July 2016, Annex III.

employment later (Article 24 (1) 6 of the AFRB).[89] The trade representation was used only as grounds for entry, on the basis of which a residence permit was issued but it did not give the individual the right to work (Article 8 (1) 2 of the ALMLM). After entering the country, many of the migrants registered a Bulgarian company so that they could exercise economic activity (as managers). However, this did not give them access to the labour market as employees. The alternative grounds for entry contained in the Act on Foreigners required the foreigner to register a firm as well as provide ten job positions for Bulgarians – an impossible condition for a small, fledgling business to meet, according to the migrants interviewed and other respondents.[90]

An alternative option for migrant workers was to use a multiple-entry visa type C to undertake short periods of work in Bulgaria and subsequently receive their remuneration in their country of origin. However, the Ukrainian participants in the focus group shared that this was neither good for their family nor for their business. Well-off Russians often used the grounds contained in Article 25 of the AFRB and invested in Bulgarian property of a value of 500,000 EUR, which gave them the opportunity to directly obtain permanent residence and secure housing. The permit also allowed them access to the labour market. Even Russian pensioners who had bought a flat whose price was not high enough to be considered an investment still needed to obtain a visa. In order to circumvent the law and obtain a residence permit, they thus registered a trade representation so that they could circulate freely and with minimal expense.[91] One of the participants in the focus groups added: 'Another entry option is to get married'.

The migrants interviewed in the focus groups described the reality in Bulgaria as 'forced circularity'. For most, being granted an EU long-term residence status was the desired option, rather than engaging in circular migration. Bulgarian migration law is based on the premise that in order to change one's status (including from visa type C to D) or renew one's permit after reaching the maximum period allowed, in most of the cases a person needs to leave Bulgaria and re-apply from another country.[92] According to one of the lawyers interviewed, this was a result of the logic of the law that '[t]he foreigner must suffer'.[93] The Russian participants in one of the focus groups shared that 'when you need to go out to change your visa from C to D and pay so much, you do not want to circulate'.

According to ALMLM's provisions, migrant workers can stay in Bulgaria for a maximum period of 3 years before they need to leave and re-apply from abroad – which means that a labour market test will have to be performed again. An interviewee commented on this provision as follows:

[89] Focus groups with Russian and Ukrainian migrants, Bulgaria, September 2016, Annex IV.

[90] Interview #18 with representatives of an IT company, Bulgaria, September 2016, Annex III.

[91] Focus groups with Russian and Ukrainian migrants, Bulgaria, September 2016, Annex IV.

[92] See also European Migration Network (2015).

[93] Interview #8 with lawyer, Bulgaria, July 2016, Annex III.

> 'If sent back home after 3 years, the migrant is discouraged to return! Everyone looses. Re-entry is not guaranteed. You can be refused visa D, which means a waste of time, money and opportunities. Who would risk coming back if he/she can go somewhere else?'[94]

The Russians who were trying to reach the 5-year threshold in order to obtain an EU long-term residence shared that they paid 350 EUR per year per person in order to renew their residence permit and their ID card,[95] and that voluntary 'circular migration would be a big risk'.[96]

Along with the restrictive entry conditions, foreigners who decide to apply for a work permit face additional hurdles as part of a cumbersome single application procedure.[97] As the European Commission stresses in its Implementation Report on the Single Permit Directive, in Bulgaria this procedure does not entail a single administrative act but duplication of submission of documents and therefore is prone to cause long delays.

6.3 Entry and Re-entry Conditions as Provided by EU Legal and Policy Instruments Implemented at the National Level

6.3.1 EU Visa & GAMM Instruments Facilitating Circular Migration

As discussed in Chap. 4, when analysing the implementation of EU's approach to circular migration, it is also important to take into consideration the instruments and initiatives developed as part of the EU Visa policy and the GAMM framework. In line with its bilateral relations in the context of the Eastern Partnership, Poland actively supports the efforts directed at visa facilitation at the EU level.[98] Both Bulgaria and Poland benefit from the already-concluded visa facilitation agreements with the Eastern Partnership countries and participate in activities under the auspices of GAMM. Both are parties to the Mobility Partnerships with Moldova (2008), Georgia (2009), Armenia (2011), Azerbaijan (2013), and Belarus (2016) respectively.

In addition, after the Local Border Traffic Regulation 1931/2006[99] was adopted at the EU level, Poland signed an agreement with Ukraine, which entered into force

---

[94] Interview #9 with lawyer, Bulgaria, July 2016, Annex III.

[95] This has been reduced as a result of the amendments to Article 10 of Tariff No. 4.

[96] Focus groups with Russian migrants, Bulgaria, September 2016, Annex IV.

[97] European Commission (2019a), pp. 4–5.

[98] European Migration Network (2012), p. 32.

[99] Regulation No. 1931/2006 of the European Parliament and of the Council of 20 December 2006 laying down rules on local border traffic at the external land borders of the Member States and amending the provisions of the Schengen Convention [2006] OJ L 405.

in July 2009,[100] and concluded another agreement with Belarus in February 2010. Both parties have ratified the latter agreement but the formal finalisation of the procedure by the Belarusian government is still pending.[101] During the Polish presidency of the Council, a local border traffic agreement for the Kaliningrad region was signed after 2 years of negotiations.[102] Such an agreement required an exception to Article 3 (2) of the Local Border Traffic Regulation 1931/2006 that would allow the entire Kaliningrad region to be considered as a border area.[103] The bilateral agreement between Poland and the Russian Federation entered into force on 27 July 2012, allowing for reciprocal visa-free entry for up to 30 days.[104] However, because of tense relations between the two countries, it was suspended in 2016 due to a NATO summit; the suspension was initially for 1 month but the agreement has not been reintroduced since.[105]

Currently the only bilateral local border traffic agreement in force between Poland and its eastern neighbours is the 2009 agreement with Ukraine. The Polish Act on Foreigners stipulates the general provisions on the crossing of the border under the local border traffic regime.[106] Since the text of the agreements that Poland concluded is analogous,[107] the information presented below also covers the agreement with the Kaliningrad region and provides some insights into the pending agreement with Belarus. According to the agreements, border area residents

[100] See Agreement between the Government of the Republic of Poland and the Cabinet of Ministers of Ukraine on local border traffic rules signed in Kiev on 28 March 2008, and Protocol signed in Warsaw on 22 December 2008 between the Government of the Republic of Poland and the Cabinet of Ministers of Ukraine on amending the Agreement between the Government of the Republic of Poland and the Cabinet of Ministers of Ukraine on local border traffic rules signed in Kiev on 28 March 2008/Umowa między Rządem Rzeczypospolitej Polskiej a Gabinetem Ministrów Ukrainy o zasadach małego ruchu granicznego, podpisana w Kijowie dnia 28 marca 2008 r., oraz Protokół, podpisany w Warszawie dnia 22 grudnia 2008 roku, między Rządem Rzeczypospolitej Polskiej a Gabinetem Ministrów Ukrainy o zmianie Umowy między Rządem Rzeczypospolitej Polskiej a Gabinetem Ministrów Ukrainy o zasadach małego ruchu granicznego, podpisanej w Kijowie dnia 28 marca 2008 r. (Dziennik Ustaw z 2009 r., poz. 858).

[101] For more information, see Znadniemna (2018).

[102] For more details, see Fomina (2011).

[103] Regulation No. 1342/2011 of the European Parliament and of the Council of 13 December 2011 amending Regulation No. 1931/2006 as regards the inclusion of the Kaliningrad oblast and certain Polish administrative districts in the eligible border area [2011] OJ L 347.

[104] OECD (2013), p. 284. Agreement between the Government of the Republic of Poland and the Government of the Russian Federation on local border traffic rules signed in Moscow on 14 December 2011/Umowa między Rządem Rzeczypospolitej Polskiej a Rządem Federacji Rosyjskiej o zasadach małego ruchu granicznego, podpisana w Moskwie dnia 14 grudnia 2011 r. (Dziennik Ustaw z 2012 r., poz. 814).

[105] Radio Poland (2016).

[106] Articles 37-48 AF.

[107] Wasilewska (2008), p. 10.

are persons with documented permanent residence in the border area[108] for a period of at least 3 years; the status includes spouses and dependent children.[109]

A permit can be issued to a border resident who holds a valid travel document, presents proof of permanent residence in the border area for at least 3 years, and has a legitimate reason for frequently crossing the respective border which, according to the legislation of the contracting parties, does not constitute gainful activity or gainful employment.[110] A local border permit holder can cross the border an unlimited number of times but can only stay in the designated border areas for a period of up to 60 days from the date of entry in the case of Ukrainian citizens and 30 days in the case of Russian citizens; in both cases, however, the total period of stay cannot exceed 90 days during any given 6 months from the date of first entry.[111]

6.3.2 *EU Visa & GAMM Instruments Facilitating Circular Migration - Implementation Dynamics*

In response to the Communication of the Commission on Mobility Partnerships and Circular Migration in 2007, the Polish government decided to expand the *Oświadczenie* procedure as part of the GAMM.[112] Therefore, this instrument possesses a dual role – a national instrument facilitating circular migration, and an initiative introduced as part of the GAMM. One interviewee claimed that the Polish representatives involved in the management of the Mobility Partnerships were asked by the European Commission to promote the *Oświadczenie* procedure as a circular migration scheme: 'they take anything that looks like it and put the label'.[113]

Initially only open to Ukraine, Belarus, and Russia, the *Oświadczenie* procedure was gradually included in the scoreboards of the countries of the Eastern Partnership that concluded Mobility Partnerships with some of the EU Member States. In practice, however, this procedure is used for circulation purposes mainly by citizens of the neighbouring areas – mostly from Ukraine, Kaliningrad, and Belarus. There are several reasons for this. First, according to the experts interviewed, the act of extending the *Oświadczenie* procedure to the Eastern Partnership countries was a matter of

[108] The Agreement with Ukraine covers the border area zone up to 30 km from the shared border. The suspended agreement with the Russian Federation provided that the local border traffic regime applies to all inhabitants on the Russian side of the Kaliningrad region and on the Polish side includes the residents of large parts of the Pomerania and Warmia-Mazury provinces respectively.

[109] See for example Article 2 (1) e and (2) of the Agreement between the Government of the Republic of Poland and the Government of the Russian federation on the Rules of Local Border Traffic.

[110] Article 3 (1) of the Agreement between the Government of the Republic of Poland and the Government of the Russian federation on the Rules of Local Border Traffic.

[111] Migrantinfo.pl (2018).

[112] Interview #18 with official, Poland, November 2016, Annex II.

[113] Interview #12 with academic, Italy, May 2013, Annex II.

foreign policy rather than a development driven by Poland's migration policy.[114] Therefore, this instrument was not advertised among the local population after the conclusion of the Mobility Partnerships with Armenia and Georgia.

Second, in practice it is mainly the nationals of the neighbouring countries who were interested in this type of migration because of the costs related to migration. Even though Ukraine is the only Eastern Partnership country that has not signed a Mobility Partnership, according to the official statistical data (see Table 6.1), Ukrainians were the main beneficiaries of the *Oświadczenie* procedure. Some of the interviewed experts claimed that it was purposefully designed as a tool to primarily attract Ukrainians.[115] An official stressed that Ukrainians started to circulate between Poland and Ukraine as early as the 1990s: 'it is not like we invented the system and they started to circulate'.[116] This circulation in the border region between Poland and Ukraine is supported by well-established informal channels for recruitment of agricultural workers through a network of bus drivers.[117] Along with its geographical proximity, Poland also attracts Ukrainians because of the historical, cultural, and linguistic proximity between the two countries.[118]

Finally, another pull factor was the higher earnings for Ukrainians from a couple of months of work in Poland that, in turn, enabled them to support their families when they returned home. As one of the interviewed employers stated, 'the ones who need some additional money would come for 1 or 1.5 months to pick apples or strawberries. A pensioner would get around 300 zloty (70 EUR) as a pension and would earn about 80-100 zloty (19-24 EUR) per day in Poland'.[119]

By way of comparison, Bulgaria has praised the GAMM,[120] supported the Eastern Partnership, and joined the Mobility Partnerships signed with five of the Eastern Partnership countries. According to the latest National Migration Strategy, this stems from the fact that Bulgaria identified the Eastern Partnership countries as its main partners within the context of the GAMM.[121] Along with the bilateral agreements with the Eastern Partnership countries concluded as part of the Mobility

[114] Interview #1 with officials, Poland, November 2016, Annex II; Interview #4 with expert, Poland, December 2016, Annex II; Interview # 18 with official, Poland, November 2016, Annex II.

[115] Interview #1 with officials, Poland, November 2016, Annex II; Interview #15 with academic, Poland, November 2016, Annex II; Interview #4 with expert, Poland, December 2016, Annex II.

[116] Interview #1 with officials, Poland, November 2016, Annex II.

[117] Interview #9 with official, Poland, November 2016, Annex II. See also Bieniecki and Pawlak (2009).

[118] Interview #10 with staff member of a private recruitment agency, Poland, October 2016, Annex II. See also Kindler et al. (2016).

[119] Interview #9 with official, Poland, November 2016, Annex II.

[120] Lessenski (2009), p. 46.

[121] National Strategy in the field of Migration, Asylum and Integration (2015–2020), p. 27; also in Interview #1 with official, Bulgaria, July 2016, Annex III.

Partnerships,[122] Bulgaria has also been interested in participating in the range of the GAMM initiatives with other migrant-sending countries to the EU.[123]

Both countries participate in the projects being implemented by the International Organization for Migration (IOM) and the International Centre for Migration Policy Development (ICMPD) in the Eastern Partnership countries, such as those carried out under the auspices of the Prague Process. In addition, as noted in Chap. 4, IOM Georgia implemented the 'Temporary Labour Migration of Georgian Workers to Poland and Estonia' project with the aim of developing operational frameworks to facilitate worker mobility from Georgia to Poland and Estonia that would, in turn, promote effective job-matching, migrant skill development, and protection of their labour and human rights.[124] According to the IOM office in Georgia, the spontaneous character of Georgians' labour migration using the *Oświadczenie* procedure did not allow the Georgian government to adequately respond to national workforce development needs and ensure better skills match to the Polish labour market's demands in the long-term.[125]

Poland also participated in another initiative aimed at facilitating labour mobility as part of the 'Ulysses' project implemented by Armenia's International Centre for Human Development (ICHD) and focused on matching European employers with Armenian job seekers. Even though the system established as a result of the project was 'very informative', it did not lead to a significant increase in the number of Armenians willing to work in Poland (see Table 6.1 and Fig. 6.1).[126] An interviewee claimed that distance was one of the main obstacles because travelling from Armenia to Poland was expensive.

One of the participants in the Russian focus group in Poland shared her experience with multiple-entry visas under the visa facilitation agreement with the Russia.[127] The participant, who was from Kaliningrad, shared that these types of agreements simplified the entry procedure and 'really facilitate it for people from Russia and Ukraine, as it is easier for them to get visa and they can avoid issuing residency card'.[128] She was eligible for a multiple-entry visa for a period of several years. After a change of the local consul, who started to demand more additional documents, and due to the changed nature of her work in Poland, which required longer periods of stay there, she decided to apply for a Pole's Card in order to be able to circulate between the two countries without encountering problems.[129]

---

[122] For more details, see Chap. 4, Sect. 4.2.3.

[123] Council of Ministers (2015), p. 27.

[124] International Organization for Migration Mission to Georgia (2017).

[125] Ibid.

[126] Interview #10 with staff member of a private recruitment agency, Poland, October 2016, Annex II.

[127] See Agreement between the European Community and the Russian Federation on the facilitation of the issuance of visas to the citizens of the European Union and the Russian Federation [2007] OJ L129.

[128] Focus group with Russian migrants, Poland, November 2016, Annex IV.

[129] Focus group with Russian migrants, Poland, November 2016, Annex IV.

6.3.3 Legal Migration Directives Aiming to Facilitate Circular Migration and Their Implementation Dynamics

6.3.3.1 Blue Card Directive

The Blue Card Directive[130] was transposed into the Polish Act on Foreigners through the introduction of a special temporary residence permit for the purposes of highly-qualified employment, referred to as the 'EU Blue Card permit', which is issued through a single administrative procedure.[131] Article 127 thereof lists the conditions that must be fulfilled in order for the permit to be granted. The foreigner needs to have concluded an employment, tolling, or civil law agreement of a minimum of 1 year that provides for an annual gross remuneration no less than the equivalent of 150% of the average monthly salary in the national economy during the preceding calendar year.[132] The applicants must fulfil the qualification requirements and other conditions in cases where they are to perform work in a regulated profession, as defined by Article 5 (4) of the Act of 22 December 2015 on the rules governing recognition of professional qualifications acquired in EU Member States.[133] Furthermore, applicants must possess the authorisation of a competent authority to hold a given position or pursue a given profession or activity, where the obligation to obtain it before entering into the agreement stems from separate regulations.[134] Additionally, foreigners must have a higher professional qualification, meaning that they have completed at least a 3-year course of study at a higher education institution or have 5 years of professional experience in a field that is compatible with the profession.[135] Finally, applicants must have health insurance or a confirmation of insurance coverage for any treatment that may be required on Polish territory.

In Bulgaria, residence and work permits of the 'EU Blue Card type' can be granted to foreigners holding a visa under Article 15 (1) or a continuous residence permit on other grounds and who are identified as highly-qualified workers pursuant to the provisions of the ALMLM (Article 33к (1) of the AFRB). It needs to be

---

[130] Council Directive 2009/50/EC of 25 May 2009 on the conditions of entry and residence of third-country nationals for the purposes of highly qualified employment [2009] OJ L 155/17.

[131] Unterschütz (2016), p. 184.

[132] Announced by the President of the Central Statistical Office in accordance with Article 20 (1) (a) of the Act of 17 December 1998 on Old-Age and Disability Pensions from the Social Insurance Fund/ Ustawa z dnia 17 grudnia 1998 r. o emeryturach i rentach z Funduszu Ubezpieczeń Społecznych (Dziennik Ustaw z 2018 r. poz. 1270, consolidated text).

[133] Ustawa z dnia 22 grudnia 2015 r. o zasadach uznawania kwalifikacji zawodowych nabytych w państwach członkowskich Unii Europejskiej (Dziennik Ustaw z 2018 r. poz. 2272, consolidated text).

[134] Article 127 (1) (e) AF.

[135] Article 127 (1) (c) AF.

stressed that before the amendment of the Act on Foreigners in December 2017, only foreigners residing in a third country could apply for this permit.[136]

The Employment Agency issues the work authorisation, which is part of the Blue Card permit, in line with the general application procedure contained in Article 7 (1) 3 and 4 of the ALMLM. Foreigners must fulfil the required professional qualification and the gross salary referred to in their labour contract should be at least 1.5 times higher than the average salary in Bulgaria, according to the data available for the last 12 months, before the conclusion of the employment contract (Article 17 (2) of the ALMLM). Even though this provision only stipulates a requirement for a higher education qualification, Article 15 (2) in conjunction with Article 2 (1) 4 of the Implementing Regulation of ALMLM, requires both higher education qualification and professional experience – which is contrary to what is stipulated in Article 2 (g) of the Blue Card Directive.[137] Furthermore, as part of the application process, the employer also needs to submit to the Employment Agency a copy of the concluded fixed labour contract with the employee that provides for the obligations of the parties concerning sickness insurance (Article 2 (1) 9 in conjunction with Article 2 (4) of the Implementing Regulation of ALMLM).

In line with Article 6 of the Directive, Poland requires the execution of a labour market test as part of the Blue Card application procedure, showing that the employer cannot 'satisfy its staffing needs with the local labour market'.[138] The procedure for conducting a labour market test follows the one described earlier and is commensurate with Article 88c of the Act of 20 April 2004 on Employment Promotion and Labour Market Institutions.[139] The labour market test is required only during the first 2 years of a foreigner's stay as a Blue Card holder in Poland.[140] In addition, Article 129 provides for several exemptions from the test, for instance, when the profession that the foreigner will perform in Poland is part of the list of professions and types of work in short supply in the local labour market, as indicated by the governor.[141] Exemptions from the labour market test are also possible if foreigners had a work permit or a residence and a work permit immediately before the filing of the application for the same employer, entrusting them to perform work in the same position or when they meet the conditions for exemption of a work permit regulated in the bylaws pursuant to Article 90 (5) of Act of 20 April 2004 on Employment Promotion and Labour Market Institutions.

In contrast, the labour market test and the 20/35% cap on the total number of third-country nationals who can be employed by a local employer as part of the general admission procedure in Bulgaria were waived for Blue Card applicants in

[136] Amendment to the AFRB from December 2017 (SG No. 97/ 5 December 2017).

[137] This also violates Article 15 of the Act on Normative Acts (SG No. 27/3 April 1973, last amendment SG No. 34/3 May 2016).

[138] Article 127 (2) AF.

[139] Article 136 AF.

[140] Article 129 (3) AF.

[141] Under Article 10 (4) 1 AEPLMI.

May 2018.[142] Before that, the labour market test was only waived if the profession was included on a List of Professions for which there was a shortage of highly-qualified specialists in line with the former Article 18 of the ALMLM. In 2016, after more than 2 years of lobbying by the IT sector in Bulgaria, the government exempted several IT positions from the labour market test in accordance with this article.[143]

The Blue Card permit in Poland is granted for a period 3 months longer than the period during which work is to be performed, but for no longer than 3 years in total.[144] This permit in Bulgaria is issued after a decision of the Employment Agency for a period of up to 4 years (Article 33k (2) of the AFRB). The possibility for extending the validity of the Blue Card permit for up to 4 years was provided through an amendment to the Act on Foreigners in late 2017.[145] In cases where the term of the labour contract is shorter, the permit is issued for the period of duration of the contract, extended by 3 months (Article 33k (2) of the AFRB).

In 2016, the European Commission launched infringement proceedings against Bulgaria for the fees charged by the Ministry of Interior for issuing residence permits to foreigners.[146] According to the Commission, the fees charged were not in compliance with the Blue Card Directive, the EU Long-term Residence Directive, the Family Reunification Directive, the Students' and Researchers' Directives, and the Single Permit Directive. The amended Tariff No. 4[147] has led to a substantial decrease in the fees charged by the Ministry of Interior. Currently, Blue Card applicants are required to pay 110 BGN (56 EUR) for the residence right and 45 BGN (23 EUR) for the issue of an ID card (Articles 10б (6) and 46 (1) of Tariff No. 4). In comparison, Blue Card applicants in Poland pay 440 PLN (103 EUR) when they file the application and 50 PLN (11,75 EUR) for the residence card.[148]

According to Article 215 (1) 3 and 4 of the Polish Act on Foreigners, former Blue Card holders who obtain an EU long-term residence permit can have it revoked if they leave the territory of Poland for a period of more than 6 years or the territory of the EU for a period of 24 consecutive months.[149] The same provision is contained in Article 40 (1) 11 of the Bulgarian Act on Foreigners.

142 With amendment of the Act on Labour Migration and Labour Mobility of 16 March 2018 (SG No. 24/16 March 2018). The amended Article 17 (2) ALMLM now exempts Blue Card applications from the previously required labour market test.

143 Interview #18 with representatives of IT company, Bulgaria, September 2016, Annex III.

144 Article 128 AFRB.

145 Amendment to the AFRB from December 2017 (SG No. 97/ 5 December 2017).

146 Infringement number 20164080, decision date 13 July 2017 concerning incorrect implementation (disproportionate charges) of Directive 2003/109/EC and other Directives by Bulgaria.

147 Tariff No. 4 on the fees collected in the system of the Ministry of Interior under the State Fees Act (adopted with Council of Ministers Decree No. 53/1998, last amendment SG No. 75/ 11 September 2018).

148 Mazowieckie For Foreigners (2019).

149 For revocation grounds, see Article 133 in conjunction with Article 101 AF.

6.3.3.2 Blue Card Directive – Implementation Dynamics

The Bulgarian administration issued 1221 Blue Card permits between January 2011 and mid-2019.[150] Most of the Blue Card permits were granted to third-country nationals from Ukraine (344 permits) and Russia (296 permits), followed by China (84) and Turkey (76).[151] Since 2016, there has been a steady rise in the number of permits issued, which surpassed 200 per year. Most of the permits were granted to IT specialists and engineers, followed by chief executives, managers, and experts.

As in Bulgaria, Ukrainian and Russian migrants are the main users of the Blue Card instrument in Poland with 3567 and 1037 permits issued, respectively.[152] However, the overall number of permits issued in the period from 2012 to October 2019 was 7090, which significantly exceeds the number issued in Bulgaria for the same period. In 2016, the number of permits in Poland surpassed 1000 per year, reaching a peak of 2046 Blue Cards in 2018. According to the Foreigners' Office data from 2017, these migrants work mainly as legal, social, and cultural professionals, business and administration professionals, as well as in the IT sector.[153]

The focus groups conducted in Poland and Bulgaria included mainly migrant workers who were supported by the IT company's relocation manager in the application process for the Blue Card permit. But they were able to share the experiences of some of their colleagues who had to apply by themselves at some point in time. Most of the participants in Poland explained that they were given a choice between the national and the Blue Card permits, and that they were advised by the recruiting company to apply for the Blue Card permit.[154] The rest had informed themselves about the advantages of the permit and even about the differences that existed among EU Member States.

The Blue Card permit appealed particularly to IT professionals because it was valid for 3 years and provided a period of 3 months to look for a new job in case of unemployment, compared to the 1 month allowed under the national permit in Poland. Furthermore, Blue Card holders' spouses were also entitled to the same duration of validity and allowed unrestricted access to the labour market without the need to apply for a work permit and free access to education for family members. Finally, this permit was part of the EU system, which offers the holder the

---

[150] The data refer to the number of permits and not to permit holders, which means that there could be more than one permit per person. The data were obtained from the Ministry of Interior in September 2019 with decision No 5364p151461/12 August 2019 of Migration Directorate of the Ministry of Interior.

[151] Ibid.

[152] The data were obtained from the Polish Office of Foreigners in October 2019.

[153] The data were obtained from the Polish Office of Foreigners in October 2017. Since 2018 the Polish administration no longer collects separate data on professions of the Blue Card holders.

[154] Focus groups with Blue Card holders from Russia, Ukraine and mixed group of Russian speakers, Poland, December 2016, Annex IV. Interview with business representatives, Poland, December 2016, Annex II.

opportunity to work in other EU Member States where many IT companies often have other branches.[155]

All applicants in Poland first applied for a national visa type D to enter the country and initially stayed on the basis of this visa. Some indicated that they had to go through a ‘probation period’ and were not offered a labour contract until a few months after their arrival, which would allow them to submit an application for the Blue Card permit.[156] According to focus group participants, it took 6 months on average from receiving the job offer to obtaining the Blue Card permit, including the 2–3 months (a maximum of up to 4 months) to receive the permit after the application.

Among the participants there were Blue Card holders who had received their permits in 2013 and considered themselves in the ‘first wave’.[157] In the beginning, they had experienced problems because the administration was still elaborating the application procedure and the relocation managers from the IT companies did not have experience handling the new system. For example, the first Blue Card holders had to go through an interview at the local authority while later applicants did not have to comply with such a requirement.

In Bulgaria, the application process initially took 7–8 months to complete, and at the time of the field research in September 2016, lasted on average 5–6 months.[158] All of the foreigners interviewed had to apply for work authorisation while still in their country of origin and then for a visa on the basis of the obtained decision for work authorisation. In order to apply for a visa, they needed to present the work authorisation decision and evidence of possessing health insurance, but it was reported that the consuls often also required a rental contract – a requirement that Blue Card applicants are exempt from. The relocation manager had to call the embassies and explain to them what the legal provisions were for this group of migrant workers, including the fact that the deadline for issuing a visa type D for Blue Card holders was 15 working days.[159] This period was not respected in practice and the visas took up to 30 days to issue. This caused delays in the already lengthy application process.

After arriving in Bulgaria, they had to first find an apartment because they needed a rental contract in order to be able to finalise the Blue Card permit application.[160] During this period, which could take up to 1 month, they could not be officially employed and work because they first had to register their personal number that was

[155] Focus groups with Blue Card holders from Russia, Ukraine and mixed group of Russian speakers, Poland, December 2016, Annex IV.

[156] Focus group with Blue Card holders: mixed group of Russian speakers, Poland, December 2016, Annex IV.

[157] Focus groups with Blue Card holders: Ukrainian and mixed group of Russian speakers, Poland, December 2016, Annex IV.

[158] Focus group with Blue Card holders from Ukraine and Russia, Bulgaria, September 2016, Annex IV.

[159] Interview #18 with representatives of an IT company, Bulgaria, September 2016, Annex III.

[160] Focus group with Blue Card holders from Russia, Bulgaria, September 2016, Annex IV.

printed on the Blue Card in the National Revenue Agency. Only afterwards could the employer register the employment contract with the same agency so that the Blue Card holder was able to start working.

The main problems highlighted by the focus group participants with regards to the application procedure in both countries were caused by the requirement to prove that the foreigner had a higher professional qualification, either through the completion of (at least) a 3-year course of study at a higher education institution or through 5 years of professional experience in a field that was compatible with the profession. One of the focus group participants in Poland said that he decided to apply for a Blue Card permit through the procedure, taking into account his diploma because his education was directly related to the job position.[161] However, because the translation of his diploma did not contain the same words as the job description, the officer did not accept his documents and he had to apply on the basis of his professional experience by translating his 'work book'.[162]

Other focus group participants shared that at first they did not know which documents would be considered as constituting solid proof so they would submit documents proving both their education and professional qualification(s).[163] Another problem in this respect was caused by the nature of work of the IT specialists in Ukraine who very often work as sole traders/self-employed.[164] This would prevent them from evidencing their professional experience, and if their diploma was not related to the job description in question, then their applications would be refused.

Currently, and in violation of the Blue Card Directive, applicants in Bulgaria are required to present both types of documents – diplomas *and* proof of professional experience. In addition, the provisions of the ALMLM do not specify how many years of experience are required and whether they need to be acquired while in the same position, which creates problems for the applicants and leaves a great deal of discretion in the hands of the administration.[165]

Other problems concerning the application process were related to the discretion of the regional authorities in Poland, which in turn led to the law being applied differently. The focus group participants shared that their company was well-known to the authorities and at the time of the data collection in late 2016, the procedure in Krakow was the fastest.[166] In other Polish cities, the waiting period could take up to 9 months, even though most of the Blue Card permits were issued in Krakow.[167] The representatives of the IT company's management who were interviewed shared that

---

[161] Focus group with Blue Card holders: mixed group of Russian speakers, Poland, December 2016, Annex IV.

[162] According to one of the respondents: '"Work book" in Russia is the official CV from the Russian employer. It comes from the Soviet Union'. From focus group with Blue Card holders: mixed group of Russian speakers, Poland, December 2016.

[163] Focus group with Blue Card holders from Ukraine, Poland, December 2016, Annex IV.

[164] Focus group with Blue Card holders from Ukraine, Poland, December 2016, Annex IV.

[165] Interview #20 with representatives of an IT company, Bulgaria, December 2016, Annex III.

[166] Focus group with Blue Card holders from Ukraine, Poland, December 2016, Annex IV.

[167] Focus group with Blue Card holders from Ukraine, Poland, December 2016, Annex IV.

they had experienced some problems with the issuing of the Blue Card permits for their employees in other cities where the company had a branch, and had to intervene and explain the regulations to the officials. For instance, in order to issue a (dependent) permit for the family member of a Blue Card holder, the authorities in one town required that the Ukrainian marriage certificate had to be recognised under Polish law; the IT company representatives stressed that this was not an explicit legal requirement.[168] By way of contrast, the authorities in Warsaw merely required a sworn translation of the marriage and birth certificates, respectively. There were also regional differences regarding the recognition of diplomas: some governors required confirmation by the Ministry of Science and Higher Education while others did not.

According to the IT company management representatives, the Polish officials did not always know the legislation in question and were afraid to issue the Blue Cards because they gave too many rights to foreigners and had to stay restrictive.[169] Therefore, the company's approach was '*to fix structural issues, not to survive*'.[170] Due to the lack of IT specialists in the Polish labour market, they pursued an active recruitment strategy and supported, both financially and logistically, their employees throughout the Blue Card permit application process. Part of their approach was to work with local authorities in order to share their best practices concerning the application of the legislation and explain how the Blue Card Directive provisions were supposed to work in practice.[171] They emphasised that the problems with the implementation of the Blue Card Directive in Poland and different authorities' broad discretion was due to the lack of a precise formulation of the requirements for granting a Blue Card permit. For instance, with regards to the qualifications, the Polish legislation was silent on what constituted professional experience; in locations where the administration did not have any experience, they did not know how to interpret this kind of provision, which consequently led to lengthy delays in the application process.[172]

For the representatives of the IT business in Bulgaria, one of the biggest problems was the labour market test as part of the Blue Card application that they were still required to perform at the time of the interview in 2016. They referred to it as an 'absolute deception'.[173] According to the interviewees, it was a redundant and formal procedure that always ended with a negative result and which led to the recruitment of a third-country candidate who had already been selected. They also said that the Labour Office officials supported them throughout this procedure because they also knew that the Bulgarian labour market lacked these types of

[168] Interview #22 with IT business management representatives, Poland, December 2016, Annex II.
[169] Interview #22 with IT business management representatives, Poland, December 2016, Annex II.
[170] Interview #22 with IT business management representatives, Poland, December 2016, Annex II.
[171] Interview #22 with IT business management representatives, Poland, December 2016, Annex II.
[172] Interview #22 with IT business management representatives, Poland, December 2016, Annex II.
[173] Interview #18 with representatives of an IT company, Bulgaria, September 2016, Annex III.

specialists and there were no other candidates that they could offer to the IT business.[174]

When asked whether they had returned to their countries of origin for work, the focus group participants in Poland and Bulgaria answering in the affirmative said that they did so mainly through the internal 'home office' system that the company offered through its network of branches in the region or simply went on business trips.[175] This type of circulation did not create any problems in relation to visas and taxation.

The interviewed Russians in both Bulgaria and Poland were not interested in returning to Russia, mainly due to political reasons.[176] Some shared that they feared for their safety, while others stated that they might get into trouble because of the applicable Russian laws on currency and tax. The Russian focus group participants also shared that it did not make much sense for them to go back home for work because their labour contracts and permits were only for 1 year and subject to renewal. In addition, most of the circulation of the Ukrainian Blue Card holders was due to personal reasons but not done very often because of the long queues at the border with Ukraine.[177] In two cases, the wives of the Blue Card holders went back and spent up to 6 months in their country of origin.

The IT company management representatives in Poland shared that they were also not interested in having their employees circulating: 'We don't need people to come here and work for only six or nine months. The key of our business is long-term engagement'.[178] However, they stressed that there were companies that focused mainly on short-term projects and which could benefit from this type of mobility. In general, they stated that what was needed was 'a more flexible idea of movement'.

6.3.3.3 Seasonal Workers' Directive

The Act of 20 April 2004 on Promotion of Employment and Labour Market Institutions was amended on 1 January 2018 in order to implement the Seasonal Workers Directive[179] into Polish law and reform the *Oświadczenie* procedure.[180] The Polish seasonal work permit is issued on the basis of Article 88 (2) when foreigners perform work in the scope of activities specified in the Regulation of the Minister of

---

[174] Interview #18 with representatives of an IT company, Bulgaria, September 2016, Annex III.

[175] Focus group with Blue Card holders from Ukraine, Bulgaria, September 2016, Annex IV. Focus group with Blue Card holders from Ukraine, Poland, December 2016, Annex IV.

[176] Focus group with Blue Card holders from Russia, Bulgaria, September 2016, Annex IV. Focus group with Blue Card holders from Russia, Poland, November 2016, Annex IV.

[177] Focus group with Blue Card holders from Ukraine, Poland, December 2016, Annex IV.

[178] Interview #22 with IT business management representatives, Poland, December 2016, Annex II.

[179] Directive 2014/36/EU of the European Parliament and of the Council of 26 February 2014 on the conditions of entry and stay of third-country nationals for the purpose of employment as seasonal workers [2014] OJ L 94.

[180] OECD (2018), p. 264.

Family, Labour and Social Policy of the 8 of December 2018[181] on the basis of contracts with entities whose registered offices or places of residence are located on the territory of Poland. The activities stipulated in this Regulation are agriculture, forestry, fishery, services, gastronomy, and hospitality related to tourism, such as short-term accommodation, camping sites, and mobile catering establishments. In line with the Seasonal Workers' Directive, this temporary permit is only issued for up to 9 months in a single calendar year.

The personal scope of the Seasonal Workers' Directive is limited to those workers who apply to be admitted to the EU from a third country[182] as well as those who have already been admitted under the terms of this EU instrument (Article 2 (1) of the Directive). However, as Peers stressed, the Directive does not explicitly ban Member States from considering in-country applications, and according to Recital 18 of its Preamble, it does not affect the right of legally resident third-country nationals to work.[183] Such applications, however, would fall outside the scope of the Directive.[184]

Polish law has used the fact that in-country applications are not explicitly banned by the Directive and has introduced two separate procedures for issuing a seasonal work permit depending on where foreigners are located at the time of application – abroad or in Poland. In the first case, referred to as the 'foreign *(Zagraniczna)* path', when foreigners apply for a visa for the purpose of performing seasonal work[185] or enter Poland on the basis of a visa-free regime, the employer submits an application to a relevant local governor, who enters the application into a seasonal work application registry and issues a certificate of entry within 7 to 30 days (see Article 88p of the AEPLMI). This certificate serves as the basis for the visa application and entry as part of the visa-free regime. A seasonal work permit is issued after the foreigner's entry and upon presentation of documents authorising the foreigner to stay in Poland, and the address of accommodation during the stay in Poland. In the second case, referred to as the 'national path', when the foreigner is staying in Poland on grounds which allow for a work permit application, the employer submits it and the local governor renders a decision within 7–30 days (Article 88pa of the AEPLMI).

According to Article 88o (1) of the Polish Act of 20 April 2004 on Promotion of Employment and Labour Market Institutions, a seasonal work permit is issued when

[181] Regulation of the Minister of Family, Labor and Social Policy of 8 December 2018 regarding activity subclasses according to the Polish Classification of Activities (PKD), in which seasonal work permits for a foreigner are issued/Rozporządzenie Ministra Rodziny, Pracy i Polityki Społecznej z dnia 7 września 2018 r. zmieniające rozporządzenie w sprawie podklas działalności według Polskiej Klasyfikacji Działalności (PKD), w których wydawane są zezwolenia na pracę sezonową cudzoziemca (Dziennik Ustaw z 2018 r., poz. 1749).

[182] See also Recital 15 in the Preamble saying that applications should only be made outside of the EU.

[183] Peers (2016), p. 384/ (footnote 546).

[184] Ibid.

[185] In line with Article 60 (1) 5a AF.

the remuneration stated in the contract between the foreigner and the employer is not lower than the remuneration of employees working for the same number of hours, performing a job of a similar type, or working in a comparable position; it also requires that the employer has attached the negative results of a labour market test issued by the local governor as part of the application. The labour market test is not required for citizens of Armenia, Belarus, Georgia, Moldova, Russia, or Ukraine (Article 80o (2) of the AEPLMI) or when the foreigner had stayed uninterruptedly in Poland for a period of 3 years preceding the application (Article 88c (8) 2 of the AEPLMI). In addition, this document is waived for foreigners who graduated from a university located within the territory of Poland, other EEA country or the Swiss Confederation within the 3 years preceding the application for the issue of a work permit, or participating in doctoral studies taking place in Poland (Article 88c (8) 1 of the AEPLMI).

The Bulgarian Act on Foreigners introduced two authorisation regimes for seasonal workers: seasonal work for up to 90 days on the basis of a short-term visa type C[186] (Article 24 л (1) of the AFRB) and seasonal work permit for no less than 90 days and no more than 9 months (Article 24к of the AFRB). The Minister of Labour and Social Policy approves a List of the economic sectors, including activities, whose implementation depends on the change of seasons. This List is drawn up after consultations with the National Council for Labour Migration and Labour Mobility (Article 25 (2) of the ALMLM), and currently covers two sectors: agriculture, forestry and fisheries; and, hotels and restaurants.[187]

In order to carry out seasonal work for up to 90 days, foreigners must have a valid visa type C for the purposes of seasonal work, where this is required, and the employment must be registered by the Employment Agency on the basis of a declaration submitted by the employer in line with the ALMLM and its Implementing Regulation (Article 24л (1) of the AFRB). In line with Article 20 of the Directive, the employer is obliged to submit evidence to the Employment Agency that the seasonal worker will be provided with appropriate accommodation, and that the health and safety requirements have been fulfilled (Article 28 of the ALMLM).

In addition, as part of the application process, the employer needs to attach a declaration stating that the work offered and the pay conditions are no less favourable than the conditions offered to Bulgarian citizens in the respective labour category (Article 32 (1) 2 in conjunction with Article 2 (1) 7 of the Implementing Regulation of the ALMLM); other required documents include a copy of the labour contract, stating amongst other things, that the transport costs and the compulsory health and social security insurance of the seasonal workers shall be borne by the employer (Article 27 (1) and (2) of the ALMLM in conjunction with Article 2 (1) 9 of the Implementing Regulation of the ALMLM). Seasonal workers coming initially for a period of up to 90 days may extend their work stay with a seasonal worker permit under Article 24 (1) of the ALMLM for up to 9 months within the

186 In case a visa is required.

187 Adopted by Order РД-01-47 /17 January 2017 of MLSP.

calendar year, starting from the date of initial registration of the employment (Article 29a (1) of the ALMLM). They are exempted from the requirement to apply for access to the Bulgarian labour market while residing outside the territory of the country (Article 5 (2) of the ALMLM).[188] However, they still need to apply for visa D, which means that in practice they have to leave the territory of Bulgaria.

In Bulgaria the 'seasonal worker permit' granting the right to continuous residence can be obtained by foreigners who meet the requirements for access to the labour market in accordance with the ALMLM[189] and who have obtained a visa type D (Article 24к (1) of the AFRB). The permit for a seasonal worker is issued upon the submission of a single application and after a positive decision has been taken by the Employment Agency. It covers the duration of the labour contract, which must be for no less than 90 days and no more than 9 months (Article 24k (2) of the AFRB). The Employment Agency issues the work authorisation decision, which forms part of the permit for the seasonal worker issued by the Ministry of Interior, within 10 days after an employer submits an application (Article 24, Para 1 (2) of the ALMLM).

With regards to circular migration facilitation, foreigners who performed work in Poland at least once in the 5 years preceding the seasonal work permit application are treated with priority when considering seasonal work permit applications, if the work will be performed on the basis of an employment contract (Article 88p (2) of the AEPLMI). In addition, Article 88q allows for seasonal multi-annual work permit applications at the request of employers for periods of up to 9 months during a calendar year falling within no more than three subsequent calendar years, when the workers are citizens of Armenia, Belarus, Georgia, Moldova, Russia, or Ukraine, have worked at least once for the employer in the period of 5 years preceding the date of the application, and the employer is not in arrears with payment of income tax advances and social security contributions (if such were required).

The Bulgarian Act on Foreigners states that the seasonal worker permit is issued on the basis of an accelerated procedure for those candidates who have worked in Bulgaria at least once as a seasonal worker at any point during the preceding 5 years (Article 24k (4) of the AFRB). This accelerated procedure was detailed on the basis of the amendments of the Implementing regulation of the ALMLM in 2018, which provided that work authorisation decisions for such seasonal workers who have already worked in Bulgaria must be issued within 10 days after an employer has submitted an application (Article 20a (1)). In addition, there is no labour market test requirement as part of the accelerated procedure (Article 20a (2) of the Implementing regulation of the ALMLM). Compared to the regular procedures for seasonal work for up to 90 days and seasonal work for no less than 90 days and no more than 9 months, this accelerated procedure does not provide any additional facilitation as

---

[188] For more details, see Article 21a of the Implementing Regulation of the ALMLM.

[189] Having been offered work and salary conditions that are not less favourable than the conditions offered to Bulgarian citizens in the respective labour category, possessing specialised knowledge, skills and professional experience that is required for the post in question in line with Article 7 (1) 3 and 4 of the ALMLM.

the regular procedure does not require a labour market test either. In both cases, the work authorisation must be issued within 90 days after an employer has submitted an application (see Article 24 (2) of the ALMLM and Article 32 (3) of Implementing regulation of the ALMLM).

As a result of the lobbying efforts of Bulgarian employers working in the ambit of tourism, the Implementing Regulation of the ALMLM was amended in June 2017 to facilitate the recruitment of foreigners to carry out seasonal work for up to 90 days.[190] The amendments stipulated that no labour market test is to be applied for this category of workers (Article 32 (1) 2 and (2) of Implementing Regulation of ALMLM) and that employers were not required to submit documents proving the education and experience of these categories of workers in line with the general application for access to the Bulgarian labour market (Article 32 (1) 2 of the Implementing Regulation of ALMLM). The labour market test was also waived for seasonal workers staying for a period of up to 9 months (Article 25 (4) the Implementing Regulation of ALMLM).[191]

6.3.3.4 Seasonal Workers' Directive – Implementation Dynamics

According to data obtained by the Bulgarian Employment Agency, the number of third-country nationals registered in the country for the purpose of exercising seasonal work for up to 90 days (in line with Article 24л of the AFRB) is rising exponentially.[192] Starting at 612 declarations in 2017, the registrations surpassed 2000 in 2018 and reached 2579 in the first half of 2019.[193] Most of these seasonal workers who came for up to 90 days were from Ukraine. By contrast, the number of seasonal work permits (in line with Article 24k of the AFRB) continues to be rather low.[194] According to data of the Ministry of Interior, in the period between 2017 and first half of 2019, a total of 162 such work permits were issued.[195] However, when compared to data for the same period obtained from the Employment Agency, this number is much higher – almost 900 seasonal work permits, which is more likely to be

[190] SG No. 48/16 June 2017, Para 5.

[191] The labour market test exemption for this category of seasonal workers is still not correctly detailed in the Implementing Regulation of ALMLM. It was amended with SG No. 43/ 25 May 2018. It needs to be stressed, however, that the new Article 20a (2) of the Implementing Regulation of ALMLM refers only to the accelerated procedure under Article 24 (4) ALMLM. Nonetheless, the Bulgarian administration is using this provision to waive the labour market exemption as part of the application procedure. Official response to an official request for information to the Employment Agency, Reg. No 10-00-16405-1/6 January 2020.

[192] Data obtained from the Employment Agency in September 2019.

[193] Ibid.

[194] Data obtained from the Employment Agency and the Ministry of Interior in September 2019.

[195] Ibid.

the correct number.[196] Here again Ukrainians are in the lead, followed by Kyrgyz and Turkish workers.

In Poland, in 2018 a total of 121,436 seasonal work permits were issued as a result of 235,294 applications for seasonal work permits that were submitted at the Local Labour Offices. This number includes 138,344 seasonal work certificates issued to foreigners staying outside Poland to enable them to apply for visas using the 'foreign path'.[197] These numbers have increased already in the first half of 2019, with 137,266 certificates and 70,027 work permits. Ukrainians, followed by Belarusians and Moldovans, dominate the seasonal work permits.[198] When it comes to seasonal work certificates, Ukrainians are in the lead, followed by Belarusian and Nepalese workers.

Poland's significant delay in transposing the Seasonal Workers' Directive, which came into force in early 2018, did not allow for empirical data collection as part of this study in order to capture the implementation of this instrument.[199] Nevertheless, since this is one of EU's legal instruments explicitly aimed at fostering circular migration, and in order to be able to compare it with the Bulgarian case, it was necessary to consider it in this chapter. Therefore, the study's research design initially envisaged only focus groups with seasonal workers in the Bulgarian tourism sector. However, during the recruitment phase in the summer of 2016, the study's informants shared that these migrants were entering on a tourist visa and, in general, worked irregularly. Interviewed tourism industry representatives later confirmed this information, which posed challenges with regards to the ethical standards employed by the study, which leave irregular migrants outside the scope of this research.[200]

Furthermore, the transposition of the Seasonal Workers' Directive allegedly did not lead to regularisation of this type of work in Bulgaria in the 2016/2017-winter season. Numerous media outlets reported that the Labour Inspectorate had found that Ukrainian and Moldovan workers were coming to Bulgaria as trainees under concluded exchange agreements between professional schools in the Eastern Partnership countries and tourist companies in Bulgaria; in reality, they were working full-time and receiving salaries without having proper labour contracts.[201]

To compensate for this empirical gap and provide more information on the implementation of this Directive in Bulgaria, two interviews with a representative of an employers' organisation in the Bulgarian tourism sector were conducted in

[196] Data obtained from the Employment Agency and the Ministry of Interior in September 2019.

[197] Ministry of Family, Labour and Social Policy, Foreigner's seasonal work permits: https://psz.praca.gov.pl/web/urzad-pracy/-/8180228-zezwolenia-na-prace-sezonowa-cudzoziemca Accessed 21 December 2019.

[198] Ibid.

[199] The transposition deadline was 30 September 2016.

[200] For more details, see Chap. 2, Sect. 2.3.

[201] Offnews (2017); Dnevnik (2017).

January and October 2017, and one with a recruiter in October 2017.[202] The representative of the employers' organisation stressed that Bulgaria needed seasonal workers from third countries because Bulgarians who once worked in this sector had left for other Member States. In January 2017, the interviewee said that he appreciated the measure of 'lifting of the restrictions' for hiring third-country nationals but complained that the procedure developed under the ALMLM was cumbersome and designed to purposefully create obstacles for the recruitment of migrant workers.[203]

The interviewee expressed his dissatisfaction with the provisions adopted and which stipulated two different procedures for seasonal work: up to 90 days and between 90 days and 9 months in line with the Seasonal Workers' Directive. He said that the authorisation of 90 days was insufficient for the needs of the tourism industry because employers needed time to train the workers and wanted them to stay during the whole summer season, which is 6 months. However, the procedure concerning seasonal work permits for up to 9 months was too expensive given the number of workers that the hotels had to recruit: up to 400 EUR per person for the visa and work permit application[204] compared to 35 EUR for workers coming for up to 90 days,[205] and even less after the visa liberalisation with Ukraine came into force.[206] The respondent claimed that the Black Sea resort, where his business operated, alone needed 15,000 workers in 2017 and that it was impossible to afford the expenses stemming from seasonal work permits. The data on permits presented above as well as research conducted in the summer of 2019 demonstrate that this trend persists and that employers consequently still prefer to recruit seasonal workers on the basis of the procedure for up to 90 days.[207]

Therefore, according to the follow-up interviews in 2017 and data collected in 2019, the tourism sector primarily hired students from Ukraine and Moldova because this was cheaper in comparison to other countries of origin due to the visa conditions and geographic proximity.[208] They worked for 90 days and were replaced by a second wave of seasonal workers also hired for 90 days in order to cover the

---

[202] Interview #11 with representative of an employers' organisation in the tourism sector, Bulgaria, January 2017, Annex III.

[203] Interview #11 with representative of an employers' organisation in the tourism sector, Bulgaria, January 2017, Annex III.

[204] Interview #22 with representative of an employers' organisation in the tourism sector, Bulgaria, October 2017, Annex III. The costs are for the work permit fee (currently 50 EUR), translation and legalisation of documents, transport, and the fee for the recruitment agency. In general, the visa fees are paid by migrant workers.

[205] Interview #11 with representative of an employers' organisation in the tourism sector, Bulgaria, January 2017, Annex III.

[206] Interview #22 with representative of an employers' organisation in the tourism sector, Bulgaria, October 2017, Annex III.

[207] Vankova and Ivanova (forthcoming).

[208] Interview #22 with representative of an employers' organisation in the tourism sector, Bulgaria, October 2017, Annex III.

rest of the summer season.[209] The recruiter interviewed in October 2017 emphasised that the cumbersome procedure for hiring seasonal workers, which caused delays at the beginning of the summer season, improved following the amendment of the Implementing Regulation of the ALMLM in 16 June 2017. These amendments were engendered as a result of petitions by different stakeholders to the administration.

The persistent problems were related to the documents required for the application for seasonal work permits for up to 9 months. The documents for the applicant's experience and qualifications had to be legalised in line with Article 4 (3) 1 of the Implementing Regulation of the ALMLM, which further added to the expenses that employers incurred in order to hire such workers. Seasonal workers who wanted to extend their stay from 90 days to 9 months faced the same hurdle, as they also needed to present the above-mentioned documents. Another problem was posed by the requirement to demonstrate experience, even for entry-level positions, such as an assistant cook, which was impossible in the majority of cases. In addition, according to the interviewee, it 'makes no sense' to require migrants to apply for work permits when they are outside Bulgaria.[210]

At time of writing, there was still no comprehensive study of the implementation of the Seasonal Workers' Directive in Poland. A 2019 report on seasonal workers entering through both the *Oświadczenie* procedure and on the basis of seasonal temporary work permits demonstrates that Poland is attractive to migrants due to, amongst other things, the possibility of legal employment and considered as offering 'the easiest option' when it comes to application documents.[211] It should be borne in mind, however, that these migrants were assisted by an intermediary agency, which facilitated the application process.

6.3.3.5 EU Long-Term Residence Directive

In line with this study's research design, the provisions of the EU Long-term Residence Directive[212] allowing individuals to be absent from the territory of the host Member State are presented in this section. As already mentioned, the European Commission planned to use these provisions as an instrument that could facilitate circular migration.[213]

The EU Long-term Residence Directive was transposed into Polish law via Chapter VI of the Polish Act on Foreigners. According to its Article 215 (1), an EU

[209] Interview #22 with representative of an employers' organisation in the tourism sector, Bulgaria, October 2017. Interview # 23 with seasonal workers recruiter in the tourism sector, Bulgaria, October 2017, Annex III.

[210] Interview #23 with seasonal workers recruiter in the tourism sector, Bulgaria, October 2017, Annex III.

[211] Pawlak and Lashchuk (forthcoming).

[212] Council Directive 2003/109/ECof 25 November 2003 concerning the status of third-country nationals who are long-term residents [2004] OJ L 16.

[213] See Chap. 3, Sects. 3.2 and 3.4 for more details.

long-term residence permit is revoked when the holder leaves the territory of Poland for a period longer than 6 years or where the holder has left the territory of the EU for a period of 12 consecutive months. The Act does not allow for longer periods of absence for specific or exceptional reasons in line with Article 9 (1) (c) of the Directive.

Article 40 (1) 6 of the Bulgarian Act on Foreigners stipulates that foreigners lose their EU long-term residence permit or national permanent residence if they are absent from the territory of Bulgaria for 12 consecutive months unless they have permits based on one of the investment grounds contained in Article 25 of the AFRB. Similar to the provisions of Polish law, according to Article 40 (4) of the Bulgarian Act on Foreigners, the permit will also be revoked in cases where the individual is absent from the territory of Bulgaria for a total period of 6 years.

6.3.3.6 EU Long-Term Residence Directive – Implementation Dynamics

Between 2011 and mid-2019, 2420 third-country nationals obtained EU long-term residence permits in Bulgaria (under Article 24r (1) of the AFRB). By contrast, 33,814 national permanent residence permits have been issued in the same period (in line with Article 25 (1) of the AFRB).[214] According to Bulgarian Ministry of Interior data, so far only two EU long-term residence permits were revoked during this period because of absences of up to 12 consecutive months (on the basis of Article 40 (1) 6 of the AFRB). In addition, no EU long-term residence permits have been rescinded on the basis of absence of up to 24 months in the case of former Blue Card holders (on the basis of Article 40 (1) 11 of the AFRB) or after a 6-year-long absence from the territory of Bulgaria (Article 40 (4) of the AFRB).

The data support the findings from the focus groups that migrant workers in Bulgaria are generally interested in achieving and keeping a secure status and not in engaging in voluntary circular migration due to risks of losing their permits. None of the interviewed EU long-term residence or national permanent residence permit holders were interested in circulation for work. In addition, at the time of the focus groups, none of the Blue Card holders in Bulgaria had stayed for more than 5 years to become eligible to apply for an EU long-term residence permit. However, judging by their attitudes toward their countries of origin, they were not planning to engage in work-related circulation outside the possibilities for business or 'home office' trips.

By way of comparison, in the same period, 13,931 EU long-term residence permits and 76,616 permanent residence permits were issued in Poland.[215] Nineteen permanent residence permits and 10 EU long-term residence permits have been revoked.

---

[214] Data obtained from the Ministry of Interior in September 2019. Despite the fact that the reference period of the official request for information filed was 2007–2019, the Ministry provided data only from 2011 onwards because this is when Article 24r was introduced in the AFRB.

[215] Data obtained from the Polish Office of Foreigners in October 2019.

During the recruitment phase for the focus groups, it was impossible to find Ukrainians and Russians who had obtained the EU long-term residence permit and were also engaged in circular movement between Poland and their country of origin. Only one participant had a circular migration trajectory, but this was before she had been granted the permit. She explained that she deliberately decided to retain her Ukrainian citizenship so she would have the possibility of moving between her country and Poland in the future.[216] The participant shared that, after 20 years, she still missed her country and would have liked to be able to spend some time there, but at the same time she had her place to live in Poland and 'a fundament to rely on – employment opportunities, my friends'.[217] Asked whether she was aware that she could lose her permit if she stayed longer than the allowed absence, she replied in the negative. This finding is in line with the latest Implementation Report of the European Commission underscoring that there was a general lack of information available about the EU long-term residence status among both migrants and national administrations.[218]

The data on permits, analysed together with the findings of the focus groups, shows that the EU long-term residence permits generally are not used for circular migration purposes.

6.3.4 *Labour Migration Directives Containing Circular Migration Elements and their Implementation Dynamics*

6.3.4.1 Intra-corporate Transferees' Directive

The Act of 24 November 2017 amending the Act on Foreigners and Certain Other Acts also introduced the Intra-corporate Transferees' Directive[219] into Polish law, which entered into force in early 2018.[220] When submitting this permit application, the foreigner must be outside the territory of the EU Member States, unless that person seeks subsequent permit extension for the purpose of work under an intra-corporate transfer. A temporary residence permit for the purpose of work as part of an intra-corporate transfer is granted when, amongst others things, applicants meet the requirements concerning professional qualifications, health insurance, accom-

[216] Focus groups with Ukrainian migrants, Poland, November 2016, Annex IV.

[217] Focus groups with Russian migrants, Poland, November 2016, Annex IV.

[218] European Commission (2019b), p. 1, p. 9.

[219] Directive 2014/66/EU of the European Parliament and of the Council of 15 May 2014 on the conditions of entry and residence of third-country nationals in the framework of an intra-corporate transfer [2014] OJ L 157.

[220] The Act of 22 February 2019 amending the Act on Foreigners and Certain Other Acts/ Ustawa z dnia 22 lutego 2019 r. o zmianie ustawy o cudzoziemcach oraz niektórych innych ustaw (Dziennik Ustaw z 2019 r., poz. 577). The amendment to the Act on Foreigners entered into force on 12 February 2018.

modation, as well as previous employment at the same enterprise in line with the Directive's provisions and possibility to transfer back to it after the stay in Poland (Article 139 of the AF). In addition, applicants need to have a contract or document specifying the period of their transfer within the enterprise, the seat of the host entity, their position in the receiving unit, remuneration,[221] and other employment conditions in the host entity.

According to Article 33п (1), foreigners can obtain a 'permit for persons transferred through intra-corporate transfer' in Bulgaria, thereby granting them the right to continuous residence so long as they meet the requirements for access to the labour market under the provisions of the ALMLM and if they possess a visa type D in accordance with Article 15 (1) of the AFRB. As in Poland, applicants need to present evidence that they have met the requirements of the Directive (Article 23 of the Implementing Regulation of the ALMLM). The employer is exempted from performing a labour market test and from complying with the 20/35% cap requirement with regards to ICT applicants (Article 31 (2) of the ALMLM). If the duration of the employment contract is less than 1 year, the permit will be issued for the duration of the employment contract (Article 33п (2) of the AFRB).

In line with Article 139b (1) of the Polish Act on Foreigners by way of a ministerial regulation, the first ICT permits granted in a given calendar year might be capped in terms of specific provinces, professions, or types of activities of the host entities. The permit may be issued for a maximum period of 3 years for managers and specialists, and 1 year for employees in training. Poland does not impose the so-called cooling off period provided for in Article 12 (2) of the Directive after the maximum period of stay has been reached, which means that ICTs can choose to circulate between Poland and their home country without the need to wait for up to 6 months before they can engage in a new transfer.

This permit is issued for a period of 1 year by the Bulgarian Ministry of Interior following a positive decision by the Employment Agency based on a single permit procedure and it can be renewed if the circumstances for its issuance have not changed. Third-country nationals can work in Bulgaria on a continuous residence permit for an intra-corporate transfer for a period of up to 3 years as employees who are managers and specialists and 1 year for trainee employees (Article 32 (2) of the ALMLM). A new work authorisation application can only be submitted after a 6-month interruption between the expiration of a third-country national's permit

---

[221] The remuneration needs to be: 1) higher than the income authorizing cash benefits from social security, mentioned in the Act of 12 March 2004 on social assistance with regard to the foreigner and each family member being subsisted by them (should exceed PLN 528 for people in the family or PLN 701 for single people);

2) Not lower than the remuneration of employees performing work comparable in type or on comparable position on the territory of the Republic of Poland in comparable working time;

3) Not lower than 70% of the average gross monthly remuneration in the national economy in the province in which host unit has the registered office, in the year preceding submission of the application for the permission; announced by the President of the Central Statistical Office on the basis of Article 30 (2) of the Act of 26 October 1995 on some forms of supporting residential construction.

and the request for a new starting period of employment (Article 32 (2) of ALMLM), which could hinder circular migration.

6.3.4.2 Intra-corporate Transferees' Directive – Implementation Dynamics

Data retrieved from the national authorities of Bulgaria and Poland demonstrate that this instrument has hardly been used. In Bulgaria, in the period 2016 to mid-2019, only 60 ICT permits were issued;[222] in Poland, from 2018 until mid-2019, only three were issued.[223] Due to the late implementation of this Directive in Polish law, the operation of this instrument could not be captured through the collection of empirical data. However, the interviews conducted in Bulgaria with representatives from the IT sector – one before the adoption of the Implementing Regulation of the ALMLM and one after – shed light on the challenges faced in relation to the implementation of this instrument.

It became clear from the interviews with representatives of the Bulgarian IT sector that the ICT Directive did not suit their business needs and they had decided to quit using this instrument.[224] The interviewees shared that they used this procedure when they started to develop the company a few years ago, but remarked that it 'turned out ineffective' due to the transfer extension conditions after the authorised 3-year period under the Directive had elapsed.[225] The fact that the worker needed to wait for 6 months before re-applying for such transfer rendered this procedure ineffective for the IT business. Initially, when the Directive was transposed into Bulgarian legislation, the waiting period had been 3 months.[226] The interviewees shared that this requirement not only interrupted the company's business and economic activities, but also the private lives of their employees:

> 'These people rent flats, buy mobile phones, buy cars, there are also procedures (...) Most of them are with their families, the children go to school and after the three-year period not only the person but also his family must leave the territory of Bulgaria and this child, being in school during the school year, should just leave and spend the remaining nine months in another school in his country.'[227]

The requirement for work experience of up to 1 year in the main company from which the worker was transferred also posed certain challenges. In addition, the longer deadlines for issuing work authorisation – 30 working days compared to 15

[222] Data obtained from the Ministry of Interior in September 2019 with Decision No. 5364p151461/12 August 2019 of Migration Directorate of the Ministry of Interior.

[223] Data obtained from the Polish Office of Foreigners in October 2019.

[224] Interview #18 with representatives of an IT company, Bulgaria, September 2016, Annex III.

[225] Interview #18 with representatives of an IT company, Bulgaria, September 2016, Annex III.

[226] Interview #20 with representatives of an IT company, Bulgaria, December 2016, Annex III.

[227] Interview #18 with representatives of an IT company, Bulgaria, September 2016, Annex III.

working days within the Blue Card procedure – was considered impractical.[228] Finally, there was 'legal nonsense' that discouraged the company from using this instrument:

> 'If a work authorisation is issued for instance on 1st of March, it becomes effective on 1st of March, the day of its issue. The applicant must submit application for visa type D with this work authorisation at the Bulgarian embassy in the country where he resides. The administrative period for issuing such a visa, which is for an intra-corporate transfer, is 30 working days. You can calculate by yourself, that it is about two months or 45 calendar days. But at this point, while I am waiting for this visa (and I will just add that they are never ready on time, so it generally takes about two months), the worker's work authorisation is elapsing and he has no right to be in the country. He is waiting for his visa in his country, he has no right to come, but his work authorisation has been issued and it needs to be issued as a required document in order to apply for a visa. This is a legal nonsense! As soon as this person receives his visa and has the right to come to the country legally, two months from his work authorisation has already expired (...). It is issued for one year, but when he comes here, he works only ten months, because in reality he was waiting for two months for a visa while there was an active work authorisation during that time. This is another signal that this procedure is not practical for us.'[229]

6.3.4.3 Students' and Researchers' Directive

At the time of empirical data collection in the period 2016–2017, Poland had still not implemented the new Students' and Researchers' Directive.[230] Directive 2005/71/EC was transposed through the Act on Amendment to the Act on Foreigners and Certain Other Acts of 24 May 2007, but its full implementation did not come about until the entry into force of the Act on Foreigners of 2013 in May 2014. The application procedure required foreigners to apply in person[231] and present agreements that have been concluded with research institutions in Poland admitting them to conduct research projects.[232] In addition, researchers also had to present a written declaration completed by the research organisation in which it committed to covering the cost of the researchers' stay in Poland as well as the costs of executing an eventual return order issued to foreigners within 6 months from the expiration of the agreement, if the grounds for issuing the return order was fulfilled by irregular stay

---

[228] Interview #20 with representatives of an IT company, Bulgaria, December 2016, Annex III.

[229] Interview #18 with representatives of an IT company, Bulgaria, September 2016, Annex III.

[230] Directive (EU) 2016/801 of the European Parliament and of the Council of 11 May 2016 on the conditions of entry and residence of third-country nationals for the purposes of research, studies, training, voluntary service, pupil exchange schemes or educational projects and au pairing [2016] OJ L 132.

[231] If the application is not submitted in person, the applicant is obliged to appear at the relevant province within 7 days of the date of submitting application in line with Article 105 (1) 2 AF.

[232] Article 151 (1) 2 AF. Only research institutions covered by the Act on the rules of financing of science of 2010 can apply for approval to sign an agreement with a foreign researcher. For all requirements towards the research organisation, see Article 151 (4) AF.

in Poland.[233] The rest of the requirements were the same as for the other temporary residence permits – i.e. obligation to have health insurance and the possession of sufficient financial resources.[234] The permit was valid for a period of up to 3 years.

The Act of 22 February 2019 amending the Act on Foreigners and Certain Other Acts[235] implemented the new Students' and Researchers' Directive into Polish law. The most significant change brought by this EU law instrument into Polish law was the opportunity for researchers to apply for a temporary residence permit for a period of 9 months thus enabling them to look for work or to start a business in Poland after completing scientific research (Article 186 (1) 7 of the AF). This category of foreigners is also exempted from the requirement to apply for a work permit pursuant to Article 87 (2) 1 of the Act of 20 April 2004 on Promotion of Employment and Labour Market Institutions.

In Bulgaria, the new Students' and Researchers' Directive was transposed in ALMLM with the amendments from 16 March 2018.[236] In line with Article 24б (1) of the AFRB, researchers are entitled to receive a continuous residence permit in cases where they have obtained a visa under Article 15 (1) and have concluded a contract for the development of a 'scientific research project' with a 'scientific research organisation' that has its seat in Bulgaria and which is entered into the register of scientific research organisations in accordance with the Article 7б (1) 1 of the Act on the Promotion of Research.[237] In order to be granted a continuous residence permit, researchers need to meet the requirements stipulated by Article 24 (2) of the AFRB, such as having secured accommodation, obligatory health insurance and social insurance, and sufficient living funds for the duration of their stay.

This permit is granted for a period not shorter than 1 year (Article 24б (2) of the AFRB). In the event that the term of development of the scientific research project is less than 1 year, the residence permit is granted for the duration of the project. Foreigners are required to submit a residence permit application first and then register with the Employment Agency through the host research organisation.[238] Third-country nationals who are admitted as researchers for the purposes of conducting a research project under a contract with a hosting research organisation in Bulgaria are exempt from the requirement to possess a work permit for the duration of their project (Article 36 (1) of the ALMLM). Upon completion of the research project, a

---

[233] Article 151 (1) 1 (c) AF.

[234] Article 151 (1) 1 (a) and (b) AF.

[235] The Act of 22 February 2019 amending the Act on Foreigners and Certain Other Acts/ Ustawa z dnia 22 lutego 2019 r. o zmianie ustawy o cudzoziemcach oraz niektórych innych ustaw (Dziennik Ustaw z 2019 r., poz. 577). The amendment entered into force on 27 April 2019.

[236] SG No. 24/ 16 March 2018 which entered into force on 23 May 2018.

[237] Закон за насърчаване на научните изследвания (SG No. 92/ 17October 2003, latest amendment SG No. 77/ 18 September 2018).

[238] The supporting documents that need to be submitted as part of the continuous permit application are stipulated in Articles 14 (1) and 29д (1) of the Implementing Regulation of the AFRB. The supporting documents that need to be submitted for the registration with the Employment Agency are stipulated in Article 33 of the Implementing Regulation of the ALMLM.

third-country national is entitled to seek and take up employment as well as benefit from the services stipulated under the Employment Promotion Act within 9 months after submitting a registration application with the Employment Agency (Article 36 (1) 3 of the ALMLM).

6.3.4.4 Students' and Researchers' Directive – Implementation Dynamics

As already mentioned, at the time of the empirical data collection in the period 2016–2017 as part of this research, neither Poland nor Bulgaria had transposed the new Students' and Researchers' Directive. This did not allow for any examination of the implementation of this instrument through focus groups with researchers or interviews with stakeholders. Data obtained from the administrations in Bulgaria and Poland, however, suggest that the use of the EU *acquis* concerning researchers, covering both the former Researchers' Directive and the Recast from 2016, is marginal in both countries.

According to the data of the Polish Office for Foreigners, 397 temporary residence permits for researchers have been issued in the past 10 years.[239] By comparison, in Bulgaria this number totals 24 for the same period.[240] Interestingly enough, the data obtained from the Bulgarian Employment Agency show that there were only three foreigners registered as researchers through a host research organisation.[241] This could mean either that host institutions are violating the requirements of the ALMLM by not registering the researchers with the Agency or that the issued residence permits were based on bilateral measures, as some of the study's research informants pointed out.

One of the interviewed Bulgarian officials shared a general problem concerning foreigners coming to the country on a scientific exchange programme as per Article 15 (2) of the Act on Foreigners.[242] This article provides for the issuance of a long-term visa with a validity of 1 year, which falls outside the scope of the Researchers' Directive. However, the use of this article as grounds for entry created problems with regard to the lack of legal possibilities to issue a personal number to a foreigner, which was required by the Ministry of Education, other public institutions, as well as private companies such as mobile phone operators.[243] According to the interviewed official: 'In such cases, after many reservations and making a compromise, we issue a visa under Article 15 (1) of the AFRB and we grant the person a permit for continuous residence, because the lack of personal number of a foreigner

239 Data obtained from the Polish Office of Foreigners in October 2019.

240 Data obtained from the Ministry of Interior in September 2019 with Decision No. 5364p151461/12 August 2019 of Migration Directorate of the Ministry of Interior.

241 Information obtained via official request to information under Decision № РД-08-2156 from 12.08.2019 of the general secretary of the Employment Agency.

242 Interview #2 with state official, Bulgaria, July 2016, Annex III.

243 Interview #2 with state official, Bulgaria, July 2016, Annex III.

does not allow for normal life in Bulgaria'.[244] It was unclear whether this provision was used only for researchers coming under national measures or whether it also covered cases under the Directive, which would constitute a violation because Member States are only required to issue residence permits.[245]

6.4 Conclusions

This chapter aimed at assessing whether the developed instruments provide options for facilitated entry for migrants from the Eastern Partnership countries and Russia, as well as circulation-friendly policies. It demonstrated that Poland facilitates entry for nationals of Eastern Partnership countries and Russia through both national and EU visa policy instruments, as well as on the basis of national mechanisms such as the *Oświadczenie* procedure and the *Karta Polaka*, which is in line with the benchmarks of this study. Poland is gradually liberalising access to its labour market based on a wide spectrum of exemptions from work permit applications and the performance of a labour market test regulated through bylaws.

Contrary to Poland, Bulgaria provides quite limited national measures for facilitated entry for the citizens of the Eastern Partnership countries. Currently the only visa facilitation concerns priority visa processing for citizens of Armenia and Moldova entering under the bilateral labour migration agreements concluded with these countries in 2018, and the outsourced Visa Application Centres for nationals of Ukraine and Russia. Citizens of the Eastern Partnership, however, can benefit from the developed EU visa facilitation instruments, visa-free regimes, and other EU legal instruments such as the Blue Card Directive and the Seasonal Workers' Directive whose transposition has been amended recently to provide liberalised access following pressure from business organisations. Furthermore, applicants can make use of the national measures targeting foreigners of Bulgarian origin that provide for facilitated access to the labour market and permanent residence, as well as a fast-track citizenship procedure in case they decide to apply for Bulgarian citizenship on the basis of their Bulgarian ethnic origin.

The migrants interviewed in Poland used both the *Karta Polaka* and the *Oświadczenie procedure* as circulation-friendly instruments as part of their changing migrant-led trajectories. In addition, the multiple-entry visa under the Visa Facilitation Agreement with Russia stands out as an EU instrument that has also been useful for circular migration purposes. On the other hand, the restrictive admission procedures in Bulgaria trigger what can be referred to as 'forced circularity', which aims to keep migrants in a temporary position. Foreign workers are supposed to stay in the country for only a limited period of time, until they or their respective

[244] Interview #2 with state official, Bulgaria, July 2016, Annex III.

[245] The Directive allowed for a transition period under Article 18, where Member States were not obliged to issue residence permits. However, this period has since expired. See also Peers (2012), p. 138.

employers train a Bulgarian or another legally-resident foreigner to fill the job in question. Therefore, currently, all categories of migrant workers, except for Blue Card holders and foreigners of Bulgarian origin, need to exit the country for a minimum period in order to obtain new permits or change their status. This policy has also been reflected in the concluded bilateral labour migration agreements and is not considered to be in line with the study's benchmarks because it does not allow for migrant-led trajectories.

Even though it provides for flexible geographic mobility, the Blue Card instrument is not used for circular migration purposes *per se*. The Blue Card holders wishing to spend time working in their countries of origin use the company's internal 'home office' mechanisms to circulate back and forth. Outside of this practice, however, they are not interested in circulating for work, mainly due to political and economic factors. The other EU instrument considered suitable for promoting circular migration, the EU long-term residence permit, also does not seem to be an attractive tool for circulation for settled migrants in Poland and Bulgaria. The few interviewed migrants with such national or EU permits used them to maintain their transnational links with their country of origin but did not consider this to be work-related circulation. In the case of Bulgaria, because migrants invest many resources and expend great effort to enter the country and stay there, they perceive voluntary circulation as being too risky and their main goal is to obtain EU long-term or permanent residence.

So far, the Researchers' and the ICTs Directives have had a marginal use in both countries mainly due to the cumbersome and ineffective procedures created as part of their transposition into Polish and Bulgarian law. In addition, apart from the *Oświadczenie* procedure in Poland and the concluded bilateral agreements in Bulgaria, the rest of the initiatives developed under the auspices of the Mobility Partnerships have not contributed much to the facilitation of entry and re-entry conditions for citizens of the Eastern Partnership countries.

Lastly, the Local Border Traffic agreement with Ukraine cannot be directly used for circular migration related to work, but can lead to the initiation of circular migration because it provides job-seeking opportunities. Nevertheless, in order to gain insights into its implementation, one needs to conduct targeted field research in the border region, which is a matter that falls outside the scope of this research.

The main problems identified in both countries concerned the protection of migrants' rights. In Poland, for example, implementation of the *Oświadczenie* procedure in many instances involved the use of fake declarations that enabled migrants to enter Poland to look for a job and obtain a new *Oświadczenie* (declaration) stating the real employment or find other ways to 'legalise their stay'. Seasonal workers, however, used the *Oświadczenie* procedure as means of gaining entry into Poland, but in most cases they did not seek to regularise their work by signing a contract to this effect. These problems naturally placed migrant workers in a vulnerable position that could lead to abuse by their employers. Poland has addressed these issues and reformed the *Oświadczenie* procedure in 2018, *inter alia*, by providing a legal basis for refusing to register a declaration and the possibility to conduct additional verification. However, the country now needs to concentrate more efforts on the enforcement of rights-based circulation and protection of migrants' rights.

On the other hand, the analysis of Bulgarian migration law and the empirical data collected demonstrated that the restrictive entry and re-entry conditions that Bulgaria has established on the basis of both national and, to some extent, the implemented EU law, in general, lead to 'forced circulation'. Due to the involuntary circulation mechanisms for change of visas and statuses, the migrants' main desire is to achieve security of residence after 5 years and no longer circulate between their country of origin and their country of destination. Seasonal workers are probably the only exception in this regard. Thus, this study suggests that the Bulgarian circular migration approach resembles the guest worker model and is not conducive to a migrant-led trajectory. Furthermore, it has created a breeding ground for abuses of migrants' rights and can lead to exploitation as it forces migrants to circumvent the applicable legislation, leads to dependency on employers, and leaves wide discretion to the administration to take decisions about migrants' lives.

This notwithstanding, EU law is the only channel that could lead to the facilitation of rights-based circular migration in Bulgaria; the visa-liberalisation regimes and the EU's desire to turn the Blue Card into a more effective instrument are expected to lead to greater opportunities for facilitated entry and re-entry. Furthermore, as a result of the infringement procedures against Bulgaria, as well as the growing demand for foreign labour and the lobbying carried out by business organisations, the incorrect and cumbersome national transposition measures are slowly being brought more in line with their *effet utile* and access to the Bulgarian market has been liberalised for certain categories of workers. On the other hand, Poland transposed the EU labour migration legislation in a rather flexible manner, for instance allowing applications and change of status to be effected both from outside and within the country. Despite the great delay in the transposition of the Seasonal Workers' Directive, the country created procedures adapted to its national labour market needs.

References

Bieniecki, M. and M. Pawlak (2009) *Strategie przetrwania. Adaptacja ukraińskich migrantów zarobkowych do polskiej rzeczywistości instytucjonalnej*. Instytut Spraw Publicznych, Warsaw.

Council of Ministers (2015) National Strategy in the field of Migration, Asylum and Integration (2015–2020), adopted with Council of Ministers Decision No. 437/12 June 2015, Sofia.

Dnevnik (2017) Hotels and restaurants camouflage the hiring of foreigners with training contracts. http://www.dnevnik.bg/biznes/turizam/2017/03/10/2932528_hoteli_i_restoranti_zamaskirat_naemaneto_na_chujdenci/ Accessed 11 October 2017.

European Commission (2019a) Report on Directive 2011/98/EU of the European Parliament and of the Council of 13 December 2011 on a single application procedure for a single permit for third-country nationals to reside and work in the territory of a Member State and on a common set of rights for third-country workers legally residing in a Member State, COM (2019) 160 final, 29 March 2019, Brussels.

European Commission (2019b) Report on the implementation of Directive 2003/109/EC concerning the status of third-country nationals who are long-term residents, COM (2019) 161 final, Brussels, 29 March 2019, Brussels.

European Migration Network (2012) Visa policy as migration channel in Poland. National Report. Prepared by the National Contact Point to the European Migration Network, Warsaw.

European Migration Network (2015) Changes in immigration status and purpose of stay: an overview of Bulgarian approaches. Report on Bulgaria.

European Migration Network (2018) Labour market integration of third-country nationals in EU Member States. Report on Bulgaria. https://ec.europa.eu/home-affairs/sites/homeaffairs/files/03a_bulgaria_labour_market_integration_final_en.pdf Accessed 12 December 2019.

Górny, A., P. Kaczmarczyk, M. Szulecka, M. Bitner, M. Okólski, U. Siedlecka and A. Stefańczyk (2018) Imigranci w Polsce w kontekście uproszczonej procedury zatrudniania cudzoziemców. Warszawa: Uniwersytet Warszawski.

Fomina, J. (2011) Local border traffic agreement for the Kaliningrad region: a success story of the Polish presidency and a trust-building exercise for Poland and Russia. *Policy Brief.* Warsaw: Stefan Batory Foundation.

International Organization for Migration Mission to Georgia (2017) *Temporary Labour Migration of Georgian Workers to Poland and Estonia.* http://iom.ge/1/labour-migration Accessed 30 April 2017.

Kindler, M., A. Kordasiewicz and M. Szulecka (2016) *Care needs and migration for domestic work: Ukraine-Poland.* Global Action Programme on Migrant Domestic Workers and their Families. An ILO/UE project. Geneva: International Labour Office.

Kozak, I., M. Tota and B. Wojcik (2014) *Long-term Residence of Foreigners in Poland in the Light of Act on Foreigners of 12 December 2013.* In T. Sieniow (ed.), Prawo I Polityka Migracyjna II. Lublin: Rule of Law Institute Foundation.

Lessenski, M. (2009) The EU New Member States as Agenda Setters in the Enlarged European Union. Country Report: Bulgaria. European Policies Initiative. Open Society Institute, Sofia.

Mazowieckie For Foreigners (2019) EU Blue Card 2019. https://mazowieckie.pl/en/for-foreigners-1/residence/temporary-residence-pe/eu-blue-card/446,What-fees-do-I-have-to-pay.html Accessed 15 December 2019.

Ministry of Foreign Affairs (2018) Visa for Bulgaria. https://www.mfa.bg/en/services-travel/consular-services/travel-bulgaria/visa-bulgaria Accessed 12 September 2019.

Migrantinfo.pl (2018) Local Border Traffic. http://www.migrant.info.pl/Local_border_traffic.html Accessed 27 April 2019.

National Assembly of the Republic of Bulgaria (2015) http://www.parliament.bg/pub/cW/20150619090636Volen-VIZI.pdf Accessed 26 November 2017.

National Contact Point to the European Migration Network (2013) Poland abolished long-term visa fees for Moldovan citizens. https://emn.gov.pl/ese/news/10543,Poland-abolished-long-term-visa-fees-for-Moldovan-citizens.html Accessed 25 April 2017.

National Council on Migration and Integration (2019) Report on the 2018 Action Plan for the Implementation of the 2015-2020 National Strategy in the field of Migration, Asylum and Integration adopted with Council of Ministers Protocol No. 24.3 of 12 June 2019, Sofia.

OECD (2013) *International Immigration Outlook 2013: Poland.* Paris: OECD Publishing.

OECD (2018) *International Immigration Outlook 2018: Poland.* Paris: OECD Publishing.

Offnews (2017) The procedure for recruitement of seasonal workers have been eased. https://offnews.bg/turizam/ulesniha-protcedurata-za-naemane-na-sezonni-rabotnitci-ot-chuzhbina-658341.html Accessed 11 October 2017.

Pawlak, M. and I. Lashchuk. (forthcoming). *Entry to the labour market, not to the state - migrant workers' perspectives in Poland.* Report commissioned by the Friedrich Ebert Foundation.

Peers, S. (2012) Admission of Researchers. In Peers, S., E. Guild and J. Tomkin (eds). *EU Immigration and Asylum Law (Text and Commentary): Second Revised Edition. Volume 1: Visas and Border Controls.* Leiden: Martinus Nijhoff Publishers, pp. 129-164.

Peers, S. (2016) *EU Justice and Home Affairs Law. Volume I: EU Immigration and Asylum Law.* Fourth Edition. Oxford: Oxford University Press.

Radio Poland (2016) Local border traffic between Poland and Russia to remain suspended, 15 September 2016. http://www.thenews.pl/1/10/Artykul/271016,Local-border-traffic-between-Poland-and-Russia-to-remain-suspended Accessed 15 May 2017.

Statistics Poland (2018) Demographic Yearbook of Poland, Warsaw. https://stat.gov.pl/files/gfx/portalinformacyjny/pl/defaultaktualnosci/5515/3/12/1/rocznik_demograficzny_2018.pdf#page=456 Accessed 1 December 2019.

Unterschütz, J. (2016) National Report on Implementation of EU Migration Directives in Poland. In R. Blanpain, F. Hendrickx and P. Olsson (ed.), *National Effects of the Implementation of EU Directives on Labour Migration from Third Countries*. Alphen aan den Rijn: Wolters Kluwer, pp. 161-192.

Vankova, Z. (2017) Implementing the EU's Circular Migration Approach: Legal and Migrant Perspectives on Entry and Re-Entry Conditions in Bulgaria and Poland. Robert Schuman Centre for Advanced Studies Research Paper No. RSCAS 2017/34. Florence: European University Institute.

Vankova, Z. and B. Ivanova (forthcoming) *Temporary home or final destination? Migrant workers' perspectives in Bulgaria*. Report commissioned by the Friedrich Ebert Foundation.

Wasilewska, O. (2008) Analysis of the visa policies of the Visegrad countries: Relative openness. Polish visa policy towards Belarus, Moldova, Russia and Ukraine. Warsaw: Stefan Batory Foundation.

Znadniemna (2018) Białoruś: Do wprowadzenia małego ruchu granicznego z Polską brakuje przejść granicznych. http://znadniemna.pl/4487/bialorus-wprowadzenia-malego-ruchu-granicznego-polska-brakuje-przejsc-granicznych/ Accessed 1 December 2019.

Chapter 7
Implementation of National Approaches to Circular Migration: Work Authorisation, Residence Status, and Social Security Coordination for Migrant Workers in Bulgaria and Poland

This chapter aims to assess the implementation of national and EU legal migration instruments conducive to circular migration against the study's benchmarks in the areas of work authorisation, residence status, and social security coordination which are key policy areas that need to be considered if this type of migration is to be facilitated. Each section commences with a legal and policy analysis of these instruments as of mid-2019 and then moves on to reveal their implementation dynamics. The implementation analysis is based on data retrieved from the Bulgarian and Polish administrations, focus groups with Ukrainian and Russian migrant workers, and interviews conducted with relevant stakeholders as part of the study as well as available data from recent studies. The chapter concludes with an assessment of the developed instruments against the study's benchmarks in these three policy areas, looking *inter alia* into the possibility for migrants to change employer and occupation, access prolonged status, and export their pensions.

7.1 Work Authorisation

Migrant workers' initial work permits very often bind them to a specific employer, sector, and region for a specified period during which they cannot be changed – something that can lead to abuse and exploitation. Therefore, the benchmarks in this policy field aim at assessing whether workers can change their employer with a maximum restriction of 2 years.[1] In addition, they examine whether loss or termination of employment constitutes a sole ground for withdrawal of a migrant worker's

[1] Based on Article 14 (a) of Convention concerning Migrations in Abusive Conditions and the Promotion of Equality of Opportunity and Treatment of Migrant Workers, C143, adopted at 60th ILC session on 24 June 1975, Geneva, entry into force on 09 December 1978 (ILO Convention No. 143); and Article 52 (3a) of the International Convention on the Protection of the Rights of All

Z. Vankova, *Circular Migration and the Rights of Migrant Workers in Central and Eastern Europe*, IMISCOE Research Series,
https://doi.org/10.1007/978-3-030-52689-4_7

authorisation of residence or work permit.[2] Finally, the benchmarks evaluate the possibility for migrant workers to find alternative work in case of loss or termination of employment[3] and the possibility for seasonal workers to take up other remunerated activities in cases where they have already been employed on the territory of the Member Sate for a significant period of time.[4] Flexible work permits allowing change of employer within the period of their validity are among the instruments that can support the implementation of these benchmarks.

7.1.1 Work Authorisation Conditions According to National Instruments in Poland and Bulgaria

According to the Polish Act of 20 April 2004 on Employment Promotion and Labour Market Institutions (AEPLMI),[5] migrant workers can perform work under an employer's declaration of commissioning work to a foreigner (referred to as the *Oświadczenie* procedure) during a period not exceeding 6 months within the subsequent 12 months in total 'regardless of the number of entities commissioning the work' to foreigners (Article 88z (2) 3 of the AEPLMI). The main requirement is that 'the Local Labour Office entered the declaration of commissioning work to the foreigner into the register of declarations before commencing work, and that the work is performed under the conditions set out in that declaration'.[6]

Even though the Act of 20 April 2004 on Employment Promotion and Labour Market Institutions does not provide for a change of employer, foreigners can register a number of declarations, allowing them to work for several employers simultaneously. This was a documented practice while the procedure was stipulated in a Ministerial regulation until early 2018[7] and used by migrant workers to avoid unemployment. Generally, unemployment would lead to violation of the work visa conditions under the *Oświadczenie* procedure, which since 2018 has been monitored through a new Central Registration Database established by the amendments to this procedure.

Furthermore, migrant workers who have worked legally for the same employer for more than 3 months under a registered declaration can apply for a work permit (through their employer) or a single permit without having to leave the territory of

Migrant Workers and Members of Their Families (ICRMW), G.A. res. 45/158, 18 December 1990, entered into force on 1 July 2003. See Vankova (2016), pp. 342–343 for more details.

[2] Based on Article 8 (1) ILO Convention No. 143 and Article 49 (2) ICRMW.

[3] Based on Article 8 (2) ILO Convention No. 143 and Article 49 (3) ICRMW.

[4] Based on Article 59 (2) ICRMW; Vankova (2016), p. 343.

[5] Ustawa o promocji zatrudnienia i instytucjach rynku pracy (Dziennik Ustaw z 2019 r., poz. 1482, consolidated text).

[6] Article 87 (3) AEPLMI.

[7] Interview #1 with officials, Poland, November 2016, Annex II. Interview #12 with civil society actor, Poland, November 2016, Annex II.

the country. This possibility is stipulated in Article 88za of the AEPLMI. A labour market test is not required in instances where a foreigner has worked for the same employer for more than 3 months and has worked in the same position throughout the duration of the *Oświadczenie* procedure directly before applying for a work permit.[8] In order to benefit from this exemption, the foreigner must present the registered declaration to the Local Labour Office along with a labour contract and documentation evidencing the payment of social security contributions.[9] Another exemption in this regard concerns nationals from the countries covered by the *Oświadczenie* procedure who provide care and nursing services or who are household staff for natural persons in a household.

Since foreigners who have obtained the *Karta Polaka* are entitled to take up employment in the territory of Poland without needing to apply for a work permit[10] and can conduct economic activity on the same conditions as those applicable to Polish citizens,[11] they do not face any restrictions regarding change of employer or occupation. In addition, the only restriction is the validity of their visas but they have facilitated access to residence permits, which additionally mitigate the potential challenges they might face. All this means that there are no consequences if they lose their jobs.

In comparison, Article 7 (2) of the Bulgarian Act on Labour Migration and Labour Mobility (ALMLM)[12] stipulates that workers who are third-country nationals and who have been granted labour market access can be employed only by a natural or legal person in a place, at a position, and for a duration matching those specified in the work authorisation. The right to access the labour market can be withdrawn if the General Labour Inspectorate establishes that the third-country national's employment does not match the one specified in the work authorisation in line with Article 7 (3) (Article 12 (1) 2 of the ALMLM). The Act does not contain a general provision for the possibility of accessing all types of employment and change of occupations subject to a maximum restriction, apart from the options derived from the specific permits based on the EU legal migration directives and the permanent residence status.

Upon the premature termination of an employment relationship with a third-country national, the employer is obliged to notify the Bulgarian Employment

---

[8] Para 3 (3) of Regulation of the Minister of Labour and Social Policy of 29 January 2009 on determining cases in which a work permit for a foreigner is issued irrespective of the detailed conditions for issuing work permit for foreigners/Rozporządzenie Ministra Pracy i Polityki Społecznej z dnia 29 stycznia 2009 r. w sprawie określenia przypadków, w których zezwolenie na pracę cudzoziemca jest wydawane bez względu na szczegółowe warunki wydawania zezwoleń na pracę cudzoziemców (Dziennik Ustaw z 2019 r., poz. 154).

[9] Ibid., Para 3 (2). When the foreigner will be employed on the basis of the so-called civil contract, they cannot make use of this provision.

[10] See Article 87(2) 6 AEPLMI.

[11] European Migration Network (2012), p. 21.

[12] Закон за трудовата миграция и трудовата мобилност (SG No 33/26 April 2016, last amendment No 34/23 April 2019).

Agency in writing within 3 days from the date of termination of the employment (Article 10 (4) of the ALMLM). In such cases, an application for a new labour market access authorisation can be submitted after a 3-month interruption between the expiry of a third-country national's permit or the premature termination and the request for a new starting period of employment (Article 7 (1) 1 and 2 of the Implementing Regulation of the ALMLM[13]). Nevertheless, migrants who are exempted from the requirement to apply for a work permit – such as foreigners of Bulgarian origin – do not face challenges concerning change of employer or occupation.

In addition, a look at the bilateral labour migration instruments concluded shows that Article 3 of the Agreements, respectively, between Bulgaria and Armenia[14] and Bulgaria and Moldova[15] stipulate that the entry, residence, and employment of nationals of the contracting parties on the territory of the other state shall be governed by the legislation of the receiving state. Since there is no explicit provision related to the change of employer and the possibility to find alternative employment, these are subject to the national law of the contracting parties.

7.1.2 Implementation Dynamics

The interviewed migrants and NGO representatives in Poland did not report any problems concerning their change of employer or any unemployment they faced during the authorised period under the *Oświadczenie* procedure. However, two other problematic issues relating to the implementation of the *Oświadczenie* procedure were identified during the fieldwork at the end of 2016. The first was the fact that the procedure was regulated by a Ministerial regulation and not by the general normative act on employment. The Ministry of Family, Labour and Social Policy issued interpretative guidelines on the application of the *Oświadczenie* procedure,

---

[13] Правилник за прилагане на Закона за трудовата трудовата миграция и трудовата мобилност (SG No. 79/ 7 October 2016, last amendment SG No. 27/ 2 April 2019).

[14] Agreement between the Republic of Bulgaria and the Republic of Armenia on labour migration management, approved by Council of Ministers Decision No. 176 of 22 March 2018, entered into force 9 October 2018/Спогодба между Република България и Република Армения за регулиране на трудовата миграция, утвърдена с Решение № 176 от 22 март 2018 г. на Министерския съвет. В сила от 9 октомври 2018 г.

[15] Agreement between the Government of the Republic of Bulgaria and the Government of the Republic of Moldova on labour migration management, approved by Council of Ministers Decision No 492 of 13 July 2018, entered into force on 11 September 2018/Спогодба между правителството на Република България и правителството на Република Молдова за регулиране на трудовата миграция, утвърдена с Решение № 492 на Министерския съвет от 13 юли 2018 г. В сила от 11 септември 2018 г.

which were in practice a non-legally binding set of recommendations.[16] On the basis of these interpretive guidelines, the Local Labour Offices were using their discretion to apply or issue their own interpretation of the Regulation that in some of the cases differed so much that it was compared to the 'invention' of new legislation.[17] The Local Labour Office in Warsaw, for example, introduced restrictions that were not provided in the Regulation, such as a cap on the number of *Oświadczenie* that one employer could obtain and a requirement that the company should have been in operation for at least 1 year. This wide discretion was limited through the amendments in 2018[18] but these did not explicitly stipulate any provisions on change of employer and sector, which continues to be a persisting problem that naturally creates legal uncertainty for the migrant workers.

The second major problem with the implementation of the *Oświadczenie* procedure was the widespread abuse of migrant workers because they often worked irregularly. In the sectors of agriculture, construction, and domestic services it was the case that even Polish nationals were working irregularly because it was too expensive to legally employ workers due to the high financial burden stemming from taxation and social security obligations, and due to onerous bureaucratic hurdles.[19] An interviewee said that it was common for employers to abuse Ukrainian workers by not paying their last salary as they knew that their work visa was expiring and would therefore have to leave the country.[20] But even more severe cases of exploitation were recorded. For example, several Ukrainian workers were forced to work in a cigarette factory without being given their wages; their passports were taken from them and they were promised payment only at the end of their stay in Poland.[21]

According to one of the interviewed NGO representatives, the Ukrainian circular migration pattern was always based on legal entry into Poland and irregular work.[22] Before Poland's accession to the Schengen area in 2017, Ukrainians did not need a visa to enter Poland. Once Poland introduced visas, the state had to invent another special instrument to allow people to enter and work by introducing new grounds for issuing a visa. The fact that this instrument was functioning for more than 10 years, and given that before the 2018 reforms there had been no political will to change it, meant that 'we are really eager to see immigrants from Ukraine working illegally in Poland because it's cheaper, it's better for Polish economy, it's better for Polish employers and it's very convenient not to interfere in this interaction, not to

[16] Interview #1 with officials, Poland, November 2016, Annex II. The guidelines contained mainly issues to be verified and the possibility to refuse registration. Interview #1 with officials, Poland, November 2016, Annex II.

[17] Interview #12 with civil society actor, Poland, November 2016, Annex II.

[18] See Chapter 5, section 5.2.1.1 for more details.

[19] Interview #5 with civil society actor, Poland, November 2016, Annex II. Interview #12 with civil society actor, Poland, November 2016, Annex II.

[20] Interview #11 with trade union representative, Poland, November 2016, Annex II.

[21] Interview #11 with trade union representative, Poland, November 2016, Annex II.

[22] Interview #12 with civil society actor, Poland, November 2016, Annex II.

assist any migrants because otherwise we could be seen as country who exploits immigrants'.[23] The latest research shows that migrants using the assistance of an employment agency are usually hired on the basis of a civil law contract (*umowa zlecenie*).[24] However, this prevents such workers from benefiting from the Labour Code protection.

By way of comparison, according to one of the interviewed Bulgarian lawyers, the labour market test makes migrants dependent on employers and is conducive to corruption, as already described in Chap. 6.[25] This is due to the fact that changing an employer means repeating the whole application process, including the labour market test, for most categories of workers. As one interviewee stressed: 'This practically leads to a legal bondage to a certain employer in 21st century. This is pure feudalism'.[26] The interviewed officials confirmed that in the majority of cases, if workers wanted to change their employer, they needed to leave the country and then re-apply after the above-mentioned interruption.[27]

The participants in the focus groups did not mention any problems in relation to changing employers because most of them stayed in Bulgaria on trade representation grounds and worked for their own companies. One of the participants in the Russian focus groups said that she had hoped to be able to apply for a Blue Card, but when she understood that the Directive was (initially) transposed in such a way that would only allow her to apply from outside the country, she decided that this would be too risky because she only had 2 years left until she would be eligible for an EU long-term residence permit.[28]

7.1.3 Work Authorisation Conditions and Implementation Dynamics According to the Legal Migration Directives Aiming to Facilitate Circular Migration

7.1.3.1 Blue Card Directive[29]

Article 133 (2) of the Polish Act on Foreigners transposed Article 13 of the Blue Card Directive, allowing Blue Card holders to retain their permits in case of unemployment for a period of 3 months within the validity of the permit. However, the period in which the foreigners remain unemployed can only occur a maximum of

---

[23] Interview #12 with civil society actor, Poland, November 2016, Annex II. Interview #11 with trade union representative, Poland, November 2016, Annex II.

[24] Pawlak and Lashchuk (forthcoming).

[25] Interview #9 with lawyer, Bulgaria, July 2016, Annex III. For more details, see Chap. 6, Sect. 6.2.

[26] Interview #9 with lawyer, Bulgaria, July 2016, Annex III.

[27] Interview #10 with officials, Bulgaria, July 2016, Annex III.

[28] Focus group with Blue Card holders from Russia, Bulgaria, September 2016, Annex IV.

[29] Council Directive 2009/50/EC of 25 May 2009 on the conditions of entry and residence of third-country nationals for the purposes of highly qualified employment [2009] OJ L 155/17.

twice and for no longer than a combined duration of 3 months[30] within the period of the permit's validity.[31] To benefit from this exception, the foreigners need to notify the competent regional governor (*voivode)* within 15 working days that they have lost their job.[32]

In the event that Blue Card holders in Bulgaria become unemployed, they are entitled to a period of 3 months to look for new employment and start working in line with the initial application procedure (Article 21 (1) of the ALMLM). The Employment Agency cannot propose the Blue Card permit's withdrawal within this period (Article 21 (3) of the ALMLM). By contrast to the Polish case, the right to look for a new employer can only be used once within the period of validity of the Blue Card (Article 21 (4) of the ALMLM).

During the first 2 years of their stay in Poland on the basis of a Blue Card permit, foreigners cannot change their employer or the position for which they are employed and they cannot receive a lower salary than the one specified in the permit without an amendment to the permit by the competent governor.[33] In line with Article 135 (1) of the Act on Foreigners, the permit can be amended by the governor at any time. After the first 2 years, foreigners are only required to inform the governor about any changes concerning position, remuneration, and other conditions specified in the permit.[34]

Likewise, in line with the Directive, during the first 2 years of their employment, Blue Card holders in Bulgaria may only change their employer after obtaining a written positive decision that has been issued by the Employment Agency (Article 20 (2) of the ALMLM). The procedure for granting the abovementioned decision is carried out in line with the criteria contained in Article 15, which outlines the initial Blue Card application process (Article 18 of the Implementing Regulation of the ALMLM). After the initial period of 2 years, in order to change employer, Blue Card holders need to restart the application process for the permit in line with Article 17 of the ALMLM (Article 20 (3) of the ALMLM).

7.1.3.2 Blue Card Directive – Implementation Dynamics

According to the migrants interviewed in Poland, one of the major advantages of this permit was that it gave them the right to change employer and be unemployed for up to 3 months.[35] Most of the focus group participants knew that single permit holders could not benefit from these favourable conditions and needed to restart the whole process if they wanted to change employer. Nonetheless, the Blue Card

[30] Unterschütz, (2016), p. 186.

[31] Article 133 (2) 2 AF.

[32] Article 134 (1) AF.

[33] Article 135 (2) AF.

[34] Unterschütz (2016), p. 186. See Article 134 (3) AF.

[35] Focus groups with Blue Card holders: mixed group of Russian speakers, Russians and Ukrainians, Poland, December 2016, Annex IV.

holders interviewed emphasised that outside their company, which provided a relocation manager to support them in the application process, their colleagues who wished to change employer had to conduct the procedure themselves and faced bureaucratic hurdles and long queues, while some had to switch to a single permit and restart the whole procedure.[36]

Another problem shared by the focus group participants from Ukraine was that changes in the permit concerning employment conditions (position, salary, another contract) with the same employer had to be notified to the governor and that the changes allowed were capped. They even gave an example of a colleague who was promoted and declined the position because he did not want to start the procedure all over again as 'there is no guarantee the second time it will also be a Blue Card'. The interviewed IT company management representatives confirmed that this requirement 'in terms of flexibility, it is a blocker'.[37] Therefore, they used some 'loopholes' in the legislation because if they followed it strictly, they would not have been able to change position within the company.

Most of the Blue Card holders who participated in the focus groups had arrived relatively recently in Bulgaria and their main problems related to the application process, the renewal of their permits, and family reunification. None indicated any plans to change employer.

7.1.3.3 Seasonal Workers' Directive[38]

In line with Article 15 (3) of the Directive, when a foreigner has entered Poland on a visa issued for the purpose of seasonal work or as part of a visa-free regime in connection with an application for a seasonal work permit,[39] the local governor may extend the seasonal work permit for the same entity entrusting work to the foreigner or to perform seasonal work for another entity entrusting the work to the foreigner (Article 88u (1) of the AEPLMI). To extend the seasonal work permit, the provisions regarding the issue of the seasonal work permit apply accordingly (Article 88v of the AEPLMI). This means that the application needs to be registered and the employer has to present to the competent local governor a copy of a valid document entitling the foreigner to stay in Poland along with the foreigner's accommodation address during their stay in the country (Article 88p (7) of the AEPLMI). The extension of the seasonal work permit is issued for a period which, together with the period of stay of the foreigner for the purpose of performing seasonal work – counted from the day of first entry into the territory of the Schengen States in a given calendar year – is not longer than 9 months during the calendar year (Article

[36] Focus group with Blue Card holders from Ukraine, Poland, December 2016, Annex IV.

[37] Interview #22 with IT business management representatives, Poland, December 2016, Annex II.

[38] Directive 2014/36/EU of the European Parliament and of the Council of 26 February 2014 on the conditions of entry and stay of third-country nationals for the purpose of employment as seasonal workers [2014] OJ L 94.

[39] Entered in the register referenced in Article 88p (1) 1 AEPLMI.

88u (3) of the AEPLMI). The Act does not limit the number of applications for extension nor does it require performance of a labour market test as part of the permit extension procedure.

A permit is extended at the written request of the employer, submitted no earlier than within 90 days and no later than within 30 days before the expiry of the permit's validity (Article 88a (1a) of the AEPLMI). If an employer has applied for an extension of a foreigner's seasonal work permit and the application does not contain formal deficiencies or formal deficiencies have been removed on time, the work of the foreigner under the conditions set out in the permit for seasonal work is considered legal from the date of the application's submission to the day on which the decision to extend the seasonal work permit becomes final[40] (Article 88u (4) of the AEPLMI). If the application for the extension of the seasonal work permit was submitted by a different employer, the work of that foreigner under the conditions specified in the application are considered legal until the date of the local governor's decision, but not longer than for a period of 30 days from the date of submission of the application that does not contain formal deficiencies (Article 88u (5) of the AEPLMI). The only way to avoid withdrawal of the seasonal worker permit under Article 88y of the AEPLMI in case of unemployment is for the worker to secure another job with a different employer before the expiry of the permit's validity since the Act does not explicitly allow for seasonal workers to look for another job within a reasonable time.[41]

Concerning the benchmark on change of occupation, seasonal workers are entitled to work only in the fields of agriculture, forestry, fishery, services, gastronomy, and hospitality (concerning short stays and high seasons) in line with Regulation of the Minister of Family, Labour and Social Policy of 8 December 2018.[42] The Act does not limit the possibility of changing seasonal occupations when a different employer applies for an extension of the seasonal work permit as long as the new seasonal occupation is specified in the work permit application and respective contract in line with Article 88a (1aa) of the AEPLMI. The only category of foreigners who have the possibility of performing work outside the list of seasonal occupations determined by the above-mentioned Regulation are the citizens of Armenia, Belarus, Georgia, Moldova, Russia, and Ukraine. They can do so for a period not longer than a total of 30 days during the validity of their permits and given that their remuneration is not lower than the one specified in their permit and that they do not work as temporary employees (Article 88 s (1) of the AEPLMI).

---

[40] The periods of legal work do not include periods of suspension of proceedings at the request of a party.

[41] See Chap. 4, Sect. 4.3 for more details.

[42] Regulation of the Minister of Family, Labor and Social Policy of 8 December 2018 regarding activity subclasses according to the Polish Classification of Activities (PKD), in which seasonal work permits for a foreigner are issued/Rozporządzenie Ministra Rodziny, Pracy i Polityki Społecznej z dnia 7 września 2018 r. zmieniające rozporządzenie w sprawie podklas działalności według Polskiej Klasyfikacji Działalności (PKD), w których wydawane są zezwolenia na pracę sezonową cudzoziemca (Dziennik Ustaw z 2018 r., poz. 1749).

According to the Bulgarian Act on Foreigners, the permit for a seasonal worker can be extended once within the respective authorised period of stay – not less than 90 days and not more than 9 months – upon a decision by the Employment Agency (Article 24к (2) of the AFRB). This provision allows seasonal workers to extend their permit with their current employer or renew it by changing their employer once (Article 29 of the ALMLM). In case they want to change their employer, seasonal workers are not bound by the requirement of Article 5 (2), namely, to reside outside the country when applying for a work permit (Article 29 (1) of the ALMLM). They have the right to continue to reside in Bulgaria until a decision on the application is made, provided that the application was submitted within the period of validity of the permit and the authorised period of stay has not expired (Article 24 k (7) of the AFRB). An opportunity to seek employment in an area other than seasonal work, for which the permit is granted, is not explicitly regulated in the provisions of the ALMLM.

7.1.3.4 Seasonal Workers' Directive – Implementation Dynamics

As mentioned in Chap. 6,[43] the delay in transposing this Directive in Poland and the fact that seasonal workers in Bulgaria were mainly working irregularly in 2016 when the study's focus groups with migrants were conducted did not allow for data collection on the implementation of this instrument. A recent study covering seasonal workers in Bulgaria,[44] as well as the annual report of Bulgaria's Main Labour Inspectorate, demonstrate that the illegal practice of presenting foreign nationals' seasonal work as internships was restricted, and currently migrant workers are generally employed on the basis of labour contracts.[45] However, most come for a stay of up to 90 days, which is the most commonly-used entry channel because it is cheaper and less cumbersome for employers. In Poland, at the time of writing, there was still no comprehensive study of the implementation of the Seasonal Workers Directive and this does not allow conclusions to be drawn regarding work authorisation.

[43] See Chap. 6, Sect. 6.3.3.4 for more details.

[44] Vankova and Ivanova (forthcoming).

[45] Executive Agency Main Labour Inspectorate (2019), p. 22.

7.1.4 *Work Authorisation Conditions and Implementation Dynamics According to the Legal Migration Directives Containing Circular Migration Elements and Flanking Rights*

7.1.4.1 Intra-corporate Transferees' Directive[46]

ICTs are entitled to perform work in Poland for a particular employer and position under the conditions stipulated in the decision to grant a temporary residence permit and their contract (in line with Articles 139a and 139h of the AF). Therefore, they cannot change their occupation and employer within the period of validity of the ICT permit as this would affect the admission conditions and lead to the permit's withdrawal (Article 139g of the AF). Furthermore, the receiving enterprise has the obligation to notify in writing within 15 working days the regional governor who granted the ICT permit of any change in circumstances affecting the conditions of the permit granted (Article 139m (1) of the AF).

In line with the ICT Directive, the relevant Bulgarian act does not legislate for a change of employer or sector: the ICT permit entitles a third-country national to work in the territory of the Republic of Bulgaria for a specific receiving enterprise or group of enterprises only and solely in the position of manager, specialist, or trainee employee (Article 31 (3) of the ALMLM). Thus, ICTs cannot change their employer and occupation within the period of validity of the ICT permit.

The receiving enterprise is obliged to notify the Bulgarian Employment Agency of any changes in relation to the conditions for granting the permit (Article 31 (4) of the ALMLM). This means that in the case of ICTs, premature termination of their employment relationship with the sending enterprise would lead to a withdrawal of their permit because this would affect the admission conditions stipulated in Article 31 of the ALMLM and Article 23 of its Implementing Regulation. Therefore, if ICT permit holders wish to look for alternative work, they need to exit the country and then apply for another permit in Bulgaria.

7.1.4.2 Intra-corporate Transferees' Directive – Implementation Dynamics

As already demonstrated in Chap. 6, the use of the ICT Directive in practice is marginal in both Bulgaria and Poland. According to the interviewed representatives of a Bulgarian IT company, the procedure developed on the basis of the ICT Directive became much more cumbersome after the adoption of the Implementing

---

[46] Directive 2014/66/EU of the European Parliament and of the Council of 15 May 2014 on the conditions of entry and residence of third-country nationals in the framework of an intra-corporate transfer [2014] OJ L 157.

Regulation of the ALMLM.[47] According to their perspective, the ICT transfer was likened to a long-term business trip. They remarked that before the adoption of this Implementing Regulation, the procedure provided for the conclusion of a labour contract with the receiving enterprise to which the workers transferred, and that this was the procedure they had followed. Currently, however, the working conditions for third-country workers with an intra-corporate transfer permit are regulated under the terms and conditions for posted or sent workers within the framework of provision of services in the territory of the Republic of Bulgaria (Article 32 (1) of the ALMLM).[48] The amendments of the procedure also raised questions with regard to the possible renewal of this permit after the expiration of the maximum allowed 3-year period.[49] This insecurity led the company to prefer making use of the other available instruments, such as the Blue Card and Single Permit Directives.

In Poland, according to a report by Deloitte, the alternative local permits on assignment or local contract basis were considered faster in terms of processing times, as well as easier and cheaper in terms of procedure, thus offering a possible explanation as to why this instrument has not been preferred by Polish business enterprises.[50]

7.1.4.3 Students' and Researchers' Directive[51]

The Polish Act on Foreigners does not legislate for any change of host research institution and the legal consequences for the researcher in case of unemployment, e.g., where the host research institution loses its authorisation. This means that in such cases, if the grounds for issuing the permit are no longer valid in line with Article 154a of the AF, the permit should be withdrawn. In the same vein, Bulgarian migration law does not legislate for the change of hosting organisation for

[47] Interview #20 with representatives of an IT company, Bulgaria, December 2016, Annex III. The interviewees were referring to the amendments made to correct the transposition of the Directive into Bulgarian law.

[48] In accordance with the Bulgarian Labour Code (Кодекс на труда, SG No. 26/1 April 1986, last amendment SG No. 92/ 6 November 2018), the sending of employees within the context of the provision of services occurs, *inter alia*, when: an undertaking that employs a worker temporarily registered under the legislation of a third country sends a worker to an undertaking in the territory of Bulgaria (Article 121a (2) 2 of the Labour Code). An employee may be posted or sent, where for the entire period of secondment/sending or posting, there is an employment relationship between the latter and the seconding or sending employer, which is one of the conditions for obtaining an ICT permit (Article 121a (3) of the Labour Code and Article 40 (2) 1 of the ALMLM in conjunction with Article 25 (1) 3 of the Implementing Regulation of the ALMLM).

[49] Interview #20 with representatives of an IT company, Bulgaria, December 2016, Annex III.

[50] Deloitte (2018), p. 42.

[51] Directive (EU) 2016/801 of the European Parliament and of the Council of 11 May 2016 on the conditions of entry and residence of third-country nationals for the purposes of research, studies, training, voluntary service, pupil exchange schemes or educational projects and au pairing [2016] OJ L 132.

researchers. There are no legal provisions regulating cases of unemployment, which means that when the hosting agreement has been terminated, the permit of the researcher would be withdrawn because the circumstances for its granting would have changed (see Article 24 (2) AFRB).

However, it should be stressed that the wording of Article 21 (5) of the Students' and Researchers' Directive implicitly sets out the possibility of changing a host institution by stating that in cases when the researcher applies for a renewal of authorisation to enter into an employment relationship with a host institution and is not continuing the employment relationship with the same host institution, Member States can apply labour market tests in this regard. Therefore, researchers should be able to change their hosting research institution if they have a hosting agreement that meets the requirements of the admission conditions specified in Article 10 of the Students' and Researchers' Directive.

Concerning unemployment, the Students' and Researchers' Directive is silent as to whether unemployment could automatically lead to withdrawal of the residence permit.[52] Article 10 (5) of the Directive states that the hosting agreement should automatically lapse when the legal relationship between the researcher and the research organisation is terminated. In addition, Article 10 (8) leaves it to the discretion of Member States to regulate in national law consequences for researchers' authorisations as a result of withdrawal of the approval or the refusal to renew the approval of the hosting agreements. Therefore, the Directive does not oppose any national rules that would allow the researchers to look for another research organisation or job. Not allowing for a reasonable period to look for another job would also take away the *effet utile* of Article 25 of Students' and Researchers' Directive allowing researchers to seek employment or set up a business for a period of at least 9 months after completion of their research.

7.1.4.4 Students' and Researchers' Directive – Implementation Dynamics

Like the ICT Directive, the Recast Students' and Researchers' Directive is only marginally used in Bulgaria and Poland as illustrated in Chap. 6.[53] The interviewed migrants in Poland who were working as researchers possessed either the *Karta Polaka* or EU long-term residence and did not report any problems associated with changing employer. This is not surprising given the fact that for the former category there are not any explicit regulations on change of employer nor any restrictions for the latter. The 2014 implementation Plan for the 'Migration Policy of Poland: current state of affairs and proposed actions' envisaged the abolition or reduction of fees for issuing a temporary residence permit for researchers arriving in Poland with the aim of conducting research.[54] This was due to the fact that many

[52] See Article 21 (1) (a) of the SRD.

[53] See Chap. 6, Sect. 6.3.4.4 for more details.

[54] Implementation Plan for the Migration Policy of Poland: current state of affairs and proposed actions/Polityka migracyjna Polski – stan obecny i postulowane działania, p. 30.

researchers preferred to reside in Poland on the basis of a long-term national visa because it was free of charge rather than on temporary residence permit, which had the effect of placing them in a precarious position. This would have had a positive impact on researchers, who would have been required to change their research institution and restart the whole permit application procedure. Nevertheless, as already emphasised, the government repealed this policy in 2016 and it is not yet clear whether there will be any measures in this regard in the future. As demonstrated in Chap. 6, the implementation of the EU *acquis* concerning researchers in Bulgaria is also not without problems.[55]

7.1.4.5 Singe Permit Directive[56]

According to Article 123 of the Polish Act on Foreigners, the single permit cannot be revoked within 30 days from the date of unemployment if the foreigners notify the competent governor of the loss of the job indicated in the permit. The notification needs to be carried out within 15 days from the loss of the job (Article 121 (1) of the AF). This provision, however, cannot apply more than once during the period of the permit's validity (Article 123 (2) of the AF). Based on this provision, foreigners can change their employer or job if working conditions are unfavourable or if they receive a better job offer elsewhere. Nevertheless, this means that migrant workers need to start a new procedure for obtaining a single permit, which additionally requires the performance of a labour market test in line with Article 125 of the Act on Foreigners.

By contrast, following Bulgaria's 'copy and paste' approach to EU law, the ALMLM is 'harmonised' with the Single Permit Directive and does not contain any explicit rules on changing employers. Furthermore, it does not provide for a period during which the migrant worker can seek alternative employment. What is clear, however, is that any interruption of a worker's employment would be grounds for refusing an extension of the single permit (Article 15 (3) of the ALMLM).

7.1.4.6 Singe Permit Directive – Implementation Dynamics

One of the Russian participants in the focus groups remarked that the introduction of single permits worsened the situation in Poland because 'they bound the employees to their employers'.[57] He explained that if one wanted to change employer, then one had to restart the whole single permit application procedure from the first step:

---

[55] See Chap. 6, Sect. 6.3.4.4 for more details.

[56] Directive 2011/98/EU of the European Parliament and of the Council of 13 December 2011 on a single application procedure for a single permit for third-country nationals to reside and work in the territory of a Member State and on a common set of rights for third-country workers legally residing in a Member [2011] OJ L 343.

[57] Focus group with Russian migrants, Poland, November 2016, Annex IV.

'but who wants to do that?' The focus group participant also emphasised another impediment: in order to get a work permit for certain professions, one's salary needed to match a set threshold.[58] Another interviewee stressed that the introduction of the single permit had 'very negative consequences' because once the migrants decided to change their job, their position within the company, the type of contract, or if their salary decreased, then they would have to apply for a new single permit.[59] The whole process could take up to 1 year from the moment of submission until the moment of receipt, 'which is already official information from the Head of the Office of foreigners in Warsaw'.[60] In case the validity of the single permit had expired, migrants from the six eligible countries could work on the basis of the *Oświadczenie* procedure but had to apply for a visa to do so.[61] However, after 2014, Ukrainian men working in Poland were generally not willing to return home and re-apply for a visa because they were sent subpoenas as part of the ongoing mobilisation to join the Ukrainian army due to the conflict in Crimea.[62]

An NGO representative shared that the transposition of the Directive itself created a 'traffic jam' in the administration and that 'it works really, really slow, sometimes unpredictable'.[63] One strategy employed by migrants who wanted to change their single permit was to apply for a separate work permit for their second/new job at the very same time as applying for a new single permit to keep their work legal while they were waiting for the single permit to come through. This costed extra money for the applicants: before the introduction of the Single Permit Directive, the procedure costed 340 złoty (80 EUR), while at the time of writing the fee was 440 złoty (104 EUR) and the employer had to pay a further 100 złoty (23.50 EUR) to obtain a work permit.

The interviewee also declared that some companies preferred to keep migrant workers out of legal employment during the period before obtaining a work permit because the employee was tied to the company: 'they are quite sure that they won't leave them until they get at least the permit and they will be working without taxes and social security contributions'.[64] Nevertheless, the foreigner was trapped:

> 'because even if he could change the employer while the procedure lasts, he still has to have somebody to employ him, to run the labour market test for him, to resubmit application and also to submit documents saying that he quitted the job because afterwards the procedure can be cancelled'.[65]

---

[58] The focus group participant was referring to the fact that in order to obtain a work permit, the remuneration specified in the contract with the foreigner could not be lower than the salaries of other employees performing work of comparable type or in a similar position.

[59] Also stressed in Interview #12 with civil society actor, Poland, November 2016, Annex II and Interview #5 with civil society actor, Poland, November 2016, Annex II.

[60] Interview #3 with civil society actor, Poland, November 2016, Annex II.

[61] See Chap. 5, Sect. 5.2.1.1. for more details.

[62] For more info, see Home Office (2016).

[63] Interview # 3 with civil society actor, Poland, November 2016, Annex II.

[64] Interview #3 with civil society actor, Poland, November 2016, Annex II.

[65] Interview #3 with civil society actor, Poland, November 2016, Annex II.

In this period the migrants were not officially unemployed because they were waiting for their new single permit, but they could not officially be employed because one needed to have a visa or a work permit to do so according to the applicable Polish legislation. They were thus working in the informal market without an official contract. According to the interviewed NGO representative, migrants preferred this mechanism rather than returning to their country of origin and waiting there for 4 months without an opportunity to plan anything before the visa was either granted or refused.[66]

In Bulgaria, as described in Chap. 6, most of the migrant workers who participated in the focus groups shared that they have avoided the single permit application due to the cumbersome entry requirements that include performance of a labour market test.

7.2 Residence Status

As already stressed in this book, access to permanent residence and circular migration seem at first like two separate and unrelated policy areas. However, 'leakage' into permanent settlement in the host country is an inherent characteristic of circular migration.[67] Therefore, this study's analytical benchmark framework aims to explore whether lawful migrants have the opportunity to qualify for a prolonged or permanent residence status[68] on the basis of permits allowing transition from a temporary to a permanent residence status.[69] In addition, this section considers whether migrants have the right to mobility and a choice of residence within the host country.[70]

7.2.1 Residence Status According to National Instruments in Poland and Bulgaria

Migrant workers in Poland who have worked legally for more than 3 months under the *Oświadczenie* procedure with the same employer under a registered declaration, can apply directly for a work permit or a single permit without having to leave the

[66] Interview #3 with civil society actor, Poland, November 2016, Annex II.

[67] Skeldon (2012), p. 53.

[68] On the basis of Article 2 of the European Convention on Establishment, ETS No. 19, 13 December 1955, entry in force 23 February 1965.

[69] Vankova (2016), p. 344.

[70] On the basis of Article 12 (1) of the International Covenant on Civil and Political Rights (ICCPR), G.A. res. 2200A (XXI), 16 December 1966, entered into force on 23 March 1976; Article 39 ICRMW; Article 2 (1) of the Fourth protocol to the ECHR, ETS No.046, 16 September 1963, entered into force on 02 May 1968.

country. Both processes provide for an accumulation of residence periods and can lead to permanent residence, depending on the individual cases, either through the national or through the EU long-term residence procedures. Holders of a valid *Karta Polaka* who plan to settle permanently in the territory of Poland are also eligible to apply for permanent residence in line with Article 195 (1) 9 of the Act on Foreigners. The application is free of charge and after the permit has been granted, the foreigners and their family members can benefit from financial assistance for a period of up to 9 months.[71]

The national permanent residence permit is issued for specific categories of foreigners such as children of foreigners with permanent residence permits or EU long-term resident permits; children or spouses of Polish citizens; foreigners of Polish origin; victims of human trafficking; tolerated residence permit holders; and, persons with granted refugee status or with subsidiary protection and humanitarian statuses.[72] Depending on the different categories, foreigners can also be required to have stayed continuously in the territory of Poland for between 1 and 10 years. Article 195 (4) of the Act on Foreigners sheds light that a stay is considered uninterrupted if none of the breaks therein was longer than 6 months and all breaks did not exceed a total of 10 months in the periods constituting the basis for granting permission for permanent residence, subject to some exceptions.[73]

Migrants in Poland need to file their permanent residence application in person, subject to some exceptions, no later than on the last day of their lawful stay in the territory of Poland.[74] They need to present, as part of the application, documentation justifying the granting of the permit such as information on foreign travel and any stays abroad that occurred within 5 years prior to filing the application and whether they were detained or sentenced to any period of imprisonment.[75] The fee for initiating the procedure for a permanent residence permit is 640 PLN (151 EUR).[76]

By way of comparison, Article 25 of the Bulgarian Act on Foreigners contains provisions for accessing national permanent residence status.[77] There are several categories of foreigners who can become eligible for this permit such as, *inter alia*,

[71] See Articles 8a-8d of the Act on Pole's Card, which were added by Act of 13 May 2016 amending the Act on the Pole's Card and certain other acts/Ustawa z dnia 13 maja 2016 r. o zmianie ustawy o Karcie Polaka oraz niektórych innych ustaw, (Dziennik Ustaw z 2016 r., poz. 753). This provision came into force on 1 January 2017.

[72] Article 195 AF.

[73] For instance, if the break was caused by performance of professional duties or performance of work outside the territory of the Republic of Poland on the basis of an agreement concluded with an employer whose office is located in Poland or a special personal situation required the presence of a foreigner outside the territory of Poland and lasted no longer than 6 months. For more details, see Article 195 (4) 1–4 AF.

[74] Article 202 AF.

[75] For detailed information on the required documents as part of the application, see Article 203 AF.

[76] MigrantInfo.pl (2019).

[77] Including also Articles 25 (a- r) AFRB.

persons of Bulgarian origin,[78] family members of Bulgarian citizens[79] or family members of permanent residence permit holders,[80] investors,[81] and foreigners who have resided, legally and continuously, in the territory of Bulgaria for the last 5 years prior to applying for permanent residence and who have not been abroad for more than 30 months during this period.[82] The latter is the only ground on which migrant workers could qualify for permanent residence, as according to the general admission framework, migrant workers need to leave the country after reaching the maximum permit validity of 3 years and reapply after a 3-month interruption between the expiration of their permit and the request for a new starting period of employment.[83] The requirements for obtaining this permit differ depending on which category of person is applying for the status. For the purposes of this study, this section focuses only on the most relevant categories.

In line with Article 34 (1) of the Implementing Regulation of the Act on Foreigners, applicants for this type of permit need to provide evidence that they have stable, regular, foreseeable, and sufficient means of subsistence without resorting to the social assistance system. The amount of the subsistence must not be less than the minimum monthly salary or the minimum pension for the country. Furthermore, they must exhibit proof of accommodation and present a criminal record certificate from their country of origin or habitual residence, a copy of their passport, and a document confirming that they have paid the state fee of 1000 BGN (500 EUR).[84]

Foreigners of Bulgarian origin who are applying for permanent residence need to fulfil all the conditions stipulated in Article 34 (1) of the Implementing Regulation of the AFRB except the subsistence requirement.[85] They also need to present a birth certificate (Article 35 (1) of the Implementing Regulation of the AFRB).[86] Foreigners who have resided legally and continuously in the territory of Bulgaria for the 5 years

[78] See Article 25 (1) 1 AFRB.

[79] See Article 25 (1) 4 and Article 25г AFRB.

[80] See Article 25 (1) 2 and 3 AFRB.

[81] Article 25 (1) 6, 7, 8, 13 and 16 AFRB.

[82] Article 25 (1) 5. According to Article 25a of the AFRB this permit can also be obtained by foreigners who have contributed to the Republic of Bulgaria in the public and economic sphere, in the sphere of national security, science, technology, culture, or sport, without having to fulfil the requirements of the AFRB.

[83] Chap. 6, Sect. 6.1.1.

[84] In accordance with Article 12 (1) of Tariff No 4 on the fees collected in the system of the Ministry of Interior under the State Fees Act/ Тарифа No 4 за таксите, които се събират в системата на Министерството на вътрешните работи по Закона за държавните такси (SG No. 27/10 March 1998, last amendment SG No. 75/11 September 2018).

[85] In case of such applicants, a notarised declaration by the owner to provide a residence address is required instead of a proof of accommodation.

[86] In the event that the applicant does not have a birth certificate, Ministry of Interior officials can substitute it with official information about the Bulgarian origin of the foreigner if the information is contained in an electronic system maintained by the State Agency for Bulgarians Abroad (Article 35 (2) of the Implementing Regulation of the AFRB).

prior to submitting the application must also present the above-mentioned documents under Article 34 (1) of the Implementing Regulation of the AFRB. In order to be granted the permanent residence permit, the Ministry of Interior's Migration Directorate has to check its databases to confirm that the individual has legally and continuously resided in Bulgaria in the previous 5 years (Article 38a of the Implementing Regulation of the AFRB).

Finally, according to the bilateral labour migration agreements concluded by Bulgaria with Moldova and Armenia, workers are required to return to the territory of their sending state upon expiration of their legal residence and personally present their passport to its consulate within 1 month of their return (Article 23 of the Agreement of Moldova/Article 13 of the Agreement with Armenia). As already mentioned above, this means that they can stay for a maximum period of 3 years, after which they are obliged to leave the country and re-apply for another authorisation.

7.2.2 *Implementation Dynamics*[87]

According to the experts interviewed in Poland, migrants working on the basis of the *Oświadczenie* procedure did not usually face any problems in obtaining a work permit or a single permit with the same employer.[88] One of the experts stressed that proof that this legal possibility was working well was the fact that in the last 2 years (2014–2016) Ukrainians who used to circulate on the basis of work visas, had begun changing their legal status to more stable single temporary permits, which then made them eligible to apply for an EU long-term or permanent residence.[89] This 'massive growth in temporary permits' was mainly due to the political situation in Ukraine at the time. Most of the migrants, who wished to switch to a work or single permit, were fleeing from western Ukraine and sought to ensure that they would be allowed to stay in Poland if the conflict expanded. They were thus more willing to pay far greater fees for the temporary permits and go through the procedure, 'which is not as easy and nice as Oświadczenie', in order to secure a more stable status in Poland.[90]

The study's focus group recruitment strategy did not aim to include migrants who were national permanent residence permit holders. Nevertheless, one of the Russian focus group participants who worked for a business corporation mentioned that she planned to apply for a permanent residence permit in the following year.[91] According to her, the most important prerequisites for obtaining the permit were for

[87] For data on permits, see Chap. 6, Sect. 6.3.3.6.

[88] Interview #12 with civil society actor, Poland, November 2016, Annex II.

[89] Interview #12 with civil society actor, Poland, November 2016, Annex II.

[90] Interview #12 with civil society actor, Poland, November 2016, Annex II.

[91] Focus group with Russian migrants, Poland, November 2016, Annex IV.

the applicant to have a work contract, preferably a permanent one, as well as to demonstrate that the applicant had paid taxes in the previous years. She emphasised that her employer had offered her a permanent contract and paid all the requisite social security contributions, as well as private medical insurance. Her plan was to apply for permanent residence as a step towards gaining Polish citizenship, which would allow her to seek employment at EU institutions where she would receive a much better salary for the same position and tasks: 'That is why permanent residence permit and Polish citizenship is some kind of breakthrough to another level'.[92]

As highlighted in Chap. 6,[93] most of the participants in the general focus groups entered Bulgaria by registering a trade representation and then seeking employment. Obtaining continuous residence on these grounds allowed them to accumulate uninterrupted periods of residence, which would then make them eligible for permanent or EU long-term residence. The challenges that they shared in this regard concerned the circumvention practices that they had to use in order to enter the country and stay, and the fact that it was too risky and too expensive to circulate on a voluntary basis since most of them wanted to stay in Bulgaria in any case.

7.2.3 Residence Status According to EU Long-Term Residence Directive

Foreigners who are not eligible for the Polish permanent residence procedure are able to rely on the EU long-term residence permit procedure. According to Article 211 of the Act on Foreigners, this permit is granted to foreigners at their request for an indefinite period of time, if they have been staying in Poland legally and uninterruptedly for a minimum of 5 years before applying and meet the following criteria: have a source of steady and regular income sufficient to cover subsistence costs for the sponsor and the dependent family members as well as health insurance[94] or a document certifying that the costs of treatment in Poland will be covered by the insurer.

The income requirement is specified in Article 114 (2) of the AF stating that the monthly income shall be higher than the amount which would entitle a person to access cash benefits from the social assistance system, as specified in the Act of 12 March 2004 on Social Assistance[95] with respect to the sponsor and each dependent family member. Foreigners must have maintained steady and regular income for

[92] Focus group with Russian migrants, Poland, November 2016, Annex IV.

[93] Sect. 6.2.

[94] Within the meaning of the Act of 27 August 2004 on health care service financed from public funds/ Ustawa z dnia 27 sierpnia 2004 r. o świadczeniach opieki zdrowotnej finansowanych ze środków publicznych, (Dziennik Ustaw z 2019 r., poz. 1373, consolidated text).

[95] Act of 12 March 2004 on Social Assistance/Ustawa z dnia 12 marca 2004 r. o pomocy społecznej (Dziennik Ustaw z 2019 r. poz. 1507, 1622, 1690, 1818, 2473, consolidated text).

3 years of stay in Poland immediately before filing the application.[96] With regards to Blue Card holders, however, only 2 years of stable income is required.[97] This requirement for steady and regular income of 3 years can be considered disproportional based on the CJEU judgment in the case *Khachab* on the Family Reunification Directive, which can also be applied by way of analogy to the EU Long-term Residence Directive since the wording of the income requirement is almost identical in both directives.[98]

The amendments of the Polish Act on Foreigners, which entered into force on 12 February 2018, made migrants' access to EU long-term residence conditional on knowledge of the Polish language.[99] According to Article 211 (3) of the Act on Foreigners, this can be evidenced through an official certificate of competency in the Polish language at least at the B1 level of language proficiency[100] or an appropriate graduation certificate.[101] Children under 16 are exempted from the language requirement.[102]

The 5-year period of stay in the territory of Poland includes the total period of legal stay in the EU for Blue Card holders, if they have resided in Poland for a minimum of 2 years immediately prior to filing the application[103] or half of the period of legal stay in Poland in cases when they were residing on the basis of visas or temporary residence permits for the purposes of obtaining higher education or vocational training.[104] The foreigners' stay, which is the grounds for granting them the EU long-term residence permit, is considered uninterrupted if none of the periods of absence were longer than 6 months and all the intervals were no longer than 10 months within the required 5-year period.[105] In the case of Blue Card holders, the 5-year period is considered uninterrupted if none of the periods of absence exceed 12 months or where all periods of absence were no longer than 18 months in total.[106]

---

[96] Article 211 (2) 2 AF.

[97] Article 211 (2) 1 AF.

[98] Case C-558/14, Khachab, ECLI:EU:C:2016:285, paras. 41–48.

[99] OECD (2018), p. 264. Currently in Article 211 (1) 3 AF.

[100] Referred to in Article 11a of the Act of 7 October 1999 on the Polish language/Ustawa z dnia 7 października 1999 r. o języku polskim (Dziennik Ustaw z 2019 r. poz. 1480, consolidated text).

[101] A certificate of completion of school in the Republic of Poland within the meaning of Article 2 (2) of the Act of 14 December 2016 on Education/Ustawa z dnia 14 grudnia 2016 r. - Prawo oświatowe (Dziennik Ustaw z 2019 r., poz. 1148, consolidated text) or higher education institutions within the meaning of the Act of 20 July 2018 on Higher Education and Science/Ustawa z dnia 20 lipca 2018 r. - Prawo o szkolnictwie wyższym i nauce (Dziennik Ustaw z 2018 poz. 1668) with Polish as the language of instruction; or a certificate of completion of a school or higher education institution with Polish as the language of instruction abroad, corresponding to the school or higher education institution within the meaning of Article 2(2) of the Act of 14 December 2016 on Education or the Act of 20 July 2018 on Higher Education and Science.

[102] Article 211 (4) AF.

[103] Article 212 (1) 1 AF.

[104] Article 212 (1) 3 AF.

[105] Article 212 (3) 1 AF.

[106] Article 212 (3) 2 AF.

These rules do not apply, however, when the intervals were due to the performance of professional duties or work outside Poland under an agreement with an employer established on the territory of Poland; where family members accompanied a foreigner performing the abovementioned activities; in cases of exceptional personal situations requiring the foreigner's presence outside Poland for up to 6 months; or, for internships or classes provided for in the course of studies by a Polish university.[107]

Foreigners who have stayed legally and without interruption in the territory of Bulgaria for 5 years are also eligible to apply for the EU long-term residence permit in accordance with Article 24г of the Act on Foreigners. In order to obtain such residence status, the applicants need to present: evidence that they have sufficient means of subsistence for themselves and their family members that is not less than the minimal salary or minimal pension, which would not require recourse to the social assistance system (Article 24г (9) of the AFRB). Furthermore, they must, as a matter of obligation, possess health insurance or insurance for the period of their stay in accordance with the applicable Bulgarian legislation.

For those who have resided exclusively on the basis of a temporary permit, such as au pairs, seasonal workers, cross-border service providers, posted workers for the purposes of cross border service provision, or where their residence permit is formally limited, they cannot have these periods of residence included in the calculation of the 5-year period (Article 24г (2) of the AFRB). EU Blue Card holders can be granted EU long-term residence status in Bulgaria if they have resided legally and uninterruptedly for a period of 5 years in a Member State as an EU Blue Card holder; two of the five last years must have been in the territory of Bulgaria (Article 33м (1) of the AFRB).

In accordance with Article 24г (8) of the Bulgarian Act on Foreigners, periods of absence from the territory of Bulgaria do not interrupt the required 5-year period and can be included in the calculation, if they are less than six consecutive months and do not exceed a total of 10 months for the 5-year period. For Blue Card holders, the 5-year period is not to be considered interrupted by periods of absence from the territory of Bulgaria, if those periods are less than 12 consecutive months and the total duration does not exceed 18 months within the 5-year period. These periods of absence must be necessary for the exercise of an economic activity as either an employed or self-employed person; due to volunteer activity; or for education purposes in the Blue Card holder's country of origin (Article 33м (2) AFRB).

7.2.4 Implementation Dynamics

The migrants in Poland with high-skilled and low-skilled occupations shared different experiences regarding the process of obtaining an EU long-term residence permit. A Ukrainian having a high-skilled occupation who participated in the focus

[107] Article 212 (4) AF.

group did not report any problems concerning the application process for the permit.[108] What she complained about was the constantly evolving legislation, which made it very hard to make an informed decision.[109]

A Russian participant in the focus groups shared that he could not choose between the EU and the national permit because he did not have 'Polish roots' nor had he done 'anything outstanding for Poland'. Therefore, he had to apply for the EU long-term residence permit, and remarked that it was more difficult to obtain this permit. The focus group participant stressed that the governor examined his monthly income from the previous 3 years on the basis of the submitted tax return declaration (*PIT*). His income could not be less than 468 złoty (110.50 EUR) per family member and at the beginning of the process his salary had been slightly less, which rendered him ineligible for the permit.

The Russian migrant worker said that the main problem with obtaining this permit was the dependency on the employer. For instance, if the employer did not pay the required taxes or social security contributions for the worker, even for a period of 2 months, this would cause the applicant to fail the permit procedure. However, in most cases, employers unlawfully demanded that migrants performing low-skilled jobs in restaurants or in the construction sector pay all social security contributions (*ZUS)* themselves.[110] The respondent emphasised that he had been working and paying social security contributions on his own for 3 years and this is how he managed to obtain an EU long-term residence status. He added: 'If it is a low-skilled work, not in a corporation, employers won't hire you or you will be required to pay them back ZUS'.

The requirement for continuous employment (*nieprzerwane zatrudnienie*) was another hurdle for those wishing to obtain the EU long-term residence permit.[111] If migrant workers became unemployed, they had to report this change in their status to the governor and were given only 1 month to find a new job. So, one could have worked for several years, striving to become eligible for an EU long-term residence permit, but if finding a new job took more than 1 month, one became unemployed and could not subsequently obtain this status. The respondent also stressed that the transposition of the Single Permit Directive had led to greater dependence on the employer due to the bureaucratic procedure for changing employment and that migrants who wished to apply for an EU long-term residence permit preferred to stay with the same employer rather than risk changing their permits and ending up with a period of unemployment.[112]

When the Russian participant in the focus group in Poland was asked whether he had circulated at any point during the 5-year period required for an EU long-term

[108] Focus group with Ukrainian migrants, Poland, November 2016, Annex IV.

[109] The participant referred to a period of more than 20 years since she was living in Poland.

[110] Migrants were paying the amount of social security contributions based on their income to employers who made the bank transfer and paperwork. Focus group with Russian migrants, Poland, November 2016, Annex IV.

[111] Focus group with Russian migrants, Poland, November 2016, Annex IV.

[112] See also Sect. 7.1.4.6 for more details.

residence status, he said that he was aware of the limits to absence from Polish territory and therefore had deliberately remained in Poland for most of the time.[113] Some of the interviewed Blue Card holders also shared this concern. Even though none of them were eligible to apply for an EU long-term residence permit at the time of the focus groups because they had only been in Poland for one-and-a-half or two years, most of them envisaged applying for this permit at some point in the future. According to the IT company management representatives, most of their employees were interested in obtaining an EU long-term residence permit in order to reduce the bureaucracy surrounding the renewal and extension of the Blue Card.[114] Furthermore, from the management's perspective, it was more convenient if workers had a more permanent and stable status because this would allow them to travel and move with greater ease to different locations as the company's needs might require. Therefore, the IT company was investing in retaining the Blue Card holders in the country by offering free Polish language classes to the Blue Card holders and their spouses, support for spouses seeking a job, and introducing a simplified procedure for obtaining a mortgage.

By way of contrast, most migrant workers in Bulgaria had difficulties accessing an EU long-term residence status. The implementation of the Single Permit Directive created challenges for migrant workers, especially in relation to Bulgaria's migration policy focus of keeping migrants in a temporary position. As already mentioned, in accordance with Article 15 (3) of the ALMLM, the overall duration of the work authorisation can be prolonged by up to 3 years if the circumstances for its issue have not changed. After this period, the employee needs to leave the country. As one of the interviewed officials said:

> 'Then [after the three-year period] he has to leave. (…) Here the legislator for me quite consciously, has foreseen this interruption so that a period for permanent residence cannot be accumulated'.[115]

Indeed, this requirement interrupts the possibility for migrant workers to extend their continuous residence permit for another year, which needs to be submitted within 14 days before the permit is due to expire. Since migrant workers can have work authorisation for only up to 3 years, unless they are Blue Card holders or workers of Bulgarian origin, they do not have grounds for applying to extend their permit. Furthermore, they cannot change the grounds on which they entered Bulgaria from inside the country, and in any case need to leave if they would like to return to Bulgaria to work for the same employer or enter on a different basis.[116] As already mentioned, this practice of 'forced circulation' creates obstacles for migrant workers in their transition to a permanent status and as a result forces them to

---

[113] Focus group with Russian migrants, Poland, November 2016, Annex IV.

[114] Interview with IT business management representative, Poland, December 2016, Annex II.

[115] Interview #2 with official, Bulgaria, July 2016, Annex III.

[116] A proposal for amendment of the AFRB is pending at the time of finalisation of this text for publication. It could partially remedy this 'forced circularity' practice. See Draft Act on Amendments to the Act on Foreigners in the Republic of Bulgaria from 22 February 2020, Para 7, Article 24и (3).

undertake circumvention practices in order to find other ways to reside and settle in Bulgaria. Furthermore, as one of the interviewed lawyers said:

> 'And by doing all this, by throwing the worker out of here and sending him back to his country of origin, after all these costs he made, let's agree, that he would not come back here again. He will go somewhere where he does not have to do this thing. Not to mention that this demotivates people, because they do not know whether they will get a visa D or not (...) because the chances of not getting it are very big. (...) No worthy specialist with a worthy amount of money will subject himself to these administrative humiliations or obstacles to have the honour of working for Bulgaria'.[117]

According to IT business representatives, the 'forced circulation' practice that is inherent in Bulgarian migration law did not conform with the company's policy of retaining their employees.[118] As stressed by one of the interviewees, the only privileged group of foreigners in Bulgaria were the Blue Card holders because they were not required to leave the country after their work authorisation period expired. However, the business representatives interviewed in 2016 stated that they were concerned whether this would be the case in practice because at the time of the interviews they did not have any Blue Card holders who had reached the threshold for permit renewal.[119] The explicit grounds for extending the validity of the Blue Card permit for up to 4 years and its renewal within the country were only introduced in 2017 with the amendments to the Act on Foreigners.[120]

7.3 Social Security Coordination[121]

Circular migrants contribute at different times to the social security and health systems of two countries, raising the question of what happens with the contributions accumulated during the circulation period and after migrants decide to settle. The study's benchmarks in the field of social security coordination aim to assess what kinds of benefits can be exported and whether the general principles of social security coordination are covered: maintenance of the acquired rights and rights in course of acquisition; totalisation of periods of insurance, employment, or residence and of assimilated periods for the purpose of the acquisition, maintenance, or recovery of rights and for the calculation of benefits; and, equality of treatment.[122] It also

[117] Interview #9 with lawyer, Bulgaria, July 2016, Annex III.

[118] Interview # 20 with IT business representatives, Bulgaria, December 2016, Annex III.

[119] Interview # 20 with IT business representatives, Bulgaria, December 2016, Annex III.

[120] See Chap. 6, Sect. 6.3.3.1 for more details.

[121] This section builds on Vankova (2018),

[122] Based on, among others, Equality of Treatment (Social Security) Convention, C118, adopted at 46th ILC session on 28 June 1962, entry into force on 25 April 1964; Equality of Treatment (Accident Compensation) Convention, C019, adopted at 7th ILC session on 05 June 1925, entry into force on 8 September 1926. See Annex V for details. For more details, see Vankova (2016), p. 338.

aims to assess whether reimbursement of social security contributions is possible[123] and whether there are any instruments that support the implementation of these standards, such as multilateral and bilateral agreements.

7.3.1 Social Security Coordination According to National Instruments in Poland and Bulgaria

Poland has concluded bilateral social security agreements with a total of eight countries, two of which have been signed with Eastern Partnership countries – Ukraine and Moldova.[124] Bulgaria, on the other hand, has concluded a total of 14 bilateral agreements. As in the case of Poland, two of the agreements were concluded with Eastern Partnership countries – Ukraine and Moldova. In addition, Bulgaria has also signed an agreement with Russia[125] and at the time of writing was negotiating an agreement with Azerbaijan.[126]

The interviewed Polish ministry official shared that the agreements concluded by Poland were always based on the fundamental principles of coordination of social security systems contained in Regulation No. 883/2004:[127] the principle of equal treatment; the principle of the export of benefits; and, the principle of the aggregation of insurance periods.[128] By way of comparison, the Bulgarian institutions use the Model Provisions for a Bilateral Social Security Agreement contained in the European Convention on Social Security when they are beginning the negotiation of an agreement.[129] Therefore the texts of the bilateral agreements concluded by both countries reflect all basic principles in social security coordination.[130]

Furthermore, as already mentioned, the Bulgarian government's policy is to conclude bilateral labour migration agreements with countries with which there is a social security coordination agreement or with which such an instrument is under

---

[123] In line with Article 27 (2) of ICRMW and Article 9 (1) of ILO Convention No. 143.

[124] The rest are with Yugoslavia (currently refers to: Bosnia and Herzegovina, Serbia and Montenegro), North Macedonia, Canada, USA, Republic of Korea, and Australia. It should be noted there is also an additional agreement with the Government of Quebec. For more details, see https://www.gov.pl/web/rodzina/dsz-wykaz-umow-o-zabezpieczeniu-spolecznym

[125] The rest are with Yugoslavia (currently refers to: Bosnia and Herzegovina), Serbia, Albania, Libya, Turkey, North Macedonia, Israel, Korea, Canada, Montenegro, and Tunisia.

[126] Interview #4 with official, Bulgaria, July 2016, Annex III. See also official information on the website of the Bulgarian Ministry of Labour and Social policy. https://www.mlsp.government.bg/index.php?section=CONTENT&I=267 Accessed 30 September 2019. For the choice of countries, see Vankova (2018), p. 191.

[127] Regulation (EC) No 883/2004 of the European Parliament and of the Council of 29 April 2004 on the coordination of social security systems [2004] OJ L 166.

[128] Interview #13 with official, Poland, December 2016, Annex II.

[129] Interview #15 with official, Bulgaria, July 2016, Annex III.

[130] Vankova (2018), p. 194.

negotiation. For instance, the bilateral labour migration agreement with Moldova is coupled with a social security agreement that was concluded in 2008 (see Article 12 of the Agreement with Moldova).[131] In line with the Bulgarian policy on social security coordination, the bilateral labour migration agreement with Armenia provides that 'until a bilateral social security agreement is concluded, the relevant legislation of the host party is applied' (Article 11). It should be stressed that Bulgaria primarily utilises the GAMM to initiate negotiations for the conclusion of bilateral social security coordination agreements.[132]

In most of the cases, the personal scope of the agreements between Poland and Bulgaria and the Eastern Partnership countries covers all persons who are or have been subject to the legislation of one or both contracting parties, as well as other persons who derive rights from such persons. The agreements between Bulgaria and Russia[133] and Bulgaria and Ukraine,[134] however, have a very limited personal scope and concern, respectively, only Russian, Ukrainian, and Bulgarian citizens.

The material scope of the two agreements with Poland cover old-age pensions, disability pensions, pensions in respect of accident at work and occupational diseases, and survivors' pensions, which are also exportable benefits.[135] In addition, the agreement between Ukraine and Poland has broader material scope that includes unemployment benefits, maternity and paternity benefits, sickness cash benefits, and death grants.[136] This agreement allows for export of maternity and paternity benefits and sickness cash benefits.[137]

Most of the social security agreements that Bulgaria has concluded with Eastern Partnership countries exclude family benefits and healthcare benefits. The agreements concluded with Ukraine, Moldova, and Russia cover sickness cash benefits, including maternity benefits, old-age pensions, disability pensions, pensions in respect of accident at work and occupational diseases, survivors' pensions, and death grants. In addition, the agreements with Moldova and Ukraine include unemployment benefits in their material scope, while the agreement with Russia includes family benefits. The agreements with Eastern Partnership countries that Bulgaria has concluded cover mainly export of pensions.[138]

---

[131] Agreement between the Republic of Bulgaria and the Republic of Moldova on social security, signed on 5 December 2008, in force since 1 September 2009/Договор между Република България и Република Молдова за социално осигуряване, подписан на 5 декември 2008 г., в сила от 1 септември 2009 г.

[132] Interview #1 with official, Bulgaria, July 2016, Annex.

[133] Article 3.

[134] See Article 3, Article (1) 1 paragraphs 4 and 6.

[135] Agreement between the Republic of Poland and the Republic of Moldova on Social Insurance, Article 2 and Article 5. Agreement between the Republic of Poland and the Republic of Ukraine on social security, Article 2 and Article 5.

[136] Article 2 of the Agreement.

[137] See European Migration Network (2014a).

[138] For more details see, European Migration Network (2014b).

According to Article 132 of the Polish Act of 17 December 1998 on Old-Age and Disability Pensions from the Social Insurance Fund,[139] even when there is no bilateral social security agreement with a particular third country, the payment of old-age and disability pensions is still possible in Poland: 'At the request of a retiree or pensioner living abroad the pension shall be received by a person authorised to receive the pension, who is domiciled in Poland or into the account of the retiree or pensioner in his/her country, unless international treaties provide otherwise'.[140] In practice this means that Polish pensions are exported to the pensioner's state of residence only if Poland has a bilateral social security agreement with that state. However, when there is no such agreement, the pensions can be paid in Poland to a person authorised by the pensioner and who resides in Poland, or to the pensioner's bank account in Poland. In cases where there is no bilateral social security agreement between Bulgaria and a third country, the Bulgarian Ordinance on Pensions and Insurance Periods[141] does not provide for the payment of pensions abroad, which was also stressed by the European Commission in its Implementation report on the Single Permit Directive.[142] Yet, according to one interviewed official, in practice pensions are transferred via the post office and the pensioners' bank accounts.[143] Thus, the person can live abroad and receive a Bulgarian pension as long as they retain a bank account in Bulgaria.

Finally, there are no legal provisions in Bulgarian or Polish social security law or any of the concluded bilateral social security agreements with the Eastern Partnership countries in question stipulating the possibility for social security contributions to be reimbursed when these are not accessible to the migrants.

7.3.2 Implementation Dynamics[144]

The data on the exported Polish pensions to Ukraine and Moldova demonstrates the marginal use of these agreements, especially when one compares them to the number of migrants working in Poland[145] or the number of foreigners registered as insured for pension at the Polish Social Security Institution.[146] The number of

[139] Ustawa z dnia 17 grudnia 1998 r. o emeryturach i rentach z Funduszu Ubezpieczeń Społecznych, (Dziennik Ustaw z 2018, poz. 1270, consolidated text).

[140] Interview #13 with official, Poland, December 2016, Annex II.

[141] Наредба за пенсиите и осигурителния стаж, (SG No. 21/17 March 2000, last amendment SG No. 62/6 August 2019)

[142] European Commission (2019), pp. 10–11.

[143] Interview #15 with official, Bulgaria, July 2016, Annex III.

[144] This section builds on Vankova (2018).

[145] See Chap. 6, Sect. 6.2.

[146] In 2018, there were 425,670 Ukrainians and 7337 Moldovans registered in ZUS. Data from Departament statystyki i prognoz aktuarialnych, Cudzoziemcy w polskim systemie ubezpieczeń społecznych, Warszawa 2019. https://www.zus.pl/documents/10182/2322024/Cudzoziemcy+w+polskim+systemie+ubezpiecze%C5%84+spo%C5%82ecznych.pdf/4498fca6-981d-a37c-3742-8e4e74e20a32#page=8 Accessed 19 November 2019.

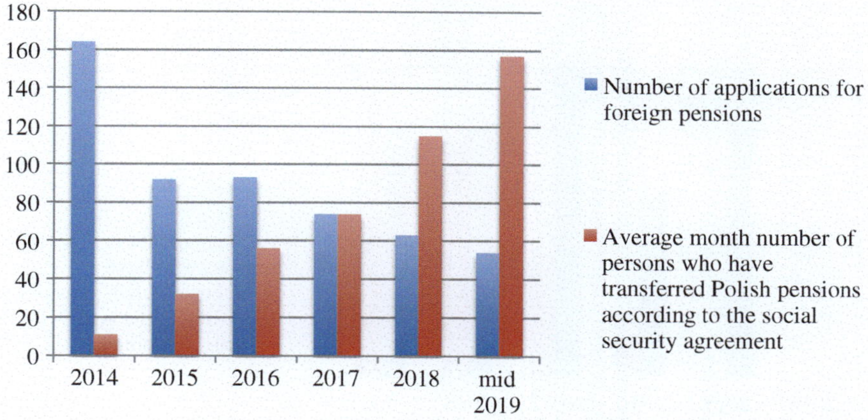

Fig. 7.1 Data on the implementation of the social security agreement between Poland and Ukraine. Source: Author's own elaboration on the basis of data of the Ministry of Family, Labour and Social Policy, Poland. (Data retrieved from an interviewed official at the Ministry of Labour and Social Policy, Poland, 7 August 2019)

persons who have transferred Polish pensions under the social security agreement with Ukraine (on the basis of a monthly average), doubled between 2017 and mid-2019, reaching a total of 157 persons in mid-2019. In addition, only five people had transferred their pensions to Moldova by mid-2019.[147] (Figs. 7.1 and 7.2).

Even though the data obtained from the Bulgarian and Polish institutions differs and cannot be directly compared, it still offers a good insight into the implementation dynamics of the social security agreements with the Eastern Partnership countries and Russia. As of mid-2019, 50 Bulgarian pensions were transferred to Ukraine on the basis of the bilateral agreement to Bulgarians who had worked both in Bulgaria and Ukraine (see Fig. 7.3).[148] On the other hand, 936 Ukrainian pensions were being paid in Bulgaria in the same reporting period. As Fig. 7.5 demonstrates, the implementation of the agreement between Moldova and Bulgaria is insignificant.

Against this background, the implementation dynamics of the agreement between Russia and Bulgaria differs drastically. The number of Russian pensions that are paid in Bulgaria as of mid-2019 is 5451 and the figure for Bulgarian pensions paid in Russia is 185 (see Fig. 7.4). One interviewed official said that in both cases the beneficiaries were mainly Russian citizens.[149] In addition, the interviewee shared that it was quite challenging to keep the data up to date because new pensioners were added or removed from the lists every month due to deaths that took place.

[147] Interview #13 with official, Poland, December 2016, Annex II.

[148] Data obtained from the National Social Security Institute in August 2019.

[149] Ibid.

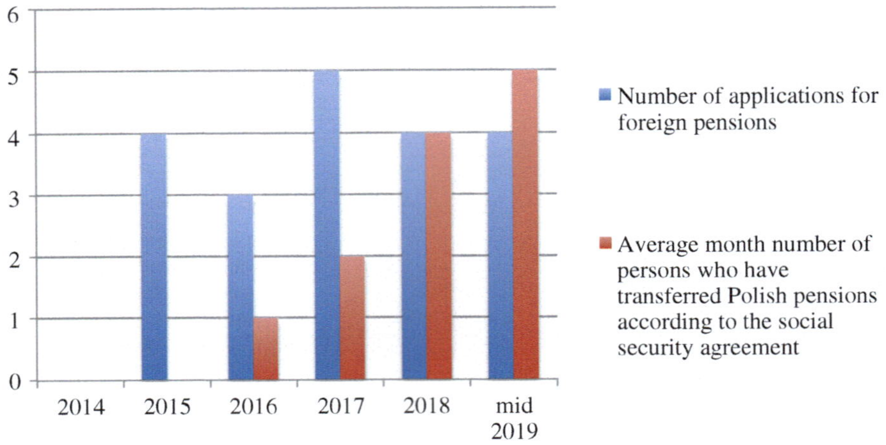

Fig. 7.2 Data on the implementation of the social security agreement between Poland and Moldova. Source: Author's own elaboration on the basis of data of the Ministry of Family, Labour and Social Policy, Poland

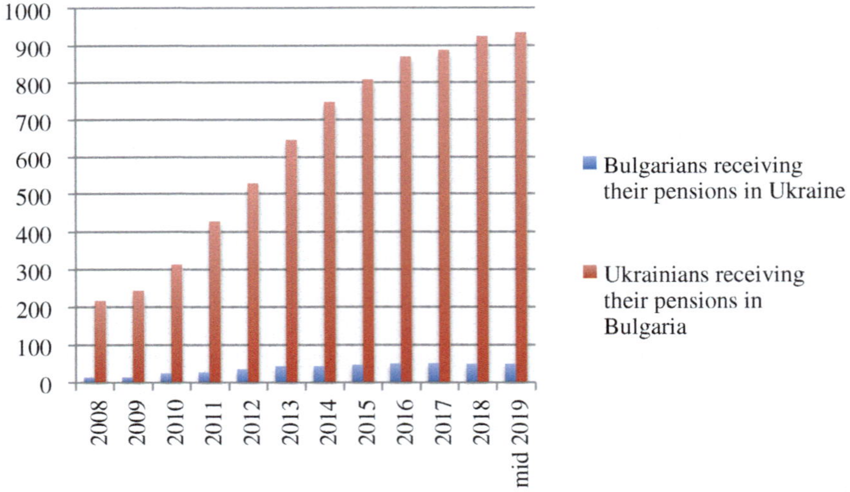

Fig. 7.3 Data on the implementation of the social security agreement between Bulgaria and Ukraine. Source: Author's own elaboration on the basis of data of the National Social Security Institute, Bulgaria

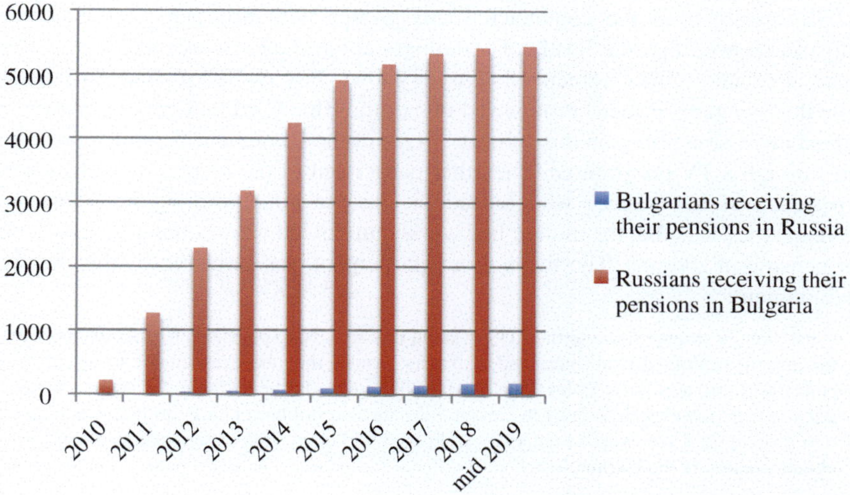

Fig. 7.4 Data on the implementation of the social security agreement between Bulgaria and Russia. Source: Author's own elaboration on the basis of data of the National Social Security Institute, Bulgaria

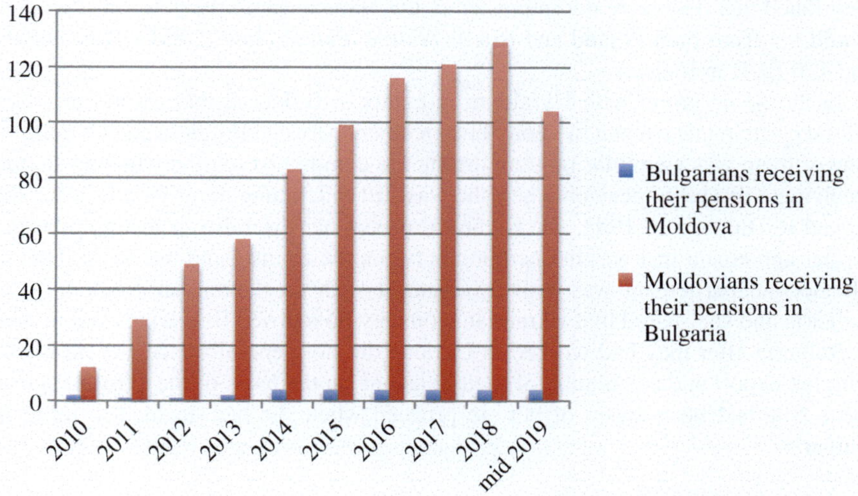

Fig. 7.5 Data on the implementation of the social security agreement between Bulgaria and Moldova. Source: Author's own elaboration on the basis of data of the National Social Security Institute, Bulgaria

The results from the conducted focus groups with migrants in Poland and Bulgaria showed that the levels of awareness about the existence and aims of the bilateral social security agreements are very low. Only two Ukrainian participants from the two general focus groups and two of the Blue Card holders interviewed in Poland knew something on this subject.[150] One of the Ukrainians had informed himself through a TV programme. The other participant in the focus group shared the experience of her mother who retired in Ukraine after working in Poland for 10 years; the years that the mother had worked in Poland were counted as a basis for her pension in Ukraine. The focus group participant was surprised by this positive development:

> 'That's really strange because the ZUS (Zakład Ubezpieczeń Społecznych, Social Security Institution) in Poland is all mess and, as far as I know, they are even unable to calculate pensions transparently for Poles, let alone for Ukrainians. In the Ukraine, for its part, I can tell you from my experience that the institutions are also stubborn in accepting documentation proving that you worked ten years in Poland and count this experience as a basis to obtain pension in the Ukraine'.

One of the interviewed lawyers from an NGO shared that since the conclusion of the agreement with Ukraine, they had very few social security cases and when they did, these mostly concerned problems with migrant workers who wanted to retire and were having trouble collecting all the documents needed to be able to benefit from the provisions of the bilateral agreement.[151]

Most of the Blue Card holders who participated in the focus groups in Poland mentioned that they were not interested in this issue because the pensions that they would get from both Poland and Russia were extremely low (250 EUR in Poland and 170 EUR in Russia).

In the focus group with Ukrainian migrants in Bulgaria, there were only two retired participants who knew about the agreement between Bulgaria and Ukraine.[152] One of them was a circular migrant during the communist regime who had started receiving a disability pension when she was still in Ukraine. Four months after she moved to Bulgaria in 1988, she began to receive her disability pension under the former agreement that was in operation between the country and the Soviet Union. The second participant was actively engaged with the Ukrainian community in Bulgaria and she shared that Ukrainian women who married Bulgarians and moved to Bulgaria after they had worked in Ukraine did not generally face any problems with the export and calculation of their pensions on the basis of the existing agreement. She had been aware of how to proceed when she had decided to retire in Bulgaria.

---

[150] Focus groups with Russian migrants and Ukrainian migrants, Poland, November 2016, Annex IV. Focus group with Blue Card holders from Ukraine, Poland, December 2016, Annex IV. Focus group with Blue Card holders from Russia, Poland, December 2016, Annex IV.

[151] Interview #13 with civil society actor, Poland, November 2016, Annex II.

[152] Focus groups with Ukrainian migrants, Sofia, September 2016, Annex IV.

There was only one pensioner in the focus groups conducted with Russian migrants.[153] She knew about the bilateral social security agreement between Russia and Bulgaria, and said that she received pensions from both countries and that she preferred to receive the Russian pension in a Russian bank because she often travelled to Russia to visit friends and family.

In Bulgaria, only two Blue Card holders who participated in the focus group with Russians knew that there was an agreement between Bulgaria and Russia.[154] They would only associate this agreement with the payment of pensions, something that was not currently on their agenda. They did not know the details of this agreement, however. Their main concern was access to healthcare. As migrants, they were entitled to medical aid only in emergency cases, and this was funded by the state budget.[155] Apart from that, unless they held EU long-term residence or permanent residence permits, they could not use the healthcare provided by the state, benefit from the Blue Card Directive or the bilateral social security agreement between Bulgaria and Russia because this was not part of the material scope of these instruments. Therefore, they all used private health insurance.

The focus group data show that mainly people who were retired or had retired relatives were aware of the existence of the concluded bilateral social security agreements. One of the factors contributing to the low awareness among the interviewed Blue Card holders was the fact that they had moved to Poland and Bulgaria relatively recently, on average between one-and-a-half and two years before the interviews were conducted. In addition, many were young people at the beginning of their careers. Finally, most of the Russian Blue Card holders did not want to return to Russia and some stressed that they did not want to have anything to do with the Russian authorities.

7.4 Conclusion

When Poland and Bulgaria's circular migration approaches are analysed and compared beyond entry and re-entry conditions for migrants, the implementation picture of the EU's approach to facilitating this type of migration becomes even more nuanced. This chapter focused on instruments considered conducive to circular migration and aimed to assess them against the study's benchmarks developed in the areas of work authorisation, residence status, and social security coordination (see Annex V).

In the field of work authorisation, Poland facilitates circular migration through both national and EU instruments. The national instruments differ in their scope insofar as they explicitly allow for a change of employer and occupation, as well as

[153] Focus groups with Russian migrants, Sofia, September 2016, Annex IV.

[154] Focus group with Blue Card holders from Russia, Bulgaria, September 2016, Annex IV.

[155] European Migration Network (2014b), p. 22. See also Huddleston et al. (2015).

the implications of unemployment on the permit status. In the Bulgarian case, however, the national regime does not contain any of the international standards employed as benchmarks of this study. Unemployment is treated as a change in migrant workers' admission grounds and leads to 'forced circularity'. The only exceptions are those introduced through EU law and for foreigners of Bulgarian origin.

Furthermore, the empirical data analysis suggests certain problems with regards to the protection of migrants' rights in both countries. In Poland this is mainly in the context of the *Oświadczenie* procedure, which despite the recent reforms still creates legal uncertainty and insufficient enforcement when it comes to obtaining employment contracts and thus places migrant workers in a vulnerable position. In Bulgaria, the 'forced circulation' logic, combined with the lack of flexible solutions to change employer and occupation as well as the risk of becoming unemployed, essentially binds migrant workers to their employers.

While transposing the EU labour migration *acquis*, both countries have established bureaucratic and cumbersome procedures, which in most cases lead to impractical solutions. For instance, the implementation of the Single Permit Directive in Poland has led to workers being bound to their employer because the transposition into Polish law has made it extremely difficult for those workers to change employer or occupation. The transposition of EU labour migration law in Bulgaria, however, is the only way through which standards in this field can become embedded in the Bulgarian legal order. Thus, it is currently only Blue Card holders and seasonal work permit holders who, on the basis of EU law, can benefit from the right to change employer. Blue Card holders can also stay in the country if they become unemployed. Neither Poland nor Bulgaria have adopted measures that allow researchers to change employer nor do they provide the possibility for researchers to find another job if they become unemployed.

When it comes to access to a more settled residence status, both case studies demonstrate that, in general, circular migration and the accumulation of continuous residence periods, which in most cases are needed in order to obtain an EU long-term residence or permanent residence status, are mutually exclusive. Migrant workers who are engaged in circulation need to either change their migration trajectory or utilise the limited permitted absences available under the EU and national permits. The empirical data analysis demonstrates that migrants who have decided to obtain an EU long-term or permanent residence status prefer not to engage in circulation for work purposes because this is perceived as being too risky.

The two case studies, however, differ regarding access to permits that allow their holders to transition from a temporary to a more permanent status. The Polish model provides some flexibility for circular migrants to transit to a single permit. However, the analysis of the implementation data in Poland highlighted a problem with the secondary transition from a single permit to a permanent or an EU long-term residence permit. The increasingly demanding requirements imposed as a result of the transposition of the EU Long-term Residence Directive into Polish law, as well as the cumbersome procedure for its implementation, make it very difficult to actually access this type of permit, especially for migrants in low-skilled occupations. Migrants shared that the main challenges while on a single permit were related to their dependency on employers – who very often required migrants in low-skilled occupations to pay their own social security contributions – as well as to the requirement for continuous employment.

By way of contrast, generally migrant workers in Bulgaria cannot transfer from one status to another, at least not while they are in the country. Furthermore, Bulgarian migration law, which establishes a practice of 'forced circulation', currently renders access to permanent residence status challenging and to EU long-term residence practically impossible for migrant workers, except for foreigners who can prove that they are Blue Card holders. This is not in line with the international standards deployed by the current study as benchmarks (see Annex V). Therefore, the circumvention practices that were documented in Bulgaria (e.g., registration of trade representation) in many cases are the only means for migrant workers to access a more secure status in Bulgaria.

In both countries, access to permanent residence status is facilitated only for certain privileged groups of migrants, which both countries try to attract based on their ethnic origin. In addition, all the instruments analysed allowed for free movement and a choice of residence within Bulgaria and Poland. None of the interviewed migrants reported any problems in this regard.

Finally, when it comes to social security coordination, this chapter has demonstrated that the bilateral agreements concluded by both Bulgaria and Poland with the Eastern Partnership countries cover all the basic social security coordination principles that served as benchmarks in this study. However, neither country has legislated for the reimbursement of social security contributions in cases where migrants and their families cannot access benefits due to waiting periods in the host country or the lack of a bilateral agreement that provides for the export of benefits. This is a provision contained in Article 27 (2) of the Convention on the Protection of the Rights of all Migrant Workers and the Members of Their Family. This standard should be especially considered in the context of the short-term circular migration of seasonal workers.[156]

With regards to their material scope, all agreements between Bulgaria and Poland and the Eastern Partnership cover pensions – old age, disability pensions, pensions in respect of accidents at work and occupational diseases, and survivors' pensions. Apart from this, the agreements with the different countries vary in other areas. Family benefits, for instance, are covered only in the agreement between Russia and Bulgaria. In addition, it seems that unemployment benefits are an important issue for Ukraine because it is foreseen in the bilateral agreements with both Poland and Bulgaria. Shedding more light on the reasons for that, however, would require additional field research covering interviews with delegation members who were part of the negotiations of these agreements.

The agreements between Bulgaria and Russia and Bulgaria and Ukraine demonstrate another potential source of vulnerability for circular migrants. Both have a very narrow personal scope insofar as they cover only nationals of the contracting

[156] A recommendation in this regard is also expressed by Verschueren (2016), p. 407.

parties and can exclude third-country nationals from benefiting from the provisions of the bilateral agreements and create additional gaps in a field that is already plagued with inconsistencies.

This study has demonstrated that the number of bilateral agreements, which are the main instruments used in the field of social security coordination, is low despite the cultural, historical, and geographic proximity of Bulgaria and Poland with the countries in the Eastern Partnership. It seems that the policy channels provided by the GAMM do not do enough to facilitate the process of starting negotiations with these countries, even though Bulgaria is actively using it.

Another problem that has been identified in both countries pertains to the implementation of these instruments. The lack of awareness among the migrants interviewed, as well as the low number of beneficiaries of the agreements, especially in the case of Poland, means that there is a need for an active information policy on these issues, which are technical and not easily understandable to migrants who do not necessarily have a good knowledge of the local language. Migrants need to be actively informed about their rights during their period of circulation. Waiting until the age of retirement could lead to the loss of entitlement or problems created by the lack of documents that need to be presented to the respective social security institution.

References

Deloitte. (2018). ICT Permit Study. *Facilitating EU Mobility for Third-country Nationals*, June 2018. https://www2.deloitte.com/content/dam/Deloitte/global/Documents/Tax/dttl-tax-ict-permit-study.pdf Accessed 19 December 2019.

European Commission. (2019). Report on Directive 2011/98/EU of the European Parliament and of the Council of 13 December 2011 on a single application procedure for a single permit for third-country nationals to reside and work in the territory of a Member State and on a common set of rights for third-country workers legally residing in a Member State, COM (2019) 160 final, 29 March 2019, Brussels.

European Migration Network. (2012). Visa policy as migration channel in Poland. National Report. Prepared by the National Contact Point to the European Migration Network, Warsaw.

European Migration Network. (2014a). Migrant access to social security in Poland. Report produced by National Contact Point to the European Migration Network in Poland, Warsaw. https://ec.europa.eu/home-affairs/sites/homeaffairs/files/what-we-do/networks/european_migration_network/reports/docs/emn-studies/21a_poland_national_report_social_security_en.pdf Accessed 19 April 2020.

European Migration Network. (2014b). Migrant access to social security and healthcare: policies and practice in Bulgaria. Ministry of Interior. https://ec.europa.eu/home-affairs/sites/homeaffairs/files/what-we-do/networks/european_migration_network/reports/docs/emn-studies/illegally-resident/03.bulgaria_national_report_social_security_en_version.pdf Accessed 19 April 2020.

Executive Agency Main Labour Inspectorate. (2019). *Activity Report of the Executive Agency Main Labour Inspectorate in 2018*, June 2019, Sofia.

Home Office. (2016). Country policy and information note. Ukraine: Military Service. Version 3.0, November 2016. https://www.refworld.org/pdfid/5890825c4.pdf Accessed 29 April 2020.

Huddleston, T., Bilgili, Ö., Joki, A., & Vankova, Z. (2015). Migrant Integration Policy Index (MIPEX). CIDOB and MPG, Brussels/Barcelona. http://www.mipex.eu/bulgaria Accessed 15 May 2015.

MigrantInfo.pl. (2019). Permanent residence permit. http://www.migrant.info.pl/permanent-residence-permit.html Accessed 19 December 2019.

OECD. (2018). *International immigration outlook 2018: Poland*. Paris: OECD Publishing.

Pawlak, M. and I. Lashchuk. (forthcoming). Entry to the labour market, not to the state - migrant workers' perspectives in Poland. Report commissioned by the Friedrich Ebert Foundation.

Skeldon, R. (2012). Going round in circles: Circular migration, poverty alleviation and marginality. *International Migration, 50*(3), 43–60.

Unterschütz, J. (2016). National report on implementation of EU migration directives in Poland. In R. Blanpain, F. Hendrickx, & P. Olsson (Eds.), *National effects of the implementation of EU directives on labour migration from third countries* (pp. 161–192). Alphen aan den Rijn: Wolters Kluwer.

Vankova, Z. (2016). EU circular migration policies: Dead or alive? Developing a rights-based benchmark framework for policy assessment. *Journal of Immigration, Asylum and Nationality Law, 30*(4), 332–352.

Vankova, Z. (2018). Poland and Bulgaria's bilateral agreements with Eastern partnership countries in the context of circular migration. *European Journal of Social Security, 20*(2), 188–203.

Vankova, Z. and B. Ivanova (forthcoming) Temporary home or final destination? Migrant workers' perspectives in Bulgaria. Report commissioned by the Friedrich Ebert Foundation.

Verschueren, H. (2016). Employment and social security rights of third-country labour migrants under EU law: An incomplete patchwork of legal protection. *European Journal of Migration and Law, 18*(4), 373–408.

Chapter 8
Flanking Rights in the Context of Circular Migration: Entry and Residence Conditions for Family Members and Recognition of Qualifications for Migrant Workers in Bulgaria and Poland

This chapter focuses on entry conditions for family members and recognition of qualifications that are considered secondary policy areas, yet could still influence migrants' willingness to engage in circular migration. The chapter commences with an analysis of the transposition of the Family Reunification Directive into Polish and Bulgarian law. It then moves on to explore its implementation dynamics through the eyes of the migrant workers who participated in the focus groups as part of this study. As a second step, the chapter examines national instruments in the field of academic and professional qualifications. It takes physicians and nurses as case studies in order to present the challenges associated with practicing regulated professions in the context of circular migration. The chapter concludes with an assessment of existing instruments against the study's benchmarks in these two policy areas, namely the conditions for family reunification and for migrant workers to have their diplomas and professional qualifications recognised.

8.1 Entry and Residence Conditions for Family Members

This section aims to assess the conditions under which circular migrants, including seasonal workers and other migrants with temporary permits, can reunite with family members under the Family Reunification Directive[1] transposed into the national legal frameworks of Bulgaria and Poland. As emphasised earlier, this benchmark[2] serves as

[1] Council Directive 2003/86/EC of 22 September 2003 on the right to family reunification [2003] OJ L 251.

[2] Benchmark developed on the basis of ILO Guidelines on Special Protective Measures for Migrant Workers in Time-bound Activities (Doc. MEIM/1997/D.4) adopted by the Tripartite Meeting of Experts on Future ILO Activities in the Field of Migration, ILO, Geneva, Annex 1, Para. 6.1. For the rest of the benchmarks, see Annex V. For more details, see Vankova (2016), p. 347.

Z. Vankova, *Circular Migration and the Rights of Migrant Workers in Central and Eastern Europe*, IMISCOE Research Series,
https://doi.org/10.1007/978-3-030-52689-4_8

a litmus of whether the EU is establishing a rights-based approach to circular migration or merely reviving the guest-worker model, which does not allow for family reunification.[3] In order to assess the legal frameworks developed at the national level, this study also takes into account two additional policy measures that can facilitate family reunion: waiting periods, which should not exceed 12 months, and housing conditions, which should not be as restrictive as to prevent family reunification.

8.1.1 Entry and Residence Conditions for Family Members According to the Family Reunification Directive

Article 159 of the Polish Act on Foreigners[4] provides for family reunification by allowing entitled family members to obtain a temporary residence permit.[5] Family reunification is only possible with a foreigner (sponsor) who is resident in Poland on the basis of specific permits or visas such as: a permanent or an EU long-term residence permit; a temporary residence permit with a duration of at least 2 years, with the last one valid for a period of not less than 1 year immediately before filing the family reunification application; a national visa or a temporary residence permit for conducting scientific research or development work (in line with Article 151 (1) of the AF) or a temporary permit after completion of scientific research or development work enabling such foreigners to look for a job (in line with Article 186 (1) 7 of the AF); and, a Blue Card or an ICT permit.[6] This means that foreigners, entering on the basis of visa, such as for work within the *Oświadczenie* procedure or on the basis of *Karta Polaka*,[7] are not eligible for family reunification. In addition, there is generally a waiting period of 2 years before the sponsor becomes eligible to apply for family reunification in Poland.

Furthermore, the sponsor needs to have health insurance within the meaning of the Act of 27 August 2004 on Health Care Services Financed from the Public Funds[8] or a document certifying that the costs of treatment in Poland will be covered by an insurer, as well as a source of steady and regular income sufficient to cover

[3] Vankova (2018), p. 170.

[4] Ustawa o cudzoziemcach z dnia 12 grudnia 2013 (Dziennik Ustaw z 2018r., poz. 2094, consolidated text).

[5] The definition of a family member covers the spouse, whereby the marriage needs to be recognised under Polish law, and minor children of the foreigner and their spouse, including an adopted child and a dependent child of whom the foreigner has actual parental custody and a minor child of the spouse, including an adopted child, dependent on him/her, of whom he/she has actual parental custody. Article 159 (3) AF.

[6] Article 159 (1) 1 AF.

[7] Unless they have obtained permanent residence permit.

[8] Ustawa z dnia 27 sierpnia 2004 r. o świadczeniach opieki zdrowotnej finansowanych ze środków publicznych (Dziennik Ustaw z 2019, poz. 1373, consolidated text).

subsistence costs for both sponsor and family members.[9] Finally, the family members need to have a guaranteed place of accommodation in Poland, which is certified by, *inter alia*, a housing rental agreement or statement from a hotel.[10]

The regional governor (*wojewoda*) must take a decision on granting a temporary residence permit within 1 month upon receipt of the documents – although this deadline can be extended in certain cases.[11] The application fees, which are payable when submitting the documents, are 340 złoty (80 EUR) plus 50 złoty (11 EUR) for the residence card.[12]

A temporary residence permit for family members of a foreigner residing in Poland can be granted either while the family members are still outside the country or if already in Poland.[13] After a positive decision from the regional governor where the sponsor is resident, if the family member granted the permit is outside Poland, they can use it to apply for visa before a Polish consul. Upon arrival in Poland, the family member needs to apply for the card and register his or her fingerprints. The permit is issued for the period of validity of the sponsor's temporary permit or for 3 years in the event that the sponsor has been granted permanent or EU long-term residence in Poland.

By way of comparison, the Bulgarian Act on Foreigners in the Republic of Bulgaria (AFRB)[14] stipulates that family members[15] can be granted a continuous residence permit in cases where: they have obtained a visa in accordance with Article 15 (1) following the approval of an application for family reunification; and, where they are joining a foreigner who holds a continuous permit with an authorised stay of at least 1 year or permanent residence permit, as well as the documents certifying family ties and the right to support are recognised or admitted for execution under Bulgarian legislation (Article 24 (1) 13 of the AFRB). Unlike Polish law, the Bulgarian legal framework provides more favourable conditions than Article 3 of the Directive by not requiring a criterion of reasonable prospects of obtaining the right of permanent residence.[16]

---

[9] Article 159 (1) 2 AF. For the determination of the income requirement, Article 114 (2) applies.

[10] Article 159 (1) 3 AF. Jacek Białas et al. (2015).

[11] See Article 109 AF.

[12] Migrant.info.pl (2019a).

[13] Article 168 AF.

[14] Закон за чужденците в Република България (SG No. 153/23 December 1998, last amendment SG No. 58/23 July 2019).

[15] The definition of a family member in Bulgarian law covers a spouse, children of the foreigner and their spouse, including any adopted children who are unmarried minors and where one of the spouses has parental custody, and the children are dependent on him/her (Article 2 (3) of the AFRB). The children of a foreigner or their spouse, who have reached the age of 18 years and who are not married are also considered to be family members, in cases where significant medical reasons mean that they require personal care or where they are otherwise unable to provide for themselves (Article 2 (4) of the AFRB).

[16] European Commission (2019), p. 2.

In order to be granted this residence permit, family members need to have met the requirements contained in Article 24 (2) of the Bulgarian Act on Foreigners, such as having accommodation, compulsory health and social insurance, and sufficient means of subsistence for the period of residence without resorting to the social assistance system.[17] Sponsors are required to submit evidence that these conditions are covered in line with Article 12 (1) of the Implementing Regulation of the AFRB.[18] Family members of researchers (Article 24б (6) of the AFRB), Blue Card holders (Article 33к (3) of the AFRB), and ICTs (Article 33р (1) of the AFRB) are also eligible to apply for a continuous residence permit under Article 24 (1) 13 of the AFRB for the same residence period as the sponsor, provided they meet the requirements of Article 24 (2) discussed above.

The Implementing Regulation of the AFRB provides that family reunification for the ICT permit holders is not linked to a residency requirement of a specified minimum period (Article 12 (9)). Unlike Poland, Bulgaria does not set any waiting period requirement before a sponsor is eligible to apply for family reunification. However, Bulgaria is one of the two Member States that have not used the derogation to Article 5 (3) of the Directive allowing family members to submit their application while already in its territory.[19]

A positive decision on the family reunification application is grounds for issuing a visa for family members in line with Article 15 (1) of the AFRB under a simplified procedure (Article 12 (8) of the Implementing Regulation of the AFRB).[20] Upon entry into the territory of the Republic of Bulgaria, and no later than 14 days before the expiration of their visa, family members need to submit a copy of their passport, a copy of their visa as per Article 15 (1) of the AFRB, and the stamp showing their last entry into the country, as well as evidence that they possess the compulsory medical insurance valid for the entire territory of Bulgaria, before the Migration Directorate or one of the Ministry of Interior's Regional Directorates (Article 13 (1) of the Implementing Regulation of the AFRB). Within 3 days after the applicants have provided these documents, and absent of grounds for revoking the right of residence for the purpose of family reunification, family members are informed of their obligation to pay a fee (generally 150 BGN (75 EUR))[21] in accordance with Tariff No. 4 on the

[17] Not less than the minimum monthly salary, the minimum stipend, or the minimum pension according to the legislation of the Republic of Bulgaria.

[18] Правилник за прилагане на Закона за трудовата трудовата миграция и трудовата мобилност (SG No. 79/ 7 October 2016, last amendment SG No. 27/ 2 April 2019).

[19] For more details, see European Commission (2019), p. 10.

[20] The family members are exempted from the requirement to present proof of secure accommodation, transport, and sufficient means of subsistence when applying for a visa on family reunification grounds (Article 8 (4) AFRB in conjunction with Article 20 (1) 2 of the Ordinance on the terms and procedures for issuing visas and determining the visa regime).

[21] Article 10б (1) of Tariff No 4 concerning family reunification under Articles 24 (1) 13 and 33р (1) of the AFRB. The fee is the same for family reunification under Article 33к (3) AFRB (Article 10б (5) of Tariff No 4). Family members of researchers are required to pay 100 BGN or 50 EUR in line with Article 10б (3) of Tariff No 4.

fees collected in the system of the Ministry of Interior under the State Fees Act[22] (Article 13 (3) of the Implementing Regulation of the AFRB). These fees were significantly reduced in 2018 as a result of an infringement proceeding against Bulgaria.[23]

In addition, the family members need to submit another application in person together with the documents proving that the requirements in terms of accommodation, income, and insurance have been met, including copies of their passports and receipt for payment of the state fee (see Article 14 (1) of the Implementing Regulation of the AFRB). The Ministry of Interior takes a final decision within 14 days after the submission of this application, which can be extended in case of 'legal and factual complexity and necessity' (Article 14 (2) of the Implementing Regulation of the AFRB).

Finally, it should be stressed that the agreements Bulgaria has concluded under the auspices of the Mobility Partnerships with Armenia (Article 3.2)[24] and Moldova (Article 3.2)[25] also provide for family reunification in line with the legislation discussed above.

8.1.2 Implementation Dynamics

One of the Ukrainian circular migrants who participated in the focus groups planned to settle in Poland only if he was able to reunite with his family.[26] He gained knowledge of the process because his friends had already gone through the application procedure. He believed that the process worked and did not expect to face any problems. The Ukrainian participant described that the family had to apply for a visa in Ukraine,[27] then move to Poland and apply for a residence permit based on the fact that he worked in Poland, 'then, the office issues a residence permit and the family starts living here – the kid goes to school as it is required by the law'.

---

[22] Tariff No. 4 on the fees collected in the system of the Ministry of Interior under the State Fees Act (adopted with Council of Ministers Decree No. 53/1998, last amendment SG No. 75/ 11 September 2018).

[23] See Chap. 6, Sect. 6.3.3.1 for more details.

[24] Agreement between the Republic of Bulgaria and the Republic of Armenia on labour migration management, approved by Council of Ministers Decision No. 176 of 22 March 2018, entered into force 9 October 2018/ Спогодба между Република България и Република Армения за регулиране на трудовата миграция, утвърдена с Решение No. 176 от 22 март 2018 г. на Министерския съвет. В сила от 9 октомври 2018 г.

[25] Agreement between the Government of the Republic of Bulgaria and the Government of the Republic of Moldova on labour migration management, approved by Council of Ministers Decision No492 of 13 July 2018, entered into force on 11 September 2018/ Спогодба между правителството на Република България и правителството на Република Молдова за регулиране на трудовата миграция, утвърдена с Решение No. 492 на Министерския съвет от 13 юли 2018 г. В сила от 11 септември 2018 г.

[26] Focus group with Ukrainian migrants, Poland, November 2016, Annex IV.

[27] The focus groups were conducted before the introduction of a visa-free regime for Ukraine.

Yet, an interviewed NGO representative described a different picture in reality.[28] The interviewee did not see a lot of Ukrainians reuniting with their families in Poland because it was not easy to obtain a family reunification permit: 'It requires money, it requires very high salary which usually they don't have. You need to prove, according to the Directive, that you can support your family, so your salary should be quite substantial and most of them, even middle level sectors, cannot do so'. The NGO representative noted that the EU directives were generally transposed into Polish law by applying the minimum possible standard: 'It's a translation, it's not really a transposition'.[29] He explained that, therefore, most of the directives were translated and inserted into Polish legislation without any assessment as to whether they created workable solutions and fitted the legal act in question.

Several of the interviewed NGO representatives stressed that because of the demanding requirements for family reunification, such as the condition of legal residence for 2 years, many of their clients were finding other ways to reunite with their families. Some reunited on the basis of a visa and afterwards stayed in Poland through a special temporary permit.[30] Usually one of the spouses had a work visa and the family member would apply for a visa based on grounds that were not specified in the Act on Foreigners in line with Article 60, Para 1 (25) thereof.[31] When the family member's visa was about to expire, the NGO advised their clients to apply for a 'temporary permit based on other circumstances' stipulated in Articles 186–187 of the Act on Foreigners. This is in fact a residence permit, which does not give its holder the right to work in Poland. However, after 2 years, the sponsor could apply for a family reunification permit.

In order to obtain the temporary permit based on other circumstances, the family members had to convince the inspector who analysed their case at the regional governor's office by explaining their situation, e.g., that the husband was working and the wife was taking care of the child and that they had enough income to support themselves: 'Then they analyse it. If they think it is enough, you can stay. It doesn't mean it always happens'; 'sometimes they give a residence permit of parents who are here. If they are old and need their children to care for them, they can stay. That is an argument'.[32]

By contrast, the interviewed Blue Card holders in Poland did not encounter any problems with the application process, which lasted on average between 2 and 4 months.[33] They used both possibilities for family reunification: while family members were already present in Poland and when their families were still in their

---

[28] Interview #12 with civil society actor, Poland, November 2016, Annex II.

[29] Interview #12 with civil society actor, Poland, November 2016, Annex II.

[30] Interview #13 with expert, Poland, November 2016, Annex II; Interview #5 with civil society actor, Poland, November 2016, Annex II; Interview #6 with civil society actor, Poland, November 2016, Annex II.

[31] Interview #13 with expert, Poland, November 2016, Annex II.

[32] Interview #13 with expert, Poland, November 2016, Annex II.

[33] Focus groups with Blue Card holders: mixed group of Russian speakers, Russians and Ukrainians, Poland, December 2016, Annex IV.

country of origin. Some brought their family members to Poland on tourist visas and submitted the family reunification application after receiving the Blue Card permit.[34] The main problem was not related to the application for the Blue Card itself but rather to the visa application. One respondent said that his spouse had to wait roughly two-and-a-half months for an interview at the Polish consulate in Ukraine.[35] According to him, this delay was caused by bribery practices as other applicants were able to buy places in the queue. Another concern of the interviewed Blue Card holders was finding work for their spouses; one Ukrainian shared that his wife was a nurse and that it would be difficult for her to find a job in Poland.[36]

In similar vein, the Ukrainian and Russian participants in the focus groups in Bulgaria, as well as the interviewed experts, shared that the biggest obstacles in relation to family reunification were the requirements laid down in Bulgarian migration law for migrants to exit the country in order to change the grounds of residence so as to switch from one visa to another.[37] Since the law did not allow family members to enter at the same time as the sponsors, most of them were utilising visa type C as an entry mechanism in order to avoid the waiting period inherent in the family reunification procedure. As a result, they had to exit the country after a 3-month stay in order to be able to then apply for a visa type D, which would enable them to re-enter and apply for a continuous residence permit.[38] This requirement significantly disrupts migrants' family life[39] as, in practice, it means that the parents had to continue paying for their children's kindergarten or school in order to keep their place in addition to all other financial resources they needed to spend as a result of this 'forced circularity'.[40] One parent shared that their child had a 'really bad year in school, just because they had to leave Bulgaria' as a result of this policy.[41]

The Blue Card permit holders raised another issue in relation to family reunification. What they found problematic was the fact that the spouses of Blue Card permit holders did not have direct access to the Bulgarian labour market; they had to find a

[34] Focus groups with Blue Card holders: mixed group of Russian speakers, Poland, December 2016, Annex IV.

[35] Focus groups with Blue Card holders: mixed group of Russian speakers, Poland, December 2016, Annex IV.

[36] See Sect. 8.2. for more details.

[37] Focus group with Blue Card holders from Russia, Bulgaria, September 2016, Annex IV; Focus group with Blue Card holders from Ukraine, Bulgaria, September 2016, Annex IV; Focus groups with Russian migrants, Sofia, September 2016, Annex IV; Interview #9 with lawyer, Bulgaria, July 2016; Interview #8 with lawyer, Bulgaria, July 2016, Annex III; Interview #12 with representative of international organisation, Bulgaria, June, 2016, Annex III.

[38] Interview #9 with lawyer, Bulgaria, July 2016, Annex III.

[39] Focus group with Blue Card holders from Russia, Bulgaria, September 2016, Annex IV; Focus group with Blue Card holders from Ukraine, Bulgaria, September 2016, Annex IV; Focus groups with Russian migrants, Sofia, September 2016, Annex IV.

[40] Focus groups with Ukrainian migrants, Sofia, September 2016, Annex IV.

[41] Focus group with Blue Card holders from Russia, Bulgaria, September 2016, Annex IV.

job with an employer willing to apply for work authorisation on their behalf.[42] According to one of the interviewed business representatives, this was seen as an anti-settlement measure.[43]

This, however, is commensurate with the provisions of the Blue Card Directive. Article 15 (6) of the Directive derogates from Article 14 (2) of the Family Reunion Directive, which has the effect of waiving the requirement of a possible waiting period for family members to access the labour market. Nevertheless, it does not waive the requirement that both sponsor and family member need to have the same level of access to employment.[44] During the first 2 years of employment, the sponsor is restricted to employment that meets the criteria of the initial admission, which means that there is not equal access to the labour market on the same conditions as nationals. Therefore, spouses of Blue Card holders are required to obtain a decision for access to the labour market from the Executive Director of the Employment Agency (Article 19 of the ALMLM). Since 2018, however, access to the Bulgarian labour market for Blue Card holders and their spouses has been further liberalised as the labour market test requirement has been waived.

One of the interviewed lawyers also stressed that the family reunification policy was extremely restrictive and 'short-sighted'.[45] She said that the more favourable visa procedure conditions that must apply to family members were not implemented in practice and that all applicants, even third-country nationals who were spouses of EU or Bulgarian nationals, had to wait for up to 40 days instead of 15. In her opinion, the main reason for this was the 'complete ignorance' of the consuls, who were appointed on nepotism-based practices and did not have even basic knowledge of the law.

The interviewee commented that she did not understand why family members of a sponsor – who had already undergone the procedure for obtaining a work permit and had thus met all required conditions related to income, insurance, and accommodation – had to face the same bureaucratic hurdle again for their family members. She found it particularly striking that sponsors could not use the same evidence for the family reunification procedure that they had already submitted for the visa application and therefore they had to request all these documents again. She said that based on her experience, the administration was allowing family reunification 'without problems' in cases where sponsors had continuous residence. One of the interviewed officials shared that family reunification was mainly refused on national security and public order grounds, as well as in cases of registered partnerships and same sex marriages, which are not recognised as valid under Bulgarian law.[46]

The family reunification procedure took an average of 2–3 months to be completed after the application was submitted by the sponsor in the territory of

---

[42] Focus group with Blue Card holders from Ukraine, Bulgaria, September 2016; Interview #20 with IT business representatives, Bulgaria, December 2016, Annex III.

[43] Interview #20 with IT business representatives, Bulgaria, December 2016, Annex III.

[44] Peers (2012), p. 61.

[45] Interview #9 with lawyer, Bulgaria, July 2016, Annex III.

[46] Interview #2 with official, Bulgaria, July 2016, Annex III.

Bulgaria.[47] Only one Blue Card holder shared that it took 4 months before his family members were able to join him.[48]

8.2 Recognition of Academic and Professional Qualifications

Another policy area that can influence the willingness of circular migrants to engage in this type of migration is the recognition of qualifications. As already mentioned, the 'triple win' proponents of circular migration claim that it enables skill transfer back to the countries of origin, which in turn supports development. Nonetheless, some evaluations of circular migration policies stress that there are cases when migrants return home and their new skills cannot be recognised or are not needed.[49] Therefore, this policy area is considered important for the purposes of this research.

The study's benchmarks in this policy area focus on the availability of provisions for recognition of occupational qualifications acquired outside the EU, including certificates and diplomas, and other means for recognising professional qualifications (see Annex V). Instruments for implementing these benchmarks include international cooperation tools and active information policies in relation to the recognition of academic qualifications that would make circular migration beneficial for the migrant worker and support both skill and knowledge transfer.[50]

8.2.1 Recognition of Academic and Professional Qualifications – Instruments at the National Level

8.2.1.1 Academic Qualifications

The adoption of the Act of 20 July 2018 on Higher Education and Science (AHES)[51] was part of Poland's reform in the field of education, research, and university governance.[52] The Act, however, did not introduce any substantial changes concerning the recognition of academic qualifications.

Recognition of academic qualifications is regulated by Chapter VIII 'Foreigners in the system of higher education and science' of the AHES. The Act provides for

[47] Focus groups with Ukrainian migrants, Sofia, September 2016, Annex IV. Focus group with Blue Card holders from Russia, Bulgaria, September 2016, Annex IV; Focus group with Blue Card holders from Ukraine, Bulgaria, September 2016, Annex IV.

[48] Focus group with Blue Card holders from Russia, Bulgaria, September 2016, Annex IV.

[49] K. Hooper and M. Sumption (2016), pp. 20–21.

[50] Vankova (2016), p. 348.

[51] Ustawa z dnia 20 lipca 2018 r. – Prawo o szkolnictwie wyższym i nauce (Dziennik Ustaw z 2018 poz. 1668).

[52] Eurydice (2018).

two different recognition regimes, that is, based on either national law or international agreements. Pursuant to Article 326 (1) thereof, there is automatic recognition of diplomas issued by an authorised higher education institution operating within the higher education system of a Member State of the EU, the OECD, or the European Free Trade Association (EFTA) – party to the Agreement on the European Economic Area.[53] This means that if a degree gives access to further studies or the right to start doctoral studies in the country where it was awarded, it also gives the same rights in Poland.[54] In case of doubt with regards to the appropriateness of the diploma or status of an issuing higher education institution, upon request of the institution concerned, the Director of the Polish Agency for Academic Exchange (NAWA) provides information about the diploma issued by a foreign university, the level of study, and the status of the university in question.[55]

In addition, pursuant to Article 327 (1) of the AHES, international agreements determining equivalence provide another option for recognition of a higher education diploma or a professional title obtained outside Poland as equivalent to the relevant Polish diploma or professional title. Concerning the Eastern Partnership countries, as of 2005 Poland had concluded bilateral agreements on the recognition of education for academic purposes with Ukraine[56] and Belarus.[57]

Finally, if a country is not covered by an international agreement, pursuant to Article 328 (4), a higher education diploma or a professional title obtained outside Poland can be recognised through the 'nostrification' procedure on the basis of Article 328 (3) of the AHES. The procedure is further detailed in a Regulation of the Minister of Science and Higher Education of 28 September 2018 on the recognition of academic degrees and degrees in the field of art awarded abroad.[58] Foreigners seeking recognition of higher education diplomas obtained outside Poland, together

[53] According to Article 326 (1) AHES provides for the equivalence at the different levels: 1) a 3-year degree programme or a first-cycle degree programme lasting at least 3 years, confirms in the Republic of Poland that its holder has a higher education at the level of the first-cycle degree programme; 2) a second-cycle degree programme confirms in the Republic of Poland that its holder has a higher education at the level of the second-cycle degree programme; 3) at least 4-year-cycle of studies confirms in the Republic of Poland that its holder has a higher education at the level of the second-cycle degree programme if it is recognized as equivalent to a diploma of completion of the second-cycle degree programme in the country in which it has been issued.

[54] For details see, Article 326 (2) AHES.

[55] Article 326 (4) AHES.

[56] Signed in April 2005, in force since 20 June 2006.

[57] Signed in April 2005, in force since 12 December 2005. For full list of the bilateral agreements, see Ministry of Science and Higher Education, 'Bilateral cooperation': https://www.gov.pl/web/nauka/wspolpraca-dwustronna Accessed 25 September 2019.

[58] Rozporządzenia Ministra Nauki i Szkolnictwa Wyższego z dnia 28 września 2018 r. w sprawie nostryfikacji stopni naukowych i stopni w zakresie sztuki nadanych za granicą (Dziennik Ustaw z 2018, poz. 1877).

with a title conferred as a result of this education, are required to submit an application initiating the nostrification procedure before a nostrification entity.[59]

During the procedure, the nostrification entity performs a formal assessment of the application[60] and can request the applicant to submit a translation into Polish of some of the documents attached to the application.[61] For instance, the institution may ask for a translation of the Master's thesis or Ph.D. dissertation into Polish, which could make the whole recognition process expensive.[62] The nostrification entity recognizes or refuses to recognize a degree as equivalent to the relevant Polish degree within 90 days from the date of submission of the application that meets the formal requirements.[63]

The applicant must pay a fee for the nostrification, regardless of the outcome. The maximum amount of the fee is specified in relation to a full professor's salary as not exceeding 50% of this rate.[64] In justified cases, the applicant may be exempted from payment. The nostrification entity itself sets the procedure for exemption from payment.

On the day of issue of a certificate, its holder acquires the right to use the Polish degree and exercise the rights arising from it. The applicant may appeal against the decisions taken by the nostrification entity pursuant to Article 127 of the Code of Administrative Procedure of 14 June 1960.[65]

In the case of secondary education (e.g., concerning the profession of technician, which is referred to as a secondary-level profession) this process is not implemented by universities but by local authorities through an administrative decision.[66]

---

[59] See Article 328 (4) AHES. Nostrification proceedings are conducted by an entity with the scientific category A+ or A in the discipline to which the application relates. According to the information of the Ministry of Science and Higher Education of 29 May 2019, until the university is parameterized (scheduled for 2021), nostrification proceedings should be conducted by units with the right to confer a post-doctoral degree of a particular field of science or a post-doctoral degree of a specific field of art in the field of scientific or artistic discipline to which the academic degree relates. https://konstytucjadlanauki.gov.pl/cudzoziemcy Accessed 3 September 2019.

[60] Para 4 (2) of the Regulation.

[61] Para 5 of the Regulation.

[62] Migrantinfo.pl (2019b).

[63] Para 7 (1) of the Regulation.

[64] Article 328 (6) and (7) AHES. The minimal remuneration of a professor is 6410 PLN according to the Regulation of the Minister of Science and Higher Education of 25 September 2018 regarding the amount of the minimum basic salary for a professor at a public university /Rozporządzenia Ministra Nauki i Szkolnictwa Wyższego z dnia 25 września 2018 r. w sprawie wysokości minimalnego miesięcznego wynagrodzenia zasadniczego dla profesora w uczelni publicznej (Dziennik Ustaw z 2018 r. poz. 1838).

[65] Ustawa z dnia 14 czerwca 1960 r. Kodeks postępowania administracyjnego Dziennik Ustaw z 2018 r., poz. 2096, consolidated text).

[66] Interview #13 with official, Poland, November 2016. For more details, see also: https://www.gov.pl/web/edukacja/recognition-of-foreign-school-certificates-and-diplomas-in-poland

By way of comparison, the recognition of academic qualifications in Bulgaria is regulated by the Act on Higher Education,[67] which stipulates that the Council of Ministers is to approve the state requirements for the recognition of higher education providing professional qualifications that have been acquired in foreign higher education institutions and the recognition of a degree acquired abroad corresponding to the educational and scientific degree 'doctor' (Article 9 (3) 9). In compliance with the Convention on Recognition of Qualifications Concerning Higher Education in the European Region (the Lisbon Recognition Convention),[68] the Council of Minsters adopted a Regulation on the State Requirements for Recognition of Higher Education Acquired or Periods of Education Completed in Foreign Higher Education Institutions.[69] It provides the general regulatory framework for recognising academic qualifications and the specific administrative procedures in relation thereto which are subject to the internal regulations of the relevant authorities.

According to the Regulation, recognition of higher education degrees acquired at a foreign higher education institution is carried out in order to provide access to further education in the higher education system, to upgrade training, and to doctoral studies (Article 6 (1)); to facilitate access to the labour market, to doctoral studies at a scientific research institution as well as for other purposes where the applicant has a legal interest (Article 6 (2)). The recognition procedure is organised by the higher education institution in the cases under Article 6 (1) of the Regulation, and by the Minister of Education and Science through the National Centre for Information and Documentation (NCID) in the cases under Article 6 (2) of the Regulation in line with the respective regulations of the higher education institutions and the Rules of Procedure of the NCID (Article 7 of the Regulation).

The Regulation stipulates the documents that applicants need to submit (Article 8 of the Regulation) and the main elements of the recognition procedure that the higher education institutions are required to follow (Article 10 of the Regulation). The procedure for recognition of higher education (Master's or Bachelor's degree)[70] that was acquired abroad must provide verification of the status of the foreign higher education institution and the diploma's authenticity; an assessment of the conformity of the data in the submitted documents with the State requirements for acquiring higher education in Bulgaria; and, finally, a decision for recognition of a higher education degree with a professional qualification, when it is indicated in the

[67] Закон за висшето образование (SG No. 112/ 27 December 1995, last amendment SG No. 98/ 27 November 2018). The Act on Pre-school and School Education/Закон за предучилищното и училищното образование (SG No. 79/13 October 2015), also contains provisions on academic recognition in the pre-school and school education. However, these matters fall outside the scope of this study.

[68] Ratified by the adoption of the Act on the Ratification of the Convention on Recognition of Qualifications Concerning Higher Education in the European Region (SG No. 51/28 March 2000).

[69] Наредба за държавните изисквания за признаване на придобито висше образование и завършени периоди на обучение в чуждестранни висши училища (adopted by a Council of Ministers Decree No. 168 of 14 August 2000, SG No. 69 /22 August 2000, last amendment SG No. 28/ 5 April 2019). In Bologna Process and European Higher Education Area, (2003), p. 4.

[70] For a doctoral degree, see Articles 13-13a of the Regulation.

diploma, a refusal to recognise it, or termination of the procedure (Article 10 of the Regulation). Amongst other things, the following indicators are taken into account: the student admissions procedure; duration of study; total tuition hours of studied subjects or acquired credits or both; and, learning outcomes of the training as a collection of the acquired knowledge, skills and competences obtained during the period of study (Article 11 (2) of the Regulation).

The recognition of higher education acquired at a foreign higher education institution is refused when there are significant differences between the submitted data and the Bulgarian state requirements for the acquisition of higher education, as well as in other cases detailed in Article 12 (1) of the Regulation. The refusal may include a recommendation stipulating the possible measures that an applicant may take in order to obtain recognition at a later stage, including sitting additional exams (Article 12 of the Regulation).

In cases where the recognition procedure is carried out to facilitate access to the labour market, to doctoral studies at a scientific research institution, or for other purposes where the applicant has a legal interest (Article 6 (2) of the Regulation), the applicant is issued a certificate by the Executive Director of NCID. In the cases under Article 6 (1) of the Regulation, the recognition of higher education is valid only for continuing education at the higher education institution that carries out the recognition and does not provide rights that can be used in relations with third countries (Article 14 (3) of the Regulation).

8.2.1.2 Professional Qualifications

Professional qualifications obtained in third countries are recognised in Poland in accordance with national legislation. How the professional qualification is recognised generally depends on whether the profession is regulated. In the case of unregulated professions, the employer takes a decision on the recognition of the foreign qualification. The latter may require confirmation of the equivalence of migrant's educational qualification with the relevant Polish award or written information on recognition of his or her degree issued by NAWA.[71] One of the officials interviewed in late 2016 said that some of the regional governors also sought confirmation of the equivalence within the Blue Card application procedure.[72]

Foreigners wishing to practice a regulated profession need to have their foreign certificate or higher education diploma recognised in the first instance. Only then can they apply for the professional rights in accordance with the regulations pertaining to the exercise of a given profession in Poland.[73] For example, the conditions for migrants to obtain the right to practice the profession of a physician are: possession of a medical diploma awarded by a state outside the EU, provided that it has been

[71] NAWA (2019).

[72] Interview #13 with official, Poland, November 2016.

[73] NAWA (2019).

recognised through a nostrification procedure at a medical school in Poland as being equivalent to a Polish medical diploma;[74] the completion of an obligatory post-graduate internship[75] lasting 13 months (*staż podyplomowy*); a positive result in the Final Medical Exam (*Lekarski Egzamin Końcowy*);[76] appropriate physical and mental health (full legal capacity); an impeccable ethical attitude;[77] and, a sufficient command of the Polish language, which is certified by an examination organised by the Polish Chamber of Physicians and Dentists in Warsaw.[78] In addition, foreigners also need to present a document certifying their right to reside in Poland.[79]

Foreigners who have their diplomas recognised, and who have passed the Polish language examination and satisfied the health and ethical requirements, can apply to the Regional Chamber of Physicians and Dentists to obtain 'a right to practice the profession awarded in order to complete the postgraduate internship' or apply to the Minster of Health for recognition of previous professional experience or an internship completed abroad that is equivalent to the Polish internship. Provided that the ethics and health conditions are still satisfied, after the completion or the recognition of the internship and a positive result in the Final Medical Exam, the foreigner can apply to the Regional Chamber of Physicians and Dentists for the right to practice the profession and can then subsequently become a member of the Regional Chamber.

The Bulgarian procedures for recognising professional qualifications likewise differ depending on the type of the profession, i.e., regulated or non-regulated. Foreigners with diplomas for non-regulated professions do not generally need to have their diplomas recognised.[80] A recognition procedure may be conducted in cases where a potential employer or a Bulgarian authority expressly requires a recognition certificate that is issued by the NCID in line with Article 14 (1) of the Regulation on the State Requirements for Recognition of Higher Education Acquired or Periods of Education Completed in Foreign Higher Education Institutions. The recognition procedure is organised by the Minister of Education and Science through the NCID in accordance with its Rules of Procedure (Article 7

---

[74] In line with Article 7 (1) 1c of the Act of 5 December 1996 on Doctors' and Dentists' Professions/ Ustawa z dnia 5 grudnia 1996 r. o zawodach lekarza i lekarza dentysty (Dziennik Ustaw, z 2019 r. poz. 537, consolidated text).

[75] Or obtained recognition of a post-graduate internship completed outside the territory of the Republic of Poland as equivalent a post-graduate internship in the Republic of Poland. See Article 5 (3) of the Act of 5 December 1996 on Doctors' and Dentists' Professions.

[76] Or an exam completing a doctor's post-graduate internship. See Article 5 (4) of the Act of 5 December 1996 on Doctors' and Dentists' Professions.

[77] See Article 7 (1) 2 and 3 of the Act of 5 December 1996 on Doctors' and Dentists' Professions.

[78] In line with the Regulation of the Minister of Health from 26 September 2012 on the post-graduate internship of a doctor and dentist/Rozporządzenie Ministra Zdrowia z dnia 26 września 2012 r. w sprawie stażu podyplomowego lekarza i lekarza dentysty (Dziennik Ustaw z 2014 r., poz. 474; z 2017 r., poz. 2194)

[79] Article 7 (1) 8 of the Act of 5 December 1996 on Doctors' and Dentists' Professions.

[80] Interview #3 with representative of NCID, Bulgaria, July 2016, Annex III.

of the Regulation) and follows the provisions of the Regulation that was described in the academic qualifications section above.

With regards to regulated professions, the Act on the Recognition of Professional Qualifications (ARPQ)[81] states that specific categories of third-country nationals who have acquired professional qualifications in a Member State will enjoy the same rights as EU nationals in having their professional qualifications recognised. These categories are family members of a Bulgarian citizen or citizen of another Member State when that citizen has moved to Bulgaria, EU long-term residence or permanent residence holders or highly-qualified workers who hold a visa under Article 15 (1) of the AFRB (Article 8 (1) of the ARPQ). In cases where Bulgaria has concluded an international agreement with a third country that includes provisions for the mutual recognition of professional qualification in a particular profession, such as in the Agreement between Israel and Bulgaria,[82] the recognition of third-country nationals' qualifications that fall within the scope of the agreement is conducted under the agreed procedure (Article 8 (2) of ARPQ).

In cases where third-country nationals do not fall within the scope of either of these two categories, the recognition of professional qualifications in a regulated profession that was acquired outside of the EU can take place under mutual recognition conditions that are established on a case-by-case basis. This may apply where the qualification complies with the Bulgarian regulatory requirements for the acquisition of the same professional qualification (Article 8 (3) of the ARPQ). In such cases, the responsible professional bodies of the regulated profession concerned have full discretion to assess the conformity of the acquired professional qualification with the current rules in Bulgaria and, if necessary, to impose compensatory measures if there are any significant differences.[83]

As with Poland, the procedure for recognising qualifications in order to obtain the right to practice the profession of a physician is taken as an example. In accordance with Article 186 (3) of the Act on Health,[84] foreigners who wish to practice this profession need to have a command of the Bulgarian language and professional terminology in the Bulgarian language, which is certified on the basis of an exam regulated in an Ordinance issued by the Minister of Education, Youth and Science and the Minister of Health.[85] Applicants also need to pass the state exams, which are

---

[81] Закон за признаване на професионални квалификации (SG No. 13/8 February 2008, last amendment SG No. 17 /26 February 2019).

[82] According to the procedure for the implementation of the Agreement, at the request of the Office for Population and Immigration at the Ministry of the Interior of the State of Israel, some or all of the pre-selected workers included in a database by the Bulgarian Employment Agency may be required to take professional exams or sit interviews or both. https://www.az.government.bg/pages/posrednichestvo-po-spogodbata-mezhdu-pravitelstvoto-na-republika-bulgaria-i-pravitelstvoto-na-darzhavata-izrael-za-posrednichestvo-i-vremenna-zaetost-na-grazhdanite-na-dvete-darzhavi/ Accessed 2 June 2019.

[83] Interview #3 with representative of NCID, Bulgaria, July 2016, Annex III.

[84] Закон за здравето (SG No. 70/ 10 August 2004, last amendment SG No. 58/23 July 2019).

[85] Наредба No. 15 от 13.12.2005 г. за установяване на владеенето на български език и професионалната терминология на български език от чужденците за упражняване на

determined in the Ordinance on the unified state requirements for the exercise of a regulated profession in the field of 'Medicine' (Article 186 (3) 3a in conjunction with Article 177 of the Act on Health).[86]

In order to be admitted to take the examination under Article 186 (3) 3a of the Act on Health, third-country nationals need to submit an application to the Minister of Health, containing amongst other things: copies of their diplomas with an officially notarised translation; an academic transcript; a certificate of command of the Bulgarian language; a medical certificate; and, documents certifying the lack of a criminal record and any disciplinary or administrative punishment that is related to the exercise of the profession.[87] The procedure for admission to an examination is organised and implemented by the Health Ministry's Directorate 'Medical Activities'.[88] The applications are examined by an Expert Commission that is designated by an order of the Minister of Health[89] under Article 79 of the Act on the Recognition of Professional Qualifications. The Commission issues a reasoned proposal to admit the applicant to the examination when the following conditions are met: the submitted documents certify a professional qualification that was acquired by the applicant in a regulated medical profession or specialty in Bulgaria; the specialty acquired by the applicant is relevant or can be assimilated to a specialty from the Bulgarian nomenclature of specialties in the healthcare system[90]; and the applicant has not received any administrative, judicial, or disciplinary penalties in relation to the exercise of the medical profession (Article 7 (2) of Regulation No. 4 of 27 May 2011).

медицинска професия в Република България/Ordinance No. 15 of 13 December 2005 on establishing the knowledge of Bulgarian language and professional terminology in Bulgarian by foreigners for practicing a medical profession in the Republic of Bulgaria (SG No. 104 /27 December 2005, last amendment SG No.96/ 20 November 2018).

[86] Currently covering five exams. For the exercise of a medical specialty, applicants are required to cover certain study programmes and successfully pass practical and theoretical exams before the State Examination Commission, which are determined by an order of the Minister of Health (Article 186 (3) 3b in conjunction with Article 180 (3) of the Act on Health).

[87] The procedure is regulated in Ordinance No. 4 of 27 May 2011 on the conditions and procedure for admission and examination of third-country nationals in line with Article 186 (3) 3 of the Act on Health, who have acquired a professional qualification in the medical profession and/or a specialty in the field of healthcare in a third country/Наредба No. 4 от 27 май 2011 г. за условията и реда за допускане и явяване на изпит по чл. 186, ал. 3, т. 3 от Закона за здравето на граждани на трети държави, придобили професионална квалификация по медицинска професия и/или специалност в областта на здравеопазването в трета държава (issued by the Minister of Health, SG No. 43 of 7 June 2011, last amendment SG No. 64/ 3 August 2018).

[88] In accordance to Article 33, Para 18 of the Rules of Procedure of the Ministry of Health/ Устройствен правилник на Министерството на здравеопазването (adopted by Council of Ministers Decree No. 55/ 28 March 2019, SG No. 26 of 29 March 2019).

[89] In accordance with Article 6 (1) of Regulation No. 4 of 27 May 2011 on the conditions and procedure for admission and examination of Article 186 (3) 3 of the Act on Health, third-country nationals who have acquired a professional qualification in a medical profession and/or a specialty in the field of healthcare in a third country.

[90] Defined by the Regulation under Article 181 (1) of the Act on Health.

The examination under Article 186 (3) 3a of the Act on Health is organised and carried out by the higher education institutions twice a year (Article 10 (1) of Regulation No. 4 of 27 May 2011). After the applicants have successfully passed the Bulgarian language exam and the state exams, in order to be allowed to practice as a physician, they also need to register with the National Register of the Bulgarian Medical Association. Along with the documents certifying that they have successfully passed the exams, they also need to present a certificate of good standing, a residence permit, and other such documents that are stipulated in Article 32 of the Act on the Professional Organisations of Doctors and Doctors of Dental Medicine.[91]

8.2.2 Implementation Dynamics

None of the participants in the focus groups in Poland were working in regulated professions. One of them had a professional qualification to work in such a profession (i.e. as a construction engineer) but she could not find a job in either Ukraine or Poland and was no longer interested in pursuing that career.[92] Apart from one participant who mentioned that recognition of her Ukrainian diploma took a very long time, none of the other interviewees reported problems in this regard.[93]

The interviews with experts involved in the process of academic and professional recognition, however, revealed the cumbersome procedures that migrants face if they wanted to work in a regulated profession. One of the main challenges was the equivalence of degrees. For example, an interviewed official stressed that the education for becoming a physician required 6 years of study and if a third-country national just had a Bachelor's degree, it would not be possible to have the qualification recognised as being equivalent to the Polish medical degree.[94] Such foreigners could be admitted to a Polish university, take additional exams, and obtain a medical degree in Poland. This did not pose a problem for recent graduates. Still, according to the interviewee's experience, migrant doctors who had exercised this profession for a longer period could find it challenging to go back to university. In addition, there were no possibilities for doctors to practice in Poland while they were still in the recognition process.

---

[91] Закон за съсловните организации на лекарите и на лекарите по дентална медицина (SG No. 83/21 July 1998, last amendment SG No. 102/11 December 2018).

[92] Focus group with Ukrainian migrants, Poland, November 2016, Annex IV.

[93] Focus group with Ukrainian migrants, Poland, November 2016, Annex IV.

[94] Interview #19 with official, Poland, November 2016, Annex II.

The interviewees shared that nurses[95] coming from the Eastern Partnership countries generally faced problems with the recognition of their academic degrees.[96] Those coming from Ukraine very often had graduated from a secondary-education school that was later upgraded to tertiary education.[97] Nevertheless, in order to practice the profession of a nurse in Poland, one had to have at least a Bachelor's level degree, which over 90% of the Ukrainians did not possess at the time of the interviews. This meant that their degree could not be recognised through a nostrification procedure and that they had to enrol in a Bachelor's program and complete their studies in either Ukraine or Poland. According to the interviewed representative of the professional body, a better option for the nurses was to complete their studies in Poland because this would exempt them from the requirement of undertaking an additional internship and the Polish language exam.[98] The professional body would recommend the latter option to the applicants, even though it was more costly, because it gave nurses a degree that would be recognised throughout the EU. When there was a possibility of having some of the subjects recognised by the university in Poland, this would have the effect of shortening the study period.[99]

Another challenge related to recognition of the academic qualifications via nostrification was the time it took to complete the process.[100] The universities sometimes did not have the capacity to finalise the procedure within the 3-month period due to the large number of foreigners applying for recognition. After the Regulation on nostrification was adopted, according to the interviewee, the universities saw many applications from Ukraine, Belarus, and other post-Soviet countries and they gained knowledge about the programmes, their differences, and knew what to expect. Thus, processing was faster for new graduates. Yet, for older qualifications, the nostrification procedure was beset with difficulties. If there were substantial

[95] The conditions for practicing the profession of a nurse are very similar tothe criteria applied for physicians and are stipulated in the Act of 15 July 2011 on Nurse and Midwife Professions/Ustawa z dnia 15 lipca 2011 r. o zawodach pielęgniarki i położnej (Dziennik Ustaw z 2019 r., poz. 576, consolidated text). The limited right to exercise the profession is granted on the basis of a command of the Polish language; a nursing certificated recognised as being equivalent to those issued in Poland; and, unlimited legal capacity and satisfactory health conditions that are conducive to the exercise of the nursing profession. After obtaining the limited right to exercise the profession, the nurse is directed by a District Chamber of Nurses to participate in an obligatory 6-month adaptation internship in a healthcare facility. Upon completion of this internship, assuming that the nurse has a residence permit in Poland, displays impeccable ethical conduct, has not been deprived of the right to practice in their country of origin, and provided that all other requirements are still satisfied, the foreigner can be granted a right to practice the profession by the District Chamber of Nurses.

[96] Interview #19 with official, Poland, November 2016, Annex II; Interview #21 with professional body representative, Poland, November 2016, Annex II; Interview #11 with trade union representative, Poland, November 2016, Annex II; Interview #5 with civil society actor, Poland, November 2016, Annex II.

[97] Interview #21 with professional body representative, Poland, November 2016, Annex II.

[98] Interview #21 with professional body representative, Poland, November 2016, Annex II.

[99] Interview #21 with professional body representative, Poland, November 2016, Annex II.

[100] Interview #19 with official, Poland, November 2016, Annex II.

differences regarding the learning outcomes and the direction of studies, the applicants had to make up for the differences by sitting an examination. The university decided what the differences were, which exams migrants had to take, and the time frame for sitting exams on a case-by-case basis.

The requirement for command of the Polish language in the academic qualification process was also a challenge identified by the interviewed official.[101] Another issue was associated with the fees charged for the process. When the first large wave of foreigners applied for recognition of their medical qualifications after Poland's accession to the EU, 'universities were charging for everything'.[102] The Regulation capped the fees that could be collected for the nostrification procedure. However, applicants had to pay for the additional courses and the language exam, which cost 400 złoty (94 EUR) for doctors. In addition, the obligatory internships were not always remunerated, placing an additional financial burden on the applicants.

These cumbersome procedures raised a logical question about the interplay of the entry and residence requirements for migrants and the process of recognition of qualifications for regulated professions. One of the interviewed lawyers stressed that these categories of professionals could not come to Poland and immediately start working.[103] Therefore, what they did instead was enter on the basis of a visa, such as for work within the *Oświadczenie* procedure or on the basis of *Karta Polaka*, for example, as a domestic worker or they used a permit for different work or one that was not work-related (i.e. for family reasons). Once they had arrived in Poland, they commenced the recognition procedure, looked for a job, or started working in most of the cases in positions that were at a level below their qualifications.[104] They also started learning the language because this was another barrier to practicing a regulated profession. The interviewed representative of the physicians' professional body confirmed that, to his knowledge, the Blue Card procedure was not used in the doctors' profession because doctors could not be hired without having obtained the right to practice, which could take between 6 months and 3 years depending on the nature of the individual case.[105]

Asked whether they had encountered any migrants who had managed to have their degrees recognised and were circulating, one of the interviewees said that nurses were investing so much time, effort, and money to obtain the right to practice that their main goal was to stay in Poland.[106] Another interviewee said that he did not think that the circular migration concept was feasible with regards to doctors.[107]

[101] Interview #19 with official, Poland, November 2016, Annex II.

[102] Interview #19 with official, Poland, November 2016, Annex II.

[103] Interview #13 with expert, Poland, November 2016, Annex II.

[104] Interview #13 with expert, Poland, November 2016, Annex II; Interview #21 with professional body representative, Poland, November 2016, Annex II; Interview #6 with lawyers, Poland, November 2016, Annex II; Interview #5 with a civil society actor, Poland, November 2016, Annex II.

[105] Interview #20 with professional body representative, Poland, December 2016, Annex II.

[106] Interview #21 with professional body representative, Poland, November 2016, Annex II.

[107] Interview #20 with professional body representative, Poland, December 2016, Annex II.

A civil society representative stressed that the few nurses whom she knew had gone through the whole recognition process successfully and obtained a Polish diploma, then ultimately moved to Germany to earn a higher salary.[108] She stressed that even nurses who were interested in circulating back and forth between Poland and Ukraine would face additional bureaucratic hurdles because the newly-obtained Polish diploma needed to be recognised anew in Ukraine due to the lack of a special agreement to facilitate this process.

The interviewed official said that existing policies for academic recognition were not working fast enough and that these were expected to change with the new act on higher education (adopted in 2018). On the other hand, the professional bodies did not plan any specific measures to foster the procedure for the recognition of professional qualifications of foreigners. An interviewee from one of the professional bodies emphasised that they still had doubts as to whether the requirements in certain Member States were met and they did not have any knowledge or control over the medical degree qualifications in countries outside the EU.[109] His experience showed, for example, that the specialist training of doctors coming from the Eastern Partnership countries was at a lower level compared to what was required in Poland thus, in practice, requiring additional exams for obtaining recognition of the relevant diploma.

The interview with a representative of the ENIC/NARIC centre in Bulgaria revealed that the centre did not have information on problems with academic recognition within the EU.[110] However, he mentioned that there were sometimes problems in relation to the recognition of diplomas between Bulgaria and third countries. The interviewee said that the 'best known' example concerned Turkey. According to him, a few years ago Turkey had ceased recognising academic qualifications for a period of time and did not give access to its higher education system. This issue was, however, resolved.

One of the Ukrainian participants in the focus groups had qualified as a physician and shared her experience with the recognition procedures.[111] The respondent moved to Bulgaria in 2015 as a family member of a migrant worker. She stressed that she was determined to start working as a doctor so she did everything to learn the Bulgarian language in order to pass the required language exam. The Ukrainian focus group participant was surprised to find out that she had to pass seven exams in medicine (five state exams and two speciality exams) and a language exam. She added:

> 'I feel insecure because these exams will take more than one year, maybe two years. We cannot work – neither as middle medical staff nor as doctors – and anyone looking for a job in a specialty cannot find anything, absolutely nothing (...) Many good specialists just go

[108] Interview #5 with a civil society actor, Poland, November 2016, Annex II.

[109] Interview #20 with professional body representative, Poland, December 2016, Annex II.

[110] Interview #3 with representative of NCID, Bulgaria, July 2016, Annex III.

[111] Focus groups with Ukrainian migrants, Bulgaria, September 2016, Annex IV.

> back to Ukraine. And all my friends who wanted to work as doctors just want to go back to Ukraine because it is an unreal situation.'[112]

She remarked, however, that although the language exam was free and very accessible, she had to pay 250 BGN (125 EUR) per state exam. Furthermore, she pointed out that she had graduated 27 years ago and the requirement to pass all these state exams was very burdensome and unreasonable because she had chosen to specialise in neurology, and if she wanted to have her professional qualifications recognised, then she would have to pass exams in areas such as gynaecology, obstetrics, surgery, and anaesthesiology.[113] Meanwhile, she had to work as a consultant for an insurance company in order to make a living. Her only 'access' to the medical profession was to work as a doctor at private medical centres at Black Sea resorts during the summer months, where Russian-speaking doctors were in demand: 'It's also unofficial, we work as students like paramedics and sometimes they say that we are translators or secretaries (...) It's just very humiliating for our specialists'.[114]

In her opinion, the current legislation complicated the situation for migrants who wanted to have their diplomas recognised and be able to work as doctors. This meant that they were looking for other opportunities to work in Western Europe, which essentially turned Bulgaria into a 'transit point'. She stressed that she could not go back to Ukraine because her home was too close to the Donbas region.[115]

One of the interviewed lawyers reiterated the problems migrants faced in having their qualifications recognised in regulated professions.[116] She stressed that unless one falls under EU legislation, there was 'a huge gap' in the field of recognition of qualifications. According to the interviewee, the equivalency exams were the only instrument that the administration used to recognise a profession and for the migrants, this meant spending additional financial resources and wasting time on a qualification that they had already acquired just so that they could practice in Bulgaria. For third-country nationals who obtained their education outside the EU, there was no other option for recognition such as, for example, to go through some brief testing and verification of the diploma that would give them the right to work in that profession in Bulgaria.

Against this background, the interview with a representative of the Bulgarian Medical Association revealed that Bulgaria was experiencing an increased emigration rate of graduates with medical qualifications; up to 90% were leaving Bulgaria upon graduation from Bulgarian medical universities.[117] In response to the question of whether the recruitment of foreign doctors was a measure that the Association could pursue, the representative answered that, currently, there was no working

[112] Focus groups with Ukrainian migrants, Bulgaria, September 2016, Annex IV.

[113] Focus groups with Ukrainian migrants, Bulgaria, September 2016, Annex IV.

[114] Focus groups with Ukrainian migrants, Bulgaria, September 2016, Annex IV.

[115] Focus groups with Ukrainian migrants, Bulgaria, September 2016, Annex IV.

[116] Interview #9 with lawyer, Bulgaria, July 2016, Annex III.

[117] Interview #21 with representative of the Bulgarian Medical Association, Bulgaria, January 2017, Annex III.

mechanism for the recognition of qualifications and the Association was not aware of the quality of the medical diplomas granted in third countries.

Most of the foreign doctors working in Bulgaria had either obtained their medical education in the country or had studied in one of the former Soviet republics. Therefore, the Bulgarian Medical Association representative concluded that it would make sense in the future to recruit doctors from the former Soviet republics because they were willing to come to Bulgaria since they had lower living standards in their countries of origin. The interviewee gave an example of good practice that was present in a former bilateral agreement between Bulgaria and Russia/USSR for the exchange of specialists in the medical field.[118]

8.3 Conclusions

Bulgaria and Poland present two different approaches when it comes to eligibility for family reunification in the context of circular migration. Bulgaria allows migrant workers holding continuous residence permits with an authorised stay of at least 1 year to reunite with family members; Poland constrains sponsors with temporary permits by imposing a 2-year delay. Polish legislation explicitly excludes migrant workers resident on the basis of a visa who therefore need to transfer to a single permit in order to be able to apply for family reunification. In the case of Bulgaria, this exclusion covers all migrants whose permit is for a period of validity of less than 1 year, such as seasonal workers. In both countries, Blue Card holders, researchers, and ICTs are the only categories of migrant workers able to benefit from some type of family reunification facilitation.

The analysis of the collected empirical data showed that the income and fee requirements for families to reunite in Poland were made difficult to satisfy for migrants employed in low-skilled sectors and middle-skilled jobs. Even though Polish legislation imposes a basic legal income requirement, the sum is required per family member; additionally migrants face relatively high application fees. No problems were raised in this regard in Bulgaria, even though before the amendments to Tariff No. 4, the family reunification fees were regarded as being disproportionately burdensome.

The restrictive and cumbersome procedures, however, do not prevent migrants from finding ways to reunite with their families. Migrants in both countries used circumvention practices in order to overcome the demanding 2-year waiting period requirement in Poland or the practice of 'forced circulation' that is inherent in Bulgarian migration law, which requires family migrants to exit the country in order to change their visas. Therefore, the legislation in both countries is not considered to be in line with the study's benchmarks in this field. The analysed transposition of

[118] Interview #21 with representative of the Bulgarian Medical Association, Bulgaria, January 2017, Annex III.

the Family Reunification Directive and the existing national legal frameworks demonstrate that it is impossible to be both a short-term circular migrant in a low-skilled occupation and at the same time be entitled to family reunification.

The analysis of the national legislation and the gathered empirical data in Bulgaria and Poland suggest that the main issues in the field of the recognition of qualifications concerns entry to regulated professions. Both countries have established long and demanding procedures for recognising qualifications in regulated professions, consisting of several stages: demonstrating proficiency in the native language and the specific terminology; undergoing a diploma recognition procedure, which might require undertaking additional studies at a university and completing an internship; passing final state exams; and, registration with a respective professional body. Neither Poland nor Bulgaria offer any facilitation with regards to access to medical professions, which was taken as an example in both countries. This leads to several challenges for migrant workers.

Firstly, migrants who seek to practice the regulated profession of a doctor or a nurse cannot enter the country on the basis of a standard work permit because they cannot have their diplomas recognised and gain the right to practice while they are still in their countries of origin. Therefore, in the case of Poland, migrants often have recourse to the *Oświadczenie* procedure, while in Bulgaria family reunification permits or other circumvention practices are used in order to to enter the country and secure residence. Furthermore, the length and complexity of the current procedure generally prevents migrants who wish to work in the medical profession from applying for a Blue Card. Once they have entered the country, and while they are preparing for the stages of the recognition procedure, they are likely to work in low-skilled occupations.

Secondly, the established procedures in both Bulgaria and Poland are time-consuming and have the effect of putting migrants' lives on hold for a couple of years, depending on the differences between the national system and the third country. As some of the respondents pointed out, these procedures are even more challenging for migrant doctors who obtained their degrees many years ago and were required to enrol in a university as part of the recognition procedure. Another challenge stems from the relatively high fees that are part of the recognition process in both countries and the Polish requirement to undertake an unpaid internship, which engenders additional financial burdens.

All these challenges result in two different options for migrant workers, neither of which involve circular migration. After investing so much time, effort, and resources in the recognition procedures and learning the language, migrants are prone to settle in the country of destination upon gaining the right to practice rather than engaging in circular migration. Furthermore, as one of the interviewees stressed, going back to their country of origin might require further recognition procedures there. An alternative option in this regard involves migrating to another Member State that has higher standards of living where the additional effort would be financially rewarded. In the case of Bulgaria, the empirical data analysis suggests that migrants choose the latter option even before commencing the recognition procedure. In the case of Poland, some medical specialists were migrating to other countries after being enrolled in an additional degree as part of the recognition procedure, which ultimately provided them with a diploma that would be recognised throughout the EU.

The research findings lead to the conclusion that circular migration and the recognition of qualifications in regulated professions are two incompatible processes within the current legislative and policy frameworks. The main reason is the lack of special measures that would allow the recognition process to be facilitated. These could be based on the conclusion of bilateral agreements between Member States and third countries using EU law provisions and mechanisms that are applied to the recognition of qualifications in regulated professions for EU citizens. However, in order to do that, countries like Bulgaria and Poland first need to articulate the need for such foreign specialists, at both the political level and the level of the professional bodies involved in the recognition process.

References

Bologna Process and European Higher Education Area (2003) National Report regarding the Bologna Process Implementation. Bulgaria. http://www.ehea.info/media.ehea.info/file/Bulgaria/34/0/National_Report_Bulgaria_2003_576340.pdf Accessed 1 March 2020.

Bialas, J., et al. (2015) How to Make It in Poland – Information Brochure for Foreigners. Residence Legalisation and Employment, Warsaw.

European Commission (2019) Report on the implementation of Directive 2003/86/EC on the right to family reunification, COM (2019) 162 final, 29 March 2019, Brussels.

Eurydice (2018) How does the new Poland law impact the Higher Education Institution's landscape? 13 September 2018. https://eacea.ec.europa.eu/national-policies/eurydice/content/how-does-new-poland-law-impact-higher-education-institutions-landscape_en Accessed 19 December 2019.

Hooper, K. and M. Sumption (2016) *Reaching a 'Fair' Deal on Talent. Emigration, Circulation and Human Capital in Countries of Origin*. Migration Policy Institute, Washington DC.

Migrant.info.pl (2019a) Temporary residence permit for family members of a foreigner residing in Poland. http://www.migrant.info.pl/temporary-residence-permit-for-family-members-of-a-foreigner-residing-in-poland.html Accessed 1 September 2019.

Migrant.info.pl (2019b) Nostrification of diplomas. http://www.migrant.info.pl/Nostrification_of_diplomas.html Accessed 3 September 2019.

NAWA (2019) Practicing Regulated Professions. https://nawa.gov.pl/en/recognition/recognition-for-professional-purposes/employment-in-higher-education-institutions Accessed 2 June 2019.

Peers, S. (2012) Family Reunion. In S. Peers, E. Guild, D. A. Arcarazo, K. Groenendijk and V. Moreno-Lax (eds), *EU Immigration and Asylum Law (Text and Commentary): Second Revised Edition. Volume 2: EU Immigration Law*. Leiden: Martinus Nijhoff Publishers, pp. 247–286.

Vankova, Z. (2016) EU Circular Migration Policies: Dead or Alive? Developing a Rights-based Benchmark Framework for Policy Assessment. Journal of Immigration, Asylum and Nationality Law 30 (4): 332–352.

Vankova, Z. (2018) The EU's Approach to Circular Migration in the Context of the Eastern Partnership Neighbourhood. In S. Carrera, L. den Hertog, D. Kostakopoulou, and M. Panizzon (eds), *EU external migration policies in an era of global mobilities: Intersecting policy universes*. Immigration and Asylum Law and Policy in Europe; Vol. 44. Leiden: Brill/Nijhoff, pp. 155–176.

Chapter 9
Conclusions and Recommendations

This book demonstrated that the EU's circular migration concept has entered the migration policy agendas of Bulgaria and Poland as part of the policy transfer driven by the process of Europeanisation. Furthermore, it became clear that the concept of circular migration, as it has been promoted by the EU, serves as an empty shell that is shaped by different EU and national instruments, depending on local contexts and labour market needs. This chapter aims to answer the main research question of this study, namely: how the EU's approach to circular migration has been implemented through the developed legal and policy instruments, and does it provide for rights-based circularity for migrant workers in the Central and Eastern European context. The chapter concludes with policy recommendations to this end.

9.1 General Conclusions

This study aims to fill a gap in existing literature by presenting a comprehensive picture of the formulation and implementation of the EU's circular migration approach in the Central and Eastern European (CEE) context. It provides a multi-level analysis of the EU legislation that facilitates circular migration, the dynamics under the aegis of the Global Approach to Migration and Mobility (GAMM), and the implementation at the national level in two CEE countries. The originality of this book lies also in the fact that it accounts for the outcomes at the individual level by employing an empirical legal research approach. Several general conclusions can thus be derived from this study.

Z. Vankova, *Circular Migration and the Rights of Migrant Workers in Central and Eastern Europe*, IMISCOE Research Series,
https://doi.org/10.1007/978-3-030-52689-4_9

More than 10 years ago, the European Commission adopted the policy idea of facilitating circular migration as part of a worldwide hype among international organisations, namely that this type of migration can benefit all parties involved. Based on several policy documents in the period 2000–2009, the EU formulated its approach towards circular migration. This study confirms the findings of other authors, such as Wickramasekara,[1] that this was done in the absence of any thorough evidence that this type of migration could indeed bring the alleged 'triple win solution' for countries of origin and destination as well as the migrant workers themselves. Furthermore, as the preceding chapters have demonstrated, the EU started to promote facilitation of this type of migration not only in relation to Member States but also to third countries without a common definition or a uniform understanding of what this policy term entails.

Against this backdrop, this study concludes that more than a decade after the EU policymakers adopted this concept, there is hardly a common approach to circular migration at the EU level. On the contrary, this is a policy notion that keeps straying into the EU Justice and Home Affairs domain and which finds a place in the Preambles of some of the EU legal migration directives and the Mobility Partnerships' annexes without a clear grasp of its meaning or tangible form. This lack of a coherent formulation leads to scattered outputs and an uneven implementation at the national level.[2]

Therefore, 'catch me if you can'[3] is a suitable phrase that captures the state of circularity of the legal and policy instruments, which are part of the EU's labour migration policy.[4] This study demonstrated that two different concepts of circular migration can be outlined under the EU approach. On the one hand, a spontaneous pattern of circularity that can be facilitated through a legislative framework such as, to a certain extent, in the context of the Blue Card Directive as well as in the Polish simplified *Oświadczenie* procedure that has been adopted under the auspices of the Mobility Partnerships with the Eastern Partnership countries; and, on the other hand, as a temporary migration scheme with a re-entry component that is regulated through the Seasonal Workers' Directive and some of the circular migration initiatives implemented under the Mobility Partnerships with the Eastern Partnership countries.[5]

Furthermore, the incoherent formulation of the EU's approach to circular migration reflects the emerging sectoral EU labour migration *acquis* as well as the dynamics of the EU external migration policy under the aegis of the GAMM. Therefore, returning to the main research question of this study on how the EU's approach to circular migration has been implemented and whether it provides rights-based circularity for migrant workers in the CEE context, the study shows

---

[1] Wickramasekara (2011).

[2] See also European Migration Network (2011).

[3] See also Nita (2016), pp. 25–27.

[4] Vankova (2018), p. 172.

[5] Ibid.

that rights come at a certain 'skill and qualifications' price in the EU.[6] The sectoral EU directives on labour migration that were analysed differentiate migrant workers on the basis of their skills and qualifications, as well as their attractiveness for the Member States' labour markets, using this as the decisive feature on which the different statuses are assigned.[7] The migrants most desired at the EU level are those with higher qualifications because they are seen as contributing the most to the Member States and the EU's economy. Blue Card holders are the only category that can benefit from rights-based circular migration that allows for a migrant-led trajectory. However, the Blue Card Directive has barely been used since it was adopted due to restrictive admission conditions, national parallel rules, and its cumbersome implementation in some Member States.

Circular migration as part of the GAMM is a vivid illustration of Member States' desire to strictly adhere to their competence under Article 79 (5) of the Treaty on the Functioning of the European Union (TFEU). Therefore, this study confirms the assessments of other authors with regards to circular migration as part of the Mobility Partnerships: behind the façade of the Mobility Partnerships' annexes containing references to circular migration initiatives, there is nothing more than several small-scale projects and there are only a few Member States that are eager to engage in this type of migration.[8] Apart from that, the only GAMM instrument that could currently make a difference for workers' rights is the Polish *Oświadczenie* procedure, which, as this study has demonstrated, is used mainly by Poland's neighbouring countries. The visa liberalisation with Georgia and Ukraine are other policy measures that can contribute to the initiation of individual circular migration projects that could not be included in the scope of this study.

This book also confirms the conclusions of other authors that circular migration, as a policy model, can provide for migrant-led trajectories and the protection of migrants' rights only if it is perceived as a spontaneous pattern of migration facilitated through the operation of flexible policies and a flexible legal framework (as in the case of Sweden and partially through the *Oświadczenie* procedure in Poland), rather than as a temporary labour migration scheme, which is redolent of the guest-worker model.[9] With the exception of Blue Card holders, such flexible solutions are only available for migrants of ethnic origin at the national level in the countries taken as case studies. In line with Newland's assessment, this study concludes that the design of circular migration policies should take as their starting point the existing spontaneous patterns of migration, such as between Poland and Ukraine.[10]

An added value of this research is that it goes one step further and provides a comprehensive picture of the implementation dynamics of the EU's approach to circular migration at the national level by adding an empirical legal research

[6] The ILO has expressed its criticism on this matter on several occations. See for instance ILO (2011), p.1 and ILO (2016), para. 106, p.34.

[7] In this regard see Fridriksdottir (2017); Wiesbrock et al. (2016), p. 960; Carrera et al. (2019).

[8] See Carrera and Hernández i Sagrera (2011); Reslow (2013); Nita (2016).

[9] Schneider and Wiesbrock, (2011); Skeldon (2012).

[10] Newland et al. (2008).

dimension. This study found that the outcomes of the implementation of the EU instruments that fall under the circular migration umbrella on the rights of migrant workers are not straightforward and depend on various factors, such as the national context, how the national migration policy has developed, the way in which EU law is transposed into domestic law, and how Member States use their margin of appreciation.[11]

In the Polish context, the entry and re-entry facilitation instruments developed are compensatory measures that aim to restore patterns of circular migration that has already been in existence for decades. All the established national and EU instruments were part of a conscious policy choice to create facilitation for the citizens of the neighbouring Commonwealth of Independent States (CIS) countries aligned with the country's foreign policy considerations. Therefore, the implementation of the EU and national instruments falling under the circular migration umbrella has provided these migrant workers in Poland with flexible possibilities for facilitated legal entry and re-entry into the country.

In the case of Bulgaria, where labour migration is regarded as a temporary solution only, the concept as such did not lead to the establishment of any new rights derived from national instruments in relation to entry and re-entry conditions due to the practice of 'forced circulation' that is inherent in Bulgarian migration law. Nonetheless, the different EU labour migration directives and the visa facilitation agreements with Eastern Partnership countries have established certain rights, which need to be further detailed and enforced by the different stakeholders: business organisations, NGOs, and migrant workers. Currently, migrants cannot fully benefit from some of the rights introduced under EU law because the directives have been inefficiently and impractically transposed into national law.

In the field of work authorisation, the EU law instruments falling under the circular migration umbrella have introduced rights for several categories of migrants with regards to the change of employer and the possibility to find alternative employment in case of unemployment. In the case of Poland, this is possible even when the EU *acquis* does not impose an obligation under the general regime, which transposed the Single Permit Directive into Polish law. However, the implementation dynamics presented by this study suggest that due to the poor 'quality' of the transposition of EU law, these rights cannot be easily enforced in practice, rendering them somewhat ineffective. In the case of Bulgaria these rights were established only for some categories of migrants as a result of the pressure to harmonise national law with EU law requirements. Therefore, in the Bulgarian case the implementation of the EU's legal instruments has led to tangible results for the rights of migrant workers.

The results of this study show that applying the minimum standards under the EU Long-term Residence Directive with regards to allowed periods of absence for circular migration facilitation is not a workable solution. Furthermore, in order to transit from permits allowing circulation to a more permanent status, migrants need

[11] On that, see also Solé et al. (2016).

to change their trajectories and plan their returns to their countries of origin within the time limits that are permitted by EU and national legislation. Therefore, except for Blue Card holders who are granted flexible geographical mobility, the remaining categories of migrant workers that fall under the EU circular migration umbrella have not gained any additional rights that could make them more prone to engage in circular migration.

The policies on social security coordination and the recognition of qualifications for third-country nationals, who fall outside the scope of the EU *acquis*, are left to the Member States to determine. The analysis of these two policy fields shows that in cases where there are no EU *acquis* requirements to be implemented, Bulgaria and Poland have not been proactive due to the embryonic development of their migration policies, the lack of any political will to attract migrant workers and create workable policies for their retention or circulation. Therefore, these policies are mainly driven by the notion of having some measures in place, rather than by incentives to create workable solutions, which would support a given policy, whether circular or not. Thus, they provide very limited rights to specific categories of migrants, and in the case of recognition of qualifications lead to a temporary 'brain waste' of a desired group of professionals such as physicians and nurses. Concerning social security coordination, Bulgaria sets a good example by pursuing an active policy for the negotiation of bilateral agreements based on the GAMM framework. However, both Bulgaria and Poland fall short of investing in information campaigns and other measures aimed at raising awareness about these rights amongst migrant workers.

The implementation of the EU's family reunification policy at the national level provides for rights-based circular migration solutions mainly with regards to highly-skilled migrants, such as Blue Card holders, researchers, and Intra-Corporate Transferees (ICTs). For the rest of the migrant workers, these two policies are incompatible, unless they decide to change their circular trajectory and settle for at least 1 year in order to be able to satisfy the eligibility criteria for family reunification. The fact that the Family Reunification Directive allows Member States to impose additional conditions before authorising family reunification is another hindrance to family reunification for circular migrants who are not regarded as being highly-skilled.

As Triandafyllidou stresses, circular migration on the ground is 'shaped by labour market dynamics, and driven by the agency of the migrants', and this does not necessarily match the EU's approach.[12] The analysis of the empirical data gathered through the conducted focus groups supports this conclusion by illustrating that migrants do not always follow the predetermined model of migration envisaged by policymakers.[13]

This implementation study shows that the EU's approach to circular migration in practice fails to accommodate the ambitious variety of migrant profiles considered as target groups for circular migration facilitation. It demonstrates that Blue Card

---

[12] Triandafyllidou (2013).

[13] Vankova (2017).

holders and EU long-term residents are generally not interested in employment-related circular migration, and that the use of the Students' and Researchers' as well as the ICTs Directives, is marginal in the CEE countries selected as case studies in this study. In line with previous studies, this book highlights seasonal workers and other migrants engaged in low-skilled occupations as the most active beneficiaries of the EU's approach to circular migration.[14]

The study's empirical data analysis suggests that flexible policies that allow migrant-led trajectories make migrants more prone to circulation and restrictive policies leave settlement as the only option, in line with other authors' conclusions.[15] This comes to show once again that the understanding of circular migration as a fluid type of movement is not necessarily reflected in current policy developments. Despite the hurdles that must be overcome in order to find jobs and the restrictive policies concerning access to family reunification or permanent residence, however, migrant workers use their agency and rely on circumvention mechanisms that allow them to continue with their own personal trajectories. Furthermore, factors such as the political situation in countries of origin, as in the case of Russian and Ukrainian migrants, also contribute to changes in migrants' trajectories and thus provide another argument for the development of rights-based labour migration policies that allow for flexibility, rather than circular migration policies.

Another important conclusion that is based on the current implementation study is that circular migration policies concern much more than just entry and re-entry conditions. The existing policies at both the EU and the national levels that are labelled as circular migration measures currently place migrant workers in a more vulnerable position. In order to distinguish the policy concept of circular migration from the guest-worker models, policymakers need to consider adopting a rights-based approach that covers other policies that concern circular migration, such as social security coordination and family reunification. Furthermore, the labour rights of such migrants need to be protected and enforced because of their vulnerable position. Even though this study did not focus on assessing the enforcement of these kinds of rights, they should nevertheless form part of any rights-based circular migration policy.

To sum up, the answer to the research question of this study on how the EU's approach to circular migration is implemented and whether it provides rights-based circularity for migrant workers in the CEE context is that the EU's approach has been driven by selectivity based on the skills and qualifications of migrants and it only rewards the most desirable migrants – the highly-qualified – with the possibility to engage in rights-based circular migration. However, circular migration approaches at the national level differ between countries due to various factors such as national context, stage of development of the national migration policy, how EU law is transposed into national law, and how Member States use their margin of appreciation. Consequently, this leads to different outcomes for the rights of migrant workers and very often to discrepancies between policymakers' predetermined models and migrants' realities.

[14] See for instance Constant and Zimmermann (2011), pp. 495–515.

[15] See for instance Massey and Pren (2012).

9.2 Policy Recommendations

In line with the general conclusions of this study, the following policy recommendations are proposed:

1. **Depart from the promotion of circular migration at the EU level and instead encourage flexible labour migration policies that allow for 'geographical mobility'.** As this study has demonstrated, migrant trajectories and circular migration patterns change over time due to different circumstances related, *inter alia*, to economic, political, and personal factors, and are very often predetermined by cultural and historical links between home and host countries. Therefore, the development of such policies needs to follow a bottom-up rather than a top-down approach from the EU level. The facilitation of this type of migration could fit the national circumstances of some Member States (e.g. Poland), but not necessarily of others. The (recast) Blue Card Directive that provides for 'geographical mobility' should serve as a model for such a flexible approach to legal migration, which is rights-based rather than utilitarian as is the case of the Seasonal Workers' Directive. This approach could facilitate spontaneous circular migration in cases where migrant workers have chosen a circular trajectory between their home and host countries, but it can also allow for a different trajectory that is related to temporary migration, settlement, or migration to another country. Such an approach is also commensurate with the realities of Member States, which also experience demographic problems along with labour market shortages.
2. **Ensure better implementation and enforcement, at the national level, of the already adopted EU *acquis* on legal migration through the establishment of a mechanism for regular monitoring and reporting to the European Commission.** Currently, this type of monitoring is done primarily through the Commission's reports, which focus on the implementation of different directives, and on the basis of signals, cases, and reports that are brought to its attention by Brussels-based or national-based NGOs. Such a mechanism could be institutionalised through the establishment of a network of independent national experts funded by the EU who report to the Commission on a regular basis, following the examples of networks such as the European network on free movement of workers within the European Union, which was coordinated by the Centre for Migration Law at the Faculty of Law of Radboud University and replaced by FreSco, as well as the Odysseus Academic Network. Furthermore, such a recommendation takes into account the attitudes of Member States, which are currently reluctant to adopt any new proposals in this respect.[16]
3. **In the long-term, EU institutions should depart from the emerging sectoral approach, which cannot encompass all categories of migrant workers**. The 2001 Commission proposal for a general framework directive on labour migra-

[16] Interview #30 with European Commission official, Belgium, November 2017, Annex I.

tion set a good example for flexible migration policy measures, such as permits allowing longer absences from the territory of the Member States, the export of pensions and benefits, and multi-entry permits that could also facilitate spontaneous circular migration. However, the 2001 proposal fell short of providing such flexible options in relation to migrants employed in low-skilled occupations. In any case, a general framework directive would provide the option to move away from the current practice favouring highly-skilled migrants with more rights and 'geographical mobility' and which discriminates against migrants in low-skilled occupations who are less attractive to the Member States.[17]

4. **The EU and the Member States should provide mechanisms that facilitate the recognition of qualifications for third-country nationals coming to work in regulated professions in the Member States, and which are not covered by the current EU legal framework.** This study has found that the current circular migration policies that aim to prevent brain drain from third countries and the recognition of qualifications are two incompatible processes. The study's empirical data analysis demonstrated that this, in fact, leads to settlement or further migration in the EU. Therefore, the EU and the Member States need to actively pursue the conclusion of multilateral and bilateral agreements to regulate this issue. As the IOM has suggested, these can be region-specific and field specific.[18] Another underused possibility at the national level, which requires the mutual cooperation of Member States' administrations and professional bodies, is to implement the (recast) Blue Card Directive, as well as the other relevant directives in this field, in such a way that it accommodates the recognition procedure for regulated professions. All of these measures, however, also need to be accompanied by reforms at the national level, which should aim to improve the current recognition procedures by making them faster and more flexible on the basis of innovative compensatory measures.[19] Additional policy measures, such as databases providing educational profiles in different sectors in third countries with which the EU cooperates, could also improve this process.
   Finally, a recommendation at the EU level is to extend the personal scope of Professional Qualifications Directive 2005/36/EC, as amended by Directive 2013/55/EU, and apply it to all third-country nationals coming to work in regulated professions. The mechanism established by these directives sets out an extensive overarching framework that provides certain procedures and safeguards.
5. **The EU and the Member States should establish family reunification policies for temporary migrants, regardless of their skill levels.** In order for migrants to engage in circular migration in a beneficial way – which, as demonstrated by this study, can encompass a variety of professions and durations of

[17] Such a recommendation was also proposed by Wiesbrock (2010), p. 734.

[18] Schuster et al. (2013), p. 255.

[19] For a list of possible compensatory measures, see ibid.

stay – they need to be able to rely on policies that minimise family disruption caused by such migration.[20] As this study has concluded, the implementation of the EU's family reunification policy at the national level currently provides for rights-based circular migration solutions only in relation to highly-skilled migrants such as Blue Card holders, researchers, and ICTs, who could also be temporary migrants staying for limited periods of time. Taking into account that the current political climate would make it impossible to justify reforms that aim to widen the personal scope of the Family Reunification Directive, Member States should establish policies that provide options for migrant workers to be accompanied by their family members. This could include frequent return options provided through longer permitted absences, long-term visas, or temporary permits allowing migrants to be joined temporarily by their family members as a measure that precedes the family reunification procedure. Furthermore, Member States need to provide an option for family members to be able to change their status (from visa type C to D, or from visa type D to a temporary permit) from inside the territory of the host country if their sponsor decides to work for a longer period of time in the EU. As this study has shown, the lack of such measures also causes family life disruptions.

6. **Member States should proactively pursue the conclusion of bilateral social security agreements and establish an accessible information policy for migrants engaged in circular migration.** This study shows that the number of such agreements, even between Member States and third countries that share cultural, economic, and historical links, is rather low. Therefore, Member States should pursue the conclusion of such agreements as part of their respective migration policies and follow the practice adopted by Bulgaria, namely to use the GAMM channels in order to negotiate such agreements. Furthermore, as the analysis of the empirical data of this research has shown, migrant workers are unaware of their rights in most cases, potentially putting them in a vulnerable position. Therefore, the Member States' administrations should pursue an active information policy, both in their territory but also, for instance, through the GAMM channels, in the home countries of migrant workers to inform them of existing possibilities for export of their benefits. Furthermore, Member States should consider the option of introducing the reimbursement of social security contributions, as has been advocated by this study.
7. **If the Member States want to pursue circular migration policies, they need to start collecting data on migrants that are engaged in this type of migration and provide regular evaluations of the policies that are in place.** The UN Economic Commission for Europe (UNECE) report *Defining and Measuring Circular Migration* provides a good starting point for policy action in this regard by proposing a statistical definition of circular migration and data sources that can be used to measure it, such as population registers, surveys, and census

[20] On that see also Hugo (2013), p. 7.

data.[21] However, this would also require a more systematic collection of data at the national level that could be complemented by additional surveys and further empirical research.[22] Only in this way can circular migration policies and their effects on the individual level be fully evaluated and understood, as well as improved.

8. **More research is needed in order to assess whether circular migration policy models based on the facilitation of spontaneous patterns could be used as complementary pathways for admission of refugees.**[23] What this study has shown, which is also commensurate with the findings of other authors, is that piloting temporary migration schemes and referring to them as 'circular' does not lead to sustainable and workable policy solutions. Therefore, there is a need for a shift in policy thinking in this respect.[24] Analysing the limited policy experience with the facilitation of spontaneous patterns of circular migration through legal and policy measures could be a good starting point for such a research exercise. What this study has demonstrated is that Ukrainians who were fleeing from the Crimea conflict in their home country used Poland's simplified *Oświadczenie* procedure rather than looking for alternative routes to enter the country irregularly or apply for asylum. They used this flexible possibility to support their migrant trajectory as it allowed them to settle if the conflict were to continue or to return to their home countries in the event that the political climate improved.
9. **New Member States such as Bulgaria and Poland are still developing their migration policies and should consider ratifying ILO Conventions No. 97 and No. 143 as well as the ICRMW.** As this study has demonstrated, Bulgaria and Poland still have not established comprehensive labour migration policies, and in both cases the policies developed did not provide adequate protection for migrant workers. Therefore, in line with the rights-based approach that has been promoted by the ILO, consistency with international standards in the field of labour migration as provided for in ILO Conventions No. 97 and No. 143, as well as the International Convention on the Protection of the Rights of All Migrant Workers and Members of Their Families (ICRMW), can support new countries of immigration, such as the ones under consideration in this research, to establish policy measures that are conducive to managing labour migration and which ensure an adequate level of protection for migrant workers.

---

[21] Conference of European Statisticians and United Nations Economic Commission for Europe (2016), pp. 20–26.

[22] On that see also Solé et al. (2016).

[23] The Global Compact on Refugees recommends that along with resettlement, countries should also offer "labour mobility opportunities for refugees, including through the identification of refugees with skills that are needed in third countries". In United Nations (2018), p. 19.

[24] In this respect, see also Newland and Agunias (2007).

References

Carrera, S. and R. Hernández i Sagrera (2011) Mobility Partnerships: 'Insecurity partnerships' for Policy Coherence and Migrant Workers' Human Rights in the EU. In R. Kunz, S. Lavenex, and M. Panizzon (eds), *Multilayered Migration Governance: The Promise of a Partnership*. London, New York: Routledge, pp. 97–117.

Carrera, S., L. Vosyliute, Z. Vankova, N. Laurentsyeva, M. Fernandes, J. Dennison, and J. Guerin (2019) The Cost of Non-Europe in the Area of Legal Migration: at the request of the European Added Value Unit of the Directorate for Impact Assessment and European Added Value within the Directorate General for Parliamentary Research Services. CEPS Paper in Liberty and Security in Europe, No 2019-01, Brussels.

Conference of European Statisticians and United Nations Economic Commission for Europe (2016) Defining and Measuring Circular Migration. Final report of the Task Force on Measuring Circular Migration, U.N doc ECE/CES/BUR/2016/FEB/14/Add.1, 20 January 2016.

Constant, A. F. and K.F. Zimmermann (2011) Circular and Repeat Migration: Counts of Exits and Years Away from the Host Country. *Population Research and Policy Review* 30 (4): 495–515.

European Migration Network (2011) Temporary and Circular Migration: empirical evidence, current policy practice and future options in EU Member States. Synthesis Report. Brussels.

Fridriksdottir, B. (2017) *What Happened to Equality? The Construction of the Right to Equal Treatment of Third-Country Nationals in Europan Union Law on Labour Migration*. Immigration and Asylum Law and Policy in Europe, Volume 43. Leiden: Brill.

Hugo, G. (2013) What We Know About Circular Migration and Enhanced Mobility. *Policy Brief* No7. Migration Policy Institute,Washington DC.

International Labour Organisation (2011) *Proposal for a Directive of the European Parliament and of the Council on the conditions of entry and residence of third-country nationals for the purposes of seasonal employment*, COM (2010) 379. Note based on International Labour Standards with reference to relevant regional standards, Brussels. http://www.ilo.org/wcmsp5/groups/public/%2D%2D-europe/%2D%2D-ro-geneva/%2D%2D-ilo-brussels/documents/genericdocument/wcms_168539.pdf Accessed 1 May 2018.

International Labour Office (2016) *International Labour Conference Report 'Promoting Fair Migration'*, ILC.105/III/1B, Geneva. https://www.ilo.org/wcmsp5/groups/public/%2D%2D-ed_norm/%2D%2D-relconf/documents/meetingdocument/wcms_453898.pdf Accessed 1 May 2018.

Schuster, A., M. V. Desiderio and G. Urso (2013) Recognition of Qualifications and Competences of Migrants. International Organization for Migration, Brussels.

Massey, D.S. and K. A. Pren (2012) Unintended consequences of US immigration policy: Explaining the post-1965 surge from Latin America. *Population and Development Review* 38 (1): 1–29.

Newland, K. and D. Agunias (2007) How can Circular migration and Sustainable Return Serve as Development tools? Background paper for Roundtable 1.4, Global Forum on Migration and Development, Brussels. https://www.gfmd.org/files/documents/gfmd_brussels07_rt1-4_en.pdf Accessed 7 March 2020.

Newland, K., D. R. Mendoza, and A. Terrazas (2008). Learning by Doing: Experiences of Circular Migration. *Migration Policy Institute Insight*, Washington DC, September 2008.

Nita, S. (2016) Circular Migration Within the EU-Moldova Mobility Partnership. In Solé, C., S. Parella, T. Sordé Martí and S. Nita (eds), *Impact of Circular Migration on Human, Political and Civil Rights. A Global Perspective*. Cham: Springer, pp. 23–44.

Reslow, N. (2013) *Partnering for mobility? Three-level Games in EU External Migration Policy'*. Datawyse/Universitaire Pers Maastricht.

Schneider, H. and A. Wiesbrock (2011) Circular migration: A triple win situation? Wishful thinking or a serious option for suitable migration policy? In D. Schiek, U. Liebert, and H. Schneider (eds), *European Economic and Social Constitutionalism after the Treaty of Lisbon*. Cambridge: Cambridge University Press, pp. 123–147.

Skeldon, R. (2012) Going Round in Circles: Circular Migration, Poverty Alleviation and Marginality. *International Migration* 50 (3): 43–60.

Solé, C., S. Parella, T. Sordé Martí and S. Nita (2016) *Impact of Circular Migration on Human, Political and Civil Rights. A Global Perspective*. Cham: Springer.

Triandafyllidou, A. (2013) Circular Migration at the Periphery of Europe: Choice, Opportunity, or Necessity? In A. Triandafyllidou (ed.), *Circular migration between Europe and its neighbourhood: choice or necessity?* Oxford: Oxford University Press, pp. 212–236.

United Nations (2018) Report of the United Nations High Commissioner for Refugees, Part II Global compact on refugees, General Assembly Official Records Seventy-third Session Supplement No. 12, (A/73/12 (Part II)), New York.

Vankova, Z. (2017) Implementing the EU's Circular Migration Approach: Legal and Migrant Perspectives on Entry and Re-Entry Conditions in Bulgaria and Poland. Robert Schuman Centre for Advanced Studies Research Paper No. RSCAS 2017/34. Florence: European University Institute.

Vankova, Z. (2018) The EU's Approach to Circular Migration in the Context of the Eastern Partnership Neighbourhood. In S. Carrera, L. den Hertog, D. Kostakopoulou, and M. Panizzon (eds), EU external migration policies in an era of global mobilities: Intersecting policy universes. Immigration and Asylum Law and Policy in Europe, Vol. 44. Leiden: Brill/Nijhoff, pp. 155–176.

Wickramasekara, P. (2011) Circular Migration: A Triple Win or a Dead End. *Discussion Paper No 15*. The Global Union Research Network.

Wiesbrock, A. (2010) *Legal Migration to the European Union*. Leiden: Brill/Nijhoff.

Wiesbrock, A., T. Jöst and A. Desmond (2016) Seasonal Workers Directive 2014/36/EU. In K. Hailbronner and D. Thym (eds), *EU Immigration and Asylum Law. A Commentary*. (2nd edition). Munchen/Oxford/Baden-Baden: C. H. Beck/Hart/Nomos, pp. 928–973.

Annexes

Annex I: Interviews with Officials from EU Institutions, Representatives of Brussels-Based NGOs, Think Tanks and International organisations in the Period 2013–2017

N	Date	Type of stakeholder	Institution	Country	Gender	Interview
1.	30.01.2013	Expert/Head of programme × 2	Think tank	Belgium	M/F	In person
2.	11.02.2013	Official	European Commission	Belgium	F	In person
3.	11.02.2013	Director	Trade Union	Belgium	F	Telephone
4.	5.02.2013	Deputy-director	Think tank	Belgium	F	In person
5.	13.02.2013	Policy officer	NGO	Belgium	F	In person
6.	19.02.2013	Head of programme	Think tank	Belgium	M	In person
7.	20.02.2013	Director	Think tank	Belgium	M	In person
8.	22.02.2013	Official	EU Council	Belgium	M	In person
9.	22.02.2013	Official	European Commission	Belgium	M	In person
10.	24.05.2013	Representative	Trade Union	Belgium	M	In person
11.	27.05.2013	Official	European Commission	Belgium	M	In person
12.	30.05.2013	Academic	University	Italy	F	Skype
13.	03.06.2013	Representative	Think-tank	Belgium	F	In person
14.	7.08.2013	Former policy officer	Think-tank	Belgium	F	In person
15.	07.08.2013	Representative	International organisation	Belgium	F	In person
16.	12.10.2013	Representative	International organisation	Switzerland	M	In person

(continued)

Z. Vankova, *Circular Migration and the Rights of Migrant Workers in Central and Eastern Europe*, IMISCOE Research Series,
https://doi.org/10.1007/978-3-030-52689-4

N	Date	Type of stakeholder	Institution	Country	Gender	Interview
17.	07.04.2014	Representative	International organisation	Switzerland	F	In person
18.	27.11.2014	State official	Council of Ministers	Georgia	M	In person
19.	24.02.2017	Representative	International organisation	Belgium	M	In person
20.	27.02.2017	Representative	International organisation	Armenia	M	Skype
21.	6.03.2017	Representative	International organisation	Austria	M	In person
22.	14.03.2017	Representative	International organisation	Austria	M	In person
23.	20.03.2017	Representative	International organisation	Austria	M	In person
24.	27.03.2017	Former MP project assistant	International NGO	Georgia	F	Skype
25.	27.03.2017	State official	Council of Ministers	Georgia	F	Skype
26.	28.03.2017	Representatives × 2	International organisation	Belarus	F/M	Skype
27.	28.03.2017	Representative	International organisation	Austria	M	In person
28.	29.03.2017	Representative	International organisation	Austria	F	In person
29.	25.05.2017	Official	European Commission	Belgium	M	In person
30.	16.11.2017	Official	European Commission	Belgium	M	In person

Annex II: Interviews with Stakeholders in Poland in the Period October – December 2016

N	Date	Type of stakeholder	Country	Institutional level	Gender	Interview
1	4.11.16	Officials × 3	Poland	Ministry	F	In person
2	3.11.16	Academic	Poland	Higher education organisation	F	In person
3	16.11.16	Civil society actor	Poland	NGO	F	In person
4	15.12.16	Expert	Poland	International organisation/ NGO	M	In person
5	10.11.16	Civil society actor	Poland	NGO	F	In person

(continued)

N	Date	Type of stakeholder	Country	Institutional level	Gender	Interview
6	24.11.16	Civil society actor/ lawyer × 2	Poland	NGO	M/F	In person
7	25.11.16	Official	Poland	Local authority (Labour office)	M	In person
8	25.11.16	Official	Poland	Border guard	M	In person
9	25.11.16	Employer/owner	Poland	Agriculture business	M	Telephone
10	25.10.16	Staff member/PR	Poland	Private recruitment agency	M	In person
11	15.11.16	Chairman	Poland	Trade union	M	In person
12	15.11.16	Civil society actor/ academic	Poland	NGO/Higher education organisation	M	In person
13	18.11.16	Expert	Poland	NGO	F	In person
14	22.11.16	Advisor	Poland	Ministry	F	In person
15	8.11.16	Academic	Poland	Higher education organisation	M	In person
16	19.12.16	Official	Poland	Ministry	F	Email
17	16.11.16	Official	Poland	Office of foreigners	F	In person
18	23.11.16	Official	Poland	Ministry	M	In person
19	18.11.16	Official	Poland	Ministry	F	In person
20	16.12.16	Lawyer	Poland	Professional body	M	In person
21	30.11.16	Lawyer/PR	Poland	Professional body	M	In person
22	7.12.16	Management representatives × 4	Poland	Business/IT company	M/M/F/F	In person

Annex III: Interviews with Stakeholders in Bulgaria, in the Period June 2016 – January 2017

N	Date	Type of stakeholder	Country	Institutional level	Gender	Interview
1	13.07.16	Official	Bulgaria	Ministry	M	In person
2	22.07.16	Official	Bulgaria	Ministry	M	In person
3	22.07.16	Official	Bulgaria	NCID	M	In person
4	14.07.16	Official	Bulgaria	Ministry	F	In person
5	14.07.16	Representatives × 3	Bulgaria	Trade Union	F	In person

(continued)

N	Date	Type of stakeholder	Country	Institutional level	Gender	Interview
6	3.08.16	Representative	Bulgaria	Employers' organisation	M	In person
7	22.07.16	Academic	Bulgaria	University	F	In person
8	11.07.16	Lawyer	Bulgaria	NGO	F	In person
9	21.07.16	Lawyer	Bulgaria	NGO	F	In person
10	8.07.16	Officials × 3	Bulgaria	Ministry	F	In person
11	4.01.17	Representative	Bulgaria	Employers' organisation	M	In person
12	26.06.16	Expert	Bulgaria	International organisation	M	In person
13	1.10.16	Official	Bulgaria	Ministry	F	Email
14	6.07.16	Advisor/Expert	Bulgaria	Council of Ministers/Political cabinet member	M	In person
15	4.11.16	Official	Bulgaria	National Social Security Institute	F	In person
16	14.10.16	Expert	Bulgaria	Employers' organisation	F	Email
17	27.09.16	Official	Bulgaria	Ministry	F	In person
18	30.09.16	Representatives	Bulgaria	IT company	M/F	In person
19	28.09.16	Foreign employer	Bulgaria	Business	F	In person
20	20.12.16	Representatives (follow up interview of # 18)	Bulgaria	IT company	M/F	In person
21	3.01.17	Representative	Bulgaria	Professional body	F	In person
22	6.10.17	Representative (follow up interview #11)	Bulgaria	Employers' organisation	M	Telephone
23	6.10.17	Seasonal workers recruiter	Bulgaria	Recruitment company	M	Telephone

Annex IV: Focus Groups in Bulgaria and Poland

Bulgaria	Number of participants	Nationality	Gender	Skill level	Working status
General focus group, 1 October 2016	5	Ukrainians/1 Russian (who wanted to be included in this group)	4 F/1 M	Mixed: Highly-skilled/ Middle-level skilled	4 Workers 1 Retiree
General focus group, 1 October 2016	5	Russians	4 F/1 M	Mostly highly-skilled	2 Workers 2 Family members of foreigners of Bulgarian origin 1 Retiree
Focus group with Blue Card holders, 30 September 2016	4	Russians	1 F/3 M	Highly-skilled	Workers
Focus group with Blue Card holders, 30 September 2016	3	Ukrainians	1 F/2 M	Highly-skilled	Workers

Poland	Number of participants	Nationality	Gender	Skill level	Working status
General focus group, 19 November 2016	5	Ukrainians	3 F/2 M	Mixed: Highly-skilled/ Middle-level skilled/ Low-skilled	5 Workers
General focus group, 19 November 2016	6	Russians	5 F/1 M	Mostly highly-skilled	6 Workers
Focus group with Blue Card holders, 6 December 2016	8	Mixed: Russians/Ukrainians/ Kazakhstanis	8 M	Highly-skilled	8 Workers
Focus group with Blue Card holders, 8 December 2016	5	Russians	2 F/3 M	Highly-skilled	5 Workers
Focus group with Blue Card holders, 8 December 2016	6	Ukrainians	2 F/4 M	Highly-skilled	6 Workers 1 applicant for a Blue Card

Annex V: Benchmark Framework for Assessment

Policy area to be addressed with circular migration	International standards	Benchmarks	Instrument/measure to implement benchmarks
Entry conditions	No general international law provisions. Admission is national prerogative of states.	–	–
Re-entry conditions	UDHR, Art. 13 ICCPR, Art. 12 ICRMW, Art. 8 ECHR Protocol No. 4, Art. 2(2) and 3(2)	Right to return to his or her own country and leave any country, including one's own country of origin.	Permits allowing periods of absence from the territory of the country of destination for long-term residents; TCNs going to their home country (also applicable to residence status)
	European Agreement on Regulations governing the Movement of Persons between Member States of the Council of Europe, Art. 1 European Convention on Establishment, Art. 1 ILO Multilateral Framework on Labour Migration, Principle 15, Guideline 15.8	Facilitated personal travel of nationals of the contracting parties without visa for visits of not more than 3 months. Facilitated entry into the territory of the contracting parties for the purpose of temporary visits. Adopting policies to encourage circular and return migration and reintegration into the country of origin, including by promoting temporary labour migration schemes and circulation-friendly visa policies.	Multiple entry visas and/or visa free regime
	ICRMW, Art. 59 (2)	The State of employment shall, subject to paragraph 1 of the present article, consider granting seasonal workers who have been employed in its territory for a significant period of time the possibility of taking up other remunerated activities and giving them priority over other workers who seek admission to that State, subject to applicable bilateral and multilateral agreements.	

Work authorisation	ECMW, Art. 8(2) ILO Migrant Workers Convention (No. 143), Art. 14 (a) ICRMW, Art. 52 (3a) ICRMW, Art. 59 (2)	Free choice of employment with maximum restriction of 1 or 2 years (with some limitations provided in the law). Contracting states shall consider granting seasonal workers who have already been employed in their territory for a significant period of time the possibility of taking up other remunerated activities.	Flexible work permits, allowing change of both sector and employer within the period of their validity
	ILO Migrant Workers Convention (No. 143), Art. 8 (1) ICRMW, Art. 49 (2) ILO Migrant Workers Convention (No. 143), Art. 8 (2) ICRMW, Art. 49 (3); Art. 51	Loss or termination of employment should not constitute a sole ground for withdrawal of migrant worker's authorization of residence or work permit. Possibility to find alternative work in case of loss or termination of employment.	
Residence status	ICCPR, Art. 12 (1), ICRMW, Art. 39, ECHR, Art. 2 (1) of the Fourth Protocol European Convention on Establishment, Art. 2	Right to free movement and choice of residence within the country, where one is lawfully resident. Contracting parties shall facilitate the prolonged or permanent residence of nationals of the other parties in its territory.	Permits allowing transit from temporary status to permanent residence rights (and thus path to citizenship)

(continued)

Policy area to be addressed with circular migration	International standards	Benchmarks	Instrument/measure to implement benchmarks
Social security coordination	ILO Equality of Treatment Convention (No. 118), Art. 5 ILO Equality of Treatment (Accident Compensation) Convention (No. 19), Art. 1	Possibility to export: Old-age pensions and benefits, Disability benefits, death grants, Benefits in respect of accidents at work and occupational diseases, survivors' benefits.	- bilateral social security agreements ILO Equality of Treatment Convention, Art. 8 Maintenance of Social Security Rights Recommendation (No. 167) European Code of Social Security, Art. 73 European Social Charter and (Revised) European Social Charter, Art. 12 -multilateral agreements ILO Equality of Treatment Convention, Art. 8 European Code of Social Security, Art. 73 European Social Charter and (Revised) European Social Charter, Art. 12
	Maintenance of Social Security Rights Convention (No. 157), Art.2 Maintenance of Social Security Rights Recommendation (No. 167) ILO Migration for Employment Convention (No. 97), Art. 6 (1) b European Code of Social Security, Art. 73	Maintenance of the acquired rights and rights in course of acquisition. Equality of treatment.	
	ILO Equality of Treatment Convention (No. 118), Art. 7 (2) ILO Migrant Workers Convention (No. 143), Art. 9(1) ICRMW, Art. 27 (2)	Totalisation of periods of insurance, employment or residence and of assimilated periods for the purpose of the acquisition, maintenance or recovery of rights and for the calculation of benefits. Reimbursement of social security contributions	
Entry and re-entry of family members	CRC, Art. 10(1) ICRMW, Art. 44 (2) ILO Migrant Workers Convention (No. 143), Art. 13(1) ILO Migrant Worker Recommendation No. 151 (Revised) European Social Charter, Art.19 (6) ILO (1997) Guidelines on Special Protective Measures for Migrant Workers in Time-bound Activities	Obligation to facilitate family reunion. Family reunion of seasonal migrants and 'special purpose workers' who are legally resident in the country.	Waiting period for family reunification shall not exceed 12 months. ECMW, Art.12 A requirement for suitable housing should not be so restrictive as to prevent family reunification. ILO Migrant Worker Recommendation No. 151

Recognition of qualifications	ILO Migrant Workers Convention (No. 143), Art. 14 c ILO Recommendation No. 151, Paragraph 6 ILO Multilateral Framework on Labour Migration. Non-binding principles and guidelines for a rights-based approach to labour migration, 2006, Guideline 12.6 Convention on the Recognition of Qualifications concerning Higher Education in the European Region Art. III.1 Art. III.3.5 Article IV.1 Section 4, 5 and 6	Members may after appropriate consultation with the representatives organizations of employers and workers, make regulations concerning recognition of occupational qualifications acquired outside its territory, including certificates and diplomas. Recognition and accreditation of migrant workers' skills and qualifications and, where that is not possible, providing a means to have their skills and qualifications recognized. Holders of qualifications issued in one country shall have adequate access to an assessment of these qualifications in another country. The responsibility to demonstrate that an application does not fulfil the relevant requirements lies with the body undertaking the assessment. Each party is required to generally recognise qualifications in regards to access to higher education, periods of study within the framework of another higher education programme or higher education degrees as similar to the corresponding qualifications in its own system, unless it can show that there are substantial differences between its own qualifications and the qualifications for which recognition is sought.	International cooperation for recognition of qualifications in both countries of origin and destination (academic and professional qualifications, non-formal and informal learning). Measures to assist migrant workers and their families on the occasion of their final return to their State of origin - information about equivalence accorded to occupational qualifications obtained abroad and any tests to be passed to secure their official recognition; equivalence accorded to educational qualifications so that migrant workers' children can be admitted to schools without downgrading. (ECMW, Art. 30) All countries are required to establish a national information centre, offering advice on the recognition of foreign qualifications to parties or persons (Article IX.2, LRC). Qualification frameworks provide for the positioning of a certain qualification in the grid of qualifications and thus facilitate recognition by supporting comparability and clarifying the learning outcomes contained in foreign degrees (Bologna process). European Credit Transfer and Accumulation System (ECTS) aims to facilitate recognition of study periods and degrees based on credits in line with the Lisbon Recognition Convention (Bologna process).

The manufacturer's authorised representative in the EU is Springer Nature Customer Service Centre GmbH, Europaplatz 3, 69115 Heidelberg, Germany. If you have any concerns regarding our products, please contact ProductSafety@springernature.com

Printed and bound by CPI Group (UK) Ltd, Croydon, CR0 4YY
30/09/2026
02225253-0007